Underwater Research

Underwater Research

edited by

E. A. DREW   *Gatty Marine Laboratory, St. Andrews, Scotland*
J. N. LYTHGOE   *MRC Vision Unit, University of Sussex, Falmer*
J. D. WOODS   *Meteorological Office, Bracknell, Berkshire*

1976

Academic Press
London   New York   San Francisco
A Subsidiary of Harcourt Brace Jovanovich Publishers

ACADEMIC PRESS INC. (LONDON) LTD.
24/28 Oval Road
London NW1

United States Edition published by
ACADEMIC PRESS INC.
111 Fifth Avenue
New York, New York 10003

Copyright © 1976 by
ACADEMIC PRESS INC. (LONDON) LTD.

All Rights Reserved

No part of this book may be reproduced in any form by photostat, microfilm, or any other means, without written permission from the publishers

Library of Congress Catalog Card Number: 74-5659
ISBN: 0 12 221950 3

PRINTED IN GREAT BRITAIN AT
THE PITMAN PRESS, BATH

Preface

Scientific diving is no longer the prerogative of a handful of students zealously adapting their sport to serve their science (or perhaps the other way around!). It has become a day-to-day research technique in almost every institution that has some special interest in ocean research.

In the hundred years before the Second World War, a handful of scientists had clambered into the unwieldy hard-hat diving suit to make their pioneering observations on underwater life. But it was not until the Cousteau-Cagnan aqualung was developed in the early 1940s that they had a diving apparatus simple and convenient enough for everyday scientific use.

In Britain, at least, underwater research had achieved considerable momentum by 1965 when five scientific diving teams, chiefly composed of undergraduates and postgraduate students, mounted expeditions to Malta, attracted there by clear water, reliable weather and cheap air fares. With so many divers on the island, all relying on the generosity of the Royal Navy for the supply of air, it was feared that chaos at the air filling station would result in the withdrawal of the air supplies. It was in response to the obvious need for our collective presence to be as little nuisance as possible to our hosts that the Underwater Association of Malta was formed.

The close contact that was maintained between the teams led to a mutual interest in each other's work, and when everyone had returned to Britain the Association organized an informal symposium to discuss the work done in Malta. This 1965 meeting was the first of the annual symposia which have since then extended to include contributions from many countries and a wide range of subject matter from sites other than Malta. The symposia of 1966–69 were published in booklet form. By 1970 it became evident that it was the interrelated subjects of diver psychology, physiology and techniques that held the common interest of all divers; the more specialized papers were best published in the appropriate scientific journals.

When a diver without artificial aids enters the water he virtually writes-off his sense of hearing, chiefly because the articulation of normal speech is impossible and he feels he cannot locate the direction of an underwater sound. Vision has always been an important sense on which a diver is often totally reliant. Yet the physical properties of natural water will certainly reduce the distance he can see to a few tens of metres and often down to nothing at all. The physics and physiology of these two senses and procedures to improve them are of such fundamental importance to all divers that the Underwater Association asked Dr Cocking to write a review chapter on the problems of underwater viewing, and Dr Hollien, Dr Rothman and Dr Feinstein to write two chapters on underwater hearing and acoustics.

Following these chapters are a selection of papers read at the symposia of the Underwater Association after 1969. These have all been up-dated where the authors deem it necessary. Again, there is a clear preoccupation with the need to understand the human reaction, both physiological and psychological, to this alien environment. The latter papers in this book show that for all its limitations diving is really the only technique we have for investigating benthic ecology where the bottom is unsuitable for the operation of the older methods of grab and trawl. It is certainly the only technique suitable for the study of sunken wrecks and cities.

Contents

Preface	v
Diver Communication H. HOLLIEN and H. B. ROTHMAN	1
Hearing in Divers H. HOLLIEN and S. FEINSTEIN	81
Improving Underwater Viewing S. J. COCKING	139
Depth Estimation by Divers HELEN E. ROSS and SAMUEL S. FRANKLIN	191
An Investigation into Colour Vision Underwater CHARLES W. FAY	199
Narcosis and Visual Attention HELEN E. ROSS and M. H. REJMAN	209
Diver Performance—Nitrogen Narcosis and Anxiety J. P. OSBORNE and F. M. DAVIS	217
Body Temperature Monitoring during Diver Performance Experiments F. M. DAVIS, J. BEVAN, J. P. OSBORNE and J. WILLIAMS	225
The Measurement of Respiration at High Ambient Pressures J. B. MORRISON	237
The Design of a Lightweight Underwater Habitat B. RAY	253
The Use of an Underwater Habitat as a Quiet Laboratory for Tests on Diver Hearing B. RAY	260
Towards the Development of a Practical Underwater Theodolite R. FARRINGTON-WHARTON	267

The Design and Application of Free-flooding Diver
Transport Vehicles   277
G. COOKE

Practical Considerations for Quantitative Estimation of
Benthos from a Submersible   285
J. F. CADDY

A Stereophotographic Method for Quantitative
Studies on Rocky-bottom Biocoenoses   299
TOMAS L. LUNDÄLV

Some Underwater Techniques for Estimating
Echinoderm Populations   303
J. K. G. DART and P. S. RAINBOW

Nocturnal Behaviour in Aggregations of
Acanthaster Planci in the Sudanese Red Sea   313
R. G. CRUMP

The Ecology of *Caryophyllia smithi* Strokes and Broderip on
South-western Coasts of the British Isles   319
K. HISCOCK and R. M. HOWLETT

Light, Zonation and Biomass of Submerged
Freshwater Macrophytes   335
D. H. N. SPENCE

Preliminary Studies on the Primary Productivity of
Macrophytes in Scottish Freshwater Lochs   347
R. M. CAMPBELL and D. H. N. SPENCE

Some Aspects of the Growth of *Posidonia oceanica* in
Malta   357
E. A. DREW and B. J. JUPP

Photosynthesis and Growth of *Laminaria hyperborea* in
British Waters   369
E. A. DREW, B. P. JUPP and W. A. A. ROBERTSON

Deposition of Calcium Carbonate Skeletons by Corals:
An Appraisal of Physiological and Ecological Evidence   381
R. K. TRENCH

Archaeological Evidence for Eustatic Sea Level Change
and Earth Movements in South West Turkey   395
N. C. FLEMING and N. M. G. CZARTORYSKA

Cape Andreas Expedition, 1969   405
J. N. GREEN

Index   413

Diver Communication

H. HOLLIEN
and
H. B. ROTHMAN

Communication Sciences Laboratory, University of Florida, Gainesville, 32601, U.S.A.

1. Speech under Normal Conditions in Air	2
2. Speech in High Pressure Air	4
3. Speech in Saturation (HeO_2/P) Diving	7
A. The effects of the environment upon communication	7
B. Techniques for unscrambling HeO_2 speech	16
C. On-line evaluation of HeO_2 speech unscramblers	24
D. Off-line evaluation of HeO_2 speech unscramblers	32
E. Discussion	37
4. SCUBA Speech	41
A. Research environments	42
B. Specialized equipment	44
C. Diver subjects	49
D. Research programmes	49
E. Equipment evaluation	58
F. Evaluation procedure	66
G. Results of diver equipment evaluation	71
H. Diver communication performance during work tasks	74
5. Summary	77
Acknowledgements	78
References	78

Divers do not enjoy adequate communication on any level—either among the individuals comprising the diving teams, between the teams or between divers and the surface. Therefore, they experience a disabling situation and cannot operate at work levels even close to their full potential. On the other hand, man is increasingly turning to the sea for scientific, military, economic, ecologic and recreational purposes; he is also diving to greater depths and, with the advent of saturation diving, he is staying at these depths for longer periods. Thus, if we are to realize our potential as diver-workers, the development of good underwater communication techniques and equipment are a must.

This chapter will discuss the available information relative to communication in both the saturation (HeO_2) and shallow water diving situations. However, before doing so, let us first re-emphasize the point that before man can become remotely as effective underwater as he is on land, he must be able to communicate, by *voice*, with his fellow-divers and colleagues on the surface. We will concede that in air, humans communicate both by written and spoken language, including gestures and facial expressions. In water, however, the usefulness of gestures and facial expressions dwindle markedly due to curtailed visual contact, and the use of life support devices such as regulators and masks. Writing is extremely limited as a method of underwater communication as are systems such as the International or Morse Codes. Since these communication approaches are both awkward and slow, speech must be utilized to solve communication problems.

The whole process of talking underwater is a much more complex one than is generally realized and before adequate speech communication can become possible, a great deal of information on this subject must become available. We need data about such issues as: (1) the interface between the life support and communication systems; (2) the effects of the various physical impingements upon the speaking mechanism; (3) the possible compensations for the resulting speech distortions; (4) predictions designed to improve existing diver communication systems; (5) evaluations of such systems as they become available and so forth.

In any case, in order to meet these needs, our group has been carrying out a broad research programme designed to answer fundamental questions about diver communication and to acquire basic knowledge about the factors that limit or enhance man's ability to communicate underwater. Our basic and applied underwater speech programmes focus on several areas of investigation: (1) studies of man's ability to produce intelligible speech under the constraints he encounters as a diver, (2) studies of underwater speech propagation and various bottom-surface thermocline, distance, filtering and masking effects as they distort speech intelligibility, (3) the analysis and appraisal of diver's communication systems and (4) development of specialized instrumentation that will permit underwater research to be carried out with a precision similar to research conducted in air. Our investigations are focused on the free SCUBA diver in relatively shallow waters as well as on saturation divers who work at greater depths. Thus, although this chapter is based primarily on our own studies, we have incorporated other relevant research into our discussion of the issues.

1. Speech under Normal Conditions in Air

Before discussing the problems encountered in speech communication under high ambient pressure we believe it is necessary to describe very briefly the nature of normal speech. Indeed, it is a uniquely complex function. In order to produce speech, one must activate a sophisticated, interacting system which involves the lungs, trachea, larynx, pharynx, nose and mouth (see Fig. 1).

Stated very simply, speech is the product of a sound source (or sources) and the resonation of the vocal cavities excited by this source. The physiological process underlying the production of the acoustic speech signal also is complex in nature. Basically stated it involves three components. These are: (1) the power source, i.e., the respiratory mechanism, consisting of the lungs, rib cage and attendant musculature

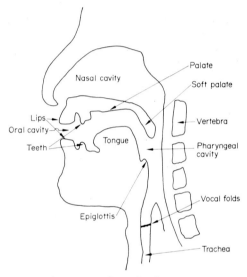

Fig. 1. Schematic view of the human speech mechanism

(not shown in Fig. 1); (2) the generator, i.e., usually the larynx, but constrictions or closures above the larynx can function as the generator and (3) the resonators, i.e., the oral, nasal and pharyngeal cavities. Using this system, humans generally are capable of generating three types of sounds, namely, (a) a periodic laryngeal tone, (b) an aperiodic random noise resulting from turbulence created within one of the available narrow passages, and (c) an aperiodic burst of noise following the sudden release of pressure. For any speech sound, only that portion of the vocal tract system superior to the sound source acts as an important resonator; essentially, the cavities below the sound source can be neglected. The nasal cavity, however, can be decoupled or coupled to the vocal cavities via the soft palate and consequently can be treated as a separate system.

The above is a simple description of a highly complex system which produces an acoustic product that is also of high complexity. One way of looking at the acoustical output of the vocal tract is to examine its resonant frequencies or formants. Formants (areas of acoustic energy maxima) are determined by the size and shape of the vocal tract and are especially important in the production of vowels. Vowels, in addition to conveying information, also provide the greatest amplitude in the speech signal (amplitude is the physical dimension most closely correlated with the perceived loudness of speech). Perhaps of greater importance are the formant transitions which represent articulatory movement from—or to—the place of consonant production to—or from—the position of the adjacent vowel. The reason why transitions are so important in speech is that (1) an articulatory gesture associated with a given sound segment characteristically varies with context; (2) acoustic segments and perceived sounds are not identical and (3) a single acoustic segment carries information in parallel about preceding and successive sounds. Indeed, much of this parallel information is carried by these transitions between sound segments. Consonants, while having less sound power than vowels, carry greater amounts of information critical to speech intelligibility. Consonants are broad band and are often transient in nature. They occur more frequently than vowels, and contain, along with their associated transitions, much important speech information.

In summary, speech is composed of an interrelated series of periodic and aperiodic waves of varying durations, amplitudes and spectra. These waves are formed into the sounds of speech by movements of the articulators within the vocal tract.

Perception of speech relies on the discrimination of various cues contained in the different sounds man is capable of producing. If these differences are masked by ambient noise or are distorted by changing or adding to the resonating structures, speech intelligibility is degraded. In the underwater situation, a great deal of ambient noise exists in the environment and even more is introduced by the diver's breathing apparatus. Further, the life support system (high pressure gas, face mask, muzzle or helmet with their restraining straps) itself introduces a restraint on the movement of the articulators and changes the resonating characteristics of the speech system. All these factors—as well as several others discussed below—contribute to the degradation of speech intelligibility in the world of the diver.

2. Speech in High Pressure Air

In normal speech, the distribution of energy in the speech spectrum is primarily determined by the shape of the vocal cavities from the vocal folds to the lips and by the velocity of sound in the exhaled gas mixture. Although increases in air pressure do not significantly effect the velocity of sound, when a diver speaks in high pressure air, his intelligibility nevertheless decreases with increasing pressure—and his speech exhibits a pronounced nasal quality. In order to investigate this effect and determine its magnitude, White (1955) judged the intelligibility of monosyllabic phonetically balanced words spoken at several simulated depths between 0 and 200 ft. Later, we studied intelligibility at approximately the same depths using similar word lists. However, our research was carried out in a substantially larger population. Table 1

TABLE 1. A comparison of mean percentages of words correct for the six depths, from Hollien and White. All recordings, with the exception of 0 ft, were made while talkers were breathing compressed air.

	N	0 ft	25 ft (7.6 m)	Depth 50 ft (15.2 m)	100 ft (30.5 m)	150 ft (45.7 m)	190/200 ft (57.9/61 m)
White[a]	4	82.2	—	79.3	78.4	70.4	57.5
Hollien	8	89.6	88.6	84.5	80.2	71.9	68.8

[a] White's subjects did not participate in the research at all depths.

presents a comparison of mean percentages of words correct from the two studies. Both sets of data show a steady decrease in intelligibility with increases in depth. The narcotic effects of nitrogen were present in both studies (manifested by obvious misarticulation or "slurring"); however, we attempted to mitigate such effects by removing obvious errors on the part of the diver/talker. Nevertheless, a 22% (residual) reduction in intelligibility occurred; it appeared to be the result of increased gas density. In any case, these two studies clearly indicate that a decrease takes place in speech intelligibility with increases in ambient pressure and such changes should be expected by the diver.

We also have studied the differential effects of high pressures on phoneme classes. Generally, we have found that the intelligibility of three phoneme classes—stops, fricatives and glides—deteriorated as depth increased from 0 ft to 190 ft. Stops showed the greatest degradation in intelligibility followed by fricatives and glides. Variation in the intelligibility of nasals—/m, n, ŋ/—was not systematic. Such variation in nasal intelligibility at different pressures may be due to the increased percept of nasal voice quality.

In an effort to determine the cause of the nasal quality found in divers' speech under conditions of high ambient pressures, Fant and associates (1964, 1968) compared vowel formant frequencies produced at atmospheric pressure to the same sounds produced at 6 ATA in air. These authors found that an increase in ambient pressure results in a non-linear shift of the lower formant frequencies (see Fig. 2). They hypothesized that the non-linear formant shift is due to the vibration

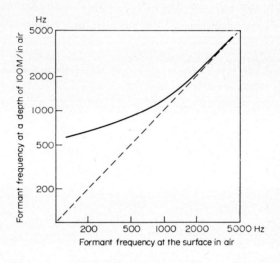

Fig. 2. Formant shift in high pressure air (after Fant and Linquist, 1968). The solid line is the measured formant shift. The dashed line is $k = 1$.

of the walls of the vocal cavity resulting from a reduction in the impedance mismatch between the air (which becomes more dense) and the cavity walls themselves (hence, the increased nasality of the signal). That is, the reduced impedance mismatch causes the cavity wall to absorb energy which results in their vibrating; hence, the cavity wall becomes a sound source itself. The additional sound source with its resulting non-linear low frequency shift constitutes a factor in the reduction of speech intelligibility because it affects both vowels and consonants and alters the relationship between concentrations of energy within a crucial portion of the spectrum. Indeed, Fant and Sonesson formulated their hypothesis after experimentally observing that the nasal cavity was not otherwise in operation at high pressures, i.e., there was no change in soft palate function due to changes in pressure. In any case, it is clear that high ambient pressures alone can substantially affect the speech of divers.

3. Speech in Saturation (HeO$_2$/P) Diving
A. The Effects of the Environment Upon Communication

As we know, in saturation diving, the life support atmosphere typically consists of mixtures composed predominantly of helium, plus oxygen and, in some cases, small percentages of nitrogen. The HeO$_2$ atmosphere (which has different sound transmission characteristics than does normal air) in combination with the high ambient pressures (for example at 1000 ft $P > 450$ psi), effectively distorts the resonant characteristics of the vocal tract with severe degradations of speech intelligibility.

A number of researchers have studied specific issues related to speech communication in the HeO$_2$ environment. Unfortunately, often their investigations have focused on narrow questions rather than on the total problem. However, taken as a whole, the data already available show that helium causes an upward shift in the formant frequencies of vowels. By itself this upward shift (a magnitude of approximately 1.8) does not materially affect intelligibility. For example, Sergeant (1967) found that breathing an 80% helium-20% oxygen mixture at atmospheric pressure did not affect general speech patterns, and that the order of difficulty in perceptually identifying phonemes in helium was similar to that found in normal speaking conditions. However, it should be noted that Sergeant's subjects were breathing the HeO$_2$ mixture through a bib device at 1 ATA. Therefore, they were talking into air with all of the unknown and complex interrelationships that would result from such a situation. Nevertheless, his data are pretty conclusive that high concentrations of helium alone affect speech intelligibility only minimally.

The shift in the frequency of vowel formants—and for that matter for all speech sounds—is very well accepted; specifically these acoustic shifts due to the helium are caused by the different transmission characteristics of that gas. In order to understand the differences between the transmission characteristics of HeO$_2$ and of air, it is helpful to think of the resonant frequencies of an acoustic tube as the function of the physical dimensions of that tube (cross sectional area and length) and the speed of sound for the gas contained in the tube. The speed of sound in a gas is related to its physical properties by the following relationship:

$$c = \sqrt{\frac{\gamma P}{\rho}}$$

where
P = pressure of the gas
ρ = density of the gas
γ = ratio of the specific heats

For most gases, P/ρ is nearly independent of pressure. Therefore, for a given gas, pressure will theoretically have no effect on **c** and, consequently, no effect on formant frequencies. However, when comparing two different gases (for example, nitrogen and helium) ρ and γ are changed (hence, **c** is changed also) and the frequencies in the acoustic signal are multiplied by a constant.

This linear frequency shift—due to the change of speed of sound in a gas—can be predicted theoretically and has been observed experimentally. For example, Fig. 3 provides data comparing observed

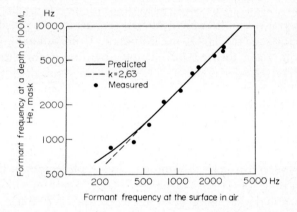

Fig. 3. Predicted and measured formant frequency transpositions for one subject when breathing a 2.5% O_2, 97.5% He mixture through a light diving mask at 100 m equivalent depth (from Fant and Lindquist, 1968).

vowel formant frequencies at 0 meters in air and 100 m in HeO_2. A linear frequency shift is apparent as is the expected slight non-linearity at the low frequencies.

In addition to the shift of vowel formant frequencies resulting from the use of HeO_2 breathing mixtures, there is an apparent loss of energy in consonants as compared to vowels in an HeO_2/P environment. In Fig. 4, Brubaker and Wurst (1968) demonstrate the comparative effects of the consonant-vowel amplitude ratio for the word "fish" spoken at 0 ft and at 300 ft (91.4 m) in an HeO_2 environment. At 300 ft there is a considerable loss of energy in the pre- and post-vocalic consonants with respect to the vowel energy. Since a high proportion of the information bearing elements necessary for intelligibility of speech are contained in the consonants and the transitions between consonants and vowels, such changes degrade the speech signal to a considerable degree. In this regard, there is some controversy concerning why consonant amplitude energy is decreased: Fant and Sonesson (1964) propose that the drop in energy is related to the physics involved in generating an air stream in the vocal tract constriction as compared to voicing produced by the vocal folds. On the other hand, Brubaker and Wurst (1968) argue that consonant suppression is the opposite effect of consonant enhancement observed in low pressure aircraft communication systems. Finally, Flower (1969) investigated several other aspects of the acoustic speech signal and reported that (1) formant bandwidths did not seem to increase with increasing depth; (2) high frequency energy seemed to decrease with increasing depth and (3) low frequency energy appeared to be increased at increasing depths. The investigations of Flower tend

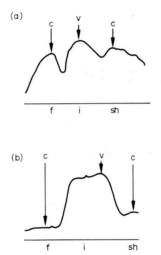

Fig. 4. Consonant-vowel amplitude display (from Brubaker and Wurst, 1968).

to support Brubaker and Wurst's contention that there is a loss of consonantal energy (primarily high frequency) relative to that of vowel energy (primarily low frequency). Since consonants carry much perceptual information, this decrease of high frequency energy at increasing depth helps to explain the concomitant reduction in speech intelligibility.

While some of the studies cited above provide much needed base information for engineers who attempt to design helium speech unscramblers, the information is of somewhat limited value since the experimenters often (1) examined the formants of vowels produced in isolation or (2) investigated only one formant. Moreover, there have been no investigations of transitions, and as we pointed out, studies of the acoustic characteristics of speech indicate that the transitions between consonants and vowels carry important information for the perception of speech.

In addition to the physical distortions of the speech signal caused by the HeO_2/P environment itself, there are behavioural factors which degrade speech intelligibility and which are difficult to quantify. Some of these behavioural distortions are introduced by the individual diver's reaction to the environment and, in particular, to his distorted speech. Some divers simply refuse to speak, others attempt to become more intelligible—often by what seems to be random modifications of their speaking patterns. In turn, the diver-introduced "distortions" may magnify some of the problems created by the HeO_2/P situation, e.g., a diver's increase in speaking intensity in order to overcome high ambient

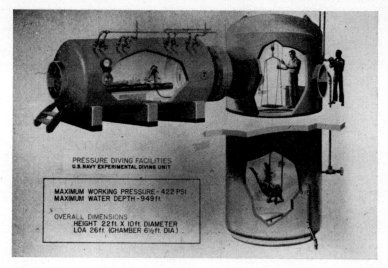

Fig. 5. An artist's view of the U.S. Navy's Experimental Diving Unit hyperbaric complex.

noise may affect the consonant-vowel ratio. Other distortions may result from the use of life-support equipment, especially when in the sea (for example, mask size and configuration). In addition, we have shown that an above-average (i.e., "good") speaker at atmospheric pressure may not necessarily be a good speaker under HeO_2/P conditions (Hollien et al., 1973).

It is evident from the above discussion that much research has been completed. However, it is also clear that much more specific research is needed in order to discover the exact nature of the distortions to speech intelligibility created by the HeO_2 environment. Hence, we are investigating the following aspects of the speech of diver/talkers in an HeO_2/P environment: (1) speaker intelligibility; (2) analysis of the talker's speech errors; (3) adaptation—or spontaneous speech improvement—by the diver over time; (4) changes in the fundamental frequency of speech caused by helium; (5) changes in the vowel formants caused by variation of helium and pressure and (6) the ways in which the talker can improve the intelligibility of his speech in the high ambient pressure and HeO_2 situation.

Very special environments and equipment are necessary in order that a research programme of the nature described may be carried out. Of fundamental importance is the hyperbaric chamber or habitat. Figure 5 provides a view of a typical chamber of this type—one of several at the U.S. Navy's Experimental Diving Unit (EDU), Washington, D.C. This particular facility is two stories high and consists of four rooms: sleeping and eating, dry work, wet work and emergency lock-in. Although structurally large, once the divers and their life support systems

Fig. 6. Photograph of a female diver reading a word list in an HeO_2 environment at 300 ft. The acoustic treatment is typical of the type utilized in this research.

are inside, these chambers are very crowded. We also work at other hyperbaric facilities—such as those at the Westinghouse Corporation's Ocean Simulation Facility, Duke University and at the Naval Coastal Systems Laboratory. It is of interest to note that we use both male and female divers in order to obtain a wider spectrum of information on diver's speech in helium; note the group seen in Fig. 6 (inside chamber) conducting a typical communication experiment. Further, it must be remembered that: (1) all of the research we report has been carried out in reasonably controlled acoustical environments (enclosures constructed of fibreglass mattresses), (2) all of our recording equipment has been calibrated at depth and (3) (when speech intelligibility is under study) all word lists read were scored by at least 10–15 listeners.

The data presented in Table 2 took us almost a year and a half to collect; the research was conducted at the EDU on aquanauts in training for Sealab-3. A total of 46 divers were subjects; 28 at sea level and 200 ft (70 m); 22 at zero and 450 ft (137.2 m) and nine at zero and 600 ft (182.9 m). The data show that intelligibility is approximately halved for every doubling of depth until, at 600 ft, it is less than 10%.* Obviously, intelligibility levels of these magnitudes constitute severe mechanically-induced speech distortion.

TABLE 2. Overall means of diver intelligibility in Helium/Oxygen. All recordings were made during Sealab 3 training at EDU. Means corrected for unequal N's.

Depth ft	Depth m	Number of Diver/Talkers	Number of Listeners	% Intelligibility
0	0	46	487	90.9
200	61.0	28	304	50.4
450	137.2	22	242	20.7
600	182.9	9	142	9.5

The four means are based on a total of 29 375 judged stimuli (words).

A question of great interest is: do aquanauts experience any spontaneous improvement of speech intelligibility? There was some suggestion from Sealab-2 that at least some of the divers gave themselves "speech correction" and hence exhibited improvement in speech intelligibility. Accordingly, we undertook a study designed to evaluate this factor. However, due to difficulty of obtaining speech samples over long periods of time, we were only able to collect data on four teams (16 divers) at 450 ft (137.2 m) over a period of two days—hardly a long enough period to permit extensive speech modification. However, Table 3 reveals that there was some trend toward speech improvement.

TABLE 3. Mean intelligibility scores of divers at 450 ft (137.2 m) in HeO_2 in the chamber at EDU. The "0" time represents the first readings immediately upon reaching depth. Subsequent times are hours elapsed from first reading at depth.

| | Cumulative Time Between Readings (h) | | | | |
	0	10	15–20	25–35	45–60
Mean	18.5	18.6	19.4	20.7	26.7
Number of lists read	16	8	28	23	19

* We now have data at 825 (251.5 m) and 1000 ft (304.8 m); they confirm the general trend noted above.

In this, as in all of our studies, there was considerable variability in the scores; hence, about half of the speakers accounted for nearly all of the improvement in speech.

Theory would predict an upward shift of vowel formants as a result of increasing concentrations of helium in breathing mixtures—and Fant has provided a model detailing these shifts. In this regard, we conducted a study (Table 4) of the vowels /u/, /ɑ/ and /æ/ spoken by five or more

TABLE 4. Mean formant frequencies for three vowels as a function of depth and helium concentration.

Condition (in ft)	Condition (in m)	/u/			/ɑ/			/æ/		
		F_1	F_2	F_3	F_1	F_2	F_3	F_1	F_2	F_3
0	0	433	1236	2307	655	1161	2235	670	1993	2712
200	61	844	2026	a	1136	2104	4672	1046	3744	4940
450	137.2	1006	2388	4718	1447	2514	a	1284	3965	5981
600	182.9	1087	2370	4444	1615	2484	4850	1546	3698	5852
825	251.5	1290	2558	4189	1749	2750	4246	1693	3656	5118

a no valid measurement available.

divers at 200, 450, 600 and 825 ft—and, as a control, at sea level. The formants of the vowels spoken in air are in reasonably good agreement with those provided by Peterson and Barney (1952) and by Fairbanks and Grubb (1961), so it can be concluded that our group of talkers is reasonably normal. As may be seen by the data in the table, the formants shift systematically with increases in helium and pressure. The increasing displacement of the formants correlates somewhat with the "severity" of the reduced speech intelligibility.

Data on consonant distortion may be observed in the next two tables. In Table 5 the manner of articulation is analyzed. From the table, it can be noted that the consonants produced normally at sea level show some involvement at 200 ft and are seriously affected at 600 ft (182.9 m). Further, the effects of depth appear greatest on the fricatives and least on the stops. Place of articulation errors are detailed in Table 6. In this case, there is a serious reduction in correct production of certain of the consonants (primarily the dentals and bilabials) at 200 ft (70 m)— and great involvement (and variability) in the place of articulation categories at 600 ft. It will be noted that at that depth, the dental and palatal consonants exhibit substantially reduced intelligibility (the palatals were the most intelligible at sea level) and that the glottals are the least affected. These findings—relating to the manner and place of consonant errors—coupled with the data on vowel distortion—have substantial implication for our work on the development of a training programme designed to improve the speech of aquanauts in the deep diving situation.

The data which suggested that divers could improve speech intelligibility *in situ*, led us to further investigate what could be done to

TABLE 5. Rank order of the intelligibility (% correct) for the phoneme categories grouped according to their manner of articulation at 0, 200 and 600 ft.

Surface		Manner of Articulation 200 ft (61 m)		600 ft (182.9 m)	
Glide	99.75	Glide	93.25	Stop	31.30
Nasal	99.69	Nasal	88.66	Nasal	22.05
Stop	99.31	Stop	87.11	Glide	19.97
Fricative	98.96	Fricative	85.38	Fricative	15.97

TABLE 6. Rank order of the intelligibility (% correct) for the phoneme categories grouped according to their place of articulation at 0, 200 and 600 ft.

Surface		Place of Articulation 200 ft (61 m)		600 ft (182.9 m)	
Palatal	99.72	Glottal	90.64	Glottal	46.62
Pre-palatal	99.24	Pre-palatal	87.39	Velar	26.43
Bilabial	99.01	Palatal	83.20	Pre-palatal	24.76
Velar	98.84	Velar	73.77	Bilabial	21.47
Glottal	98.71	Dental	68.33	Dental	9.09
Dental	98.36	Bilabial	62.84	Palatal	5.97

improve divers' intelligibility. Specifically, we believe that a major factor in good speech communication underwater will ultimately depend on the diver himself. That is, if a diver can modify his speech to compensate for some or all of the distortions created by the environment, good communications will become more likely. Accordingly, we designed an investigation to study the procedures a diver may develop and utilize in order to become more intelligible when communicating in an HeO_2/P environment. In order to do so, we placed phoneticians and/or trained talkers in the HeO_2 milieu with instructions to produce carefully controlled utterances. Twelve CSL divers who had training in phonetics and speech research and who had all served as diver/talkers in previous experiments were chosen for this purpose. In addition, all divers were trained and had demonstrated that they could produce the required speech modes with reasonable precision.

The twelve divers (six males and six females) descended to 300 ft (91.4 m) at the Westinghouse Ocean Research and Engineering Center in a gas mixture of approximately 86% helium. They were divided into equal groups; the first produced the speech with normal sidetone, the second wore TDH-39 earphones into which was fed a 95 dB noise signal which eliminated any auditory feedback. This procedure permitted us to make comparisons between talkers who were able to immediately attempt to enhance their speech intelligibility levels with those who could not.

All talkers read eight 50-word Griffiths (1967) lists (equated for difficulty) in seven different speech modes with the sequence counterbalanced to avoid "order effects". The speaking modes are as follows (in pairs): (1) normal articulation—most intelligible, (2) high and low fundamental frequency, (3) high and low vocal intensity and (4) slow and fast speaking rates. Subjects received extensive training in using each mode while keeping all others constant (except for the first pair). Fundamental frequency was controlled by means of our Fundamental Frequency Indicator (FFI); during training sessions (and the dive itself) intensity was controlled via a sound level meter, and rate was controlled by a stop watch. Flashcards were used at portholes to caution talkers who were not meeting these rigid protocols. Recordings were made via a calibrated ElectroVoice 664 microphone coupled to an Ampex 601 tape recorder.

A preliminary analysis of the data indicates that, of the seven voice parameters used, three were found to enhance speech intelligibility. These are: (1) low fundamental frequency (f_0), (2) slow rate of speech and (3) high speaking intensity. A further investigation will contrast the above three parameters with those found to result in poorest intelligibility (i.e., high f_0, fast rate and low intensity). At the conclusion of the projected study we will determine if a significant difference exists between the contrasting parameters.

To summarize this section, we have found the following:
(1) Speech intelligibility is approximately halved for every doubling of depth.
(2) Some aquanauts can improve their speech intelligibility spontaneously.

(3) Vowel formants shift systematically with increases in helium and pressure.
(4) Dental and palatal consonants were most affected at greater depths. When grouped according to manner of articulation the fricatives were most affected.
(5) Divers, trained to modify a particular speech parameter while holding others constant, were able to enhance their speech intelligibility.

These results have considerable implication for improving the speech intelligibility of aquanauts in the deep diving situation. Most important of these is the ability of trained divers to modify their speech in such a way as to overcome to a degree, the distorting effects of the environment. Further, the effects of these distortions to vowels, consonants and transitions must be studied and specified in greater detail. Indeed, we believe that, ultimately, good speech can result almost totally from the efforts of the diver. In the interim, however, electronic devices have been built which are designed to "unscramble" speech distorted by the HeO_2/P milieu.

B. Techniques for Unscrambling HeO_2 Speech

Since it will be some time before divers will have the ability to improve their speech to acceptable levels in the HeO_2 atmosphere, it is currently necessary to rely on helium speech "unscramblers" in order to mitigate the previously described problems to diver communication that are induced by this environment. These devices consist of electronic circuits designed to specifically improve the intelligibility of speech which has been distorted by the breathing of helium/oxygen mixtures under pressure. A number of techniques have been proposed as attempts to handle this problem; currently, however, only about half of these approaches have been implemented by the fabrication and testing of actual hardware. This section will discuss the philosophies and theories behind those techniques; at least those that are known to us. Indeed, it is often the case that information about how a particular unscrambler works is considered proprietary—and hence, we find it unavailable. In these cases, however, we usually have been able to deduce the principles upon which the unit was designed. (For further details see Giordano *et al.*, 1972.)

As of this writing, we deduce that the primary function of any of the available unscramblers is to linearly shift the vowel formant frequencies down to their "normal" positions. This linear downward shift, of course, does not take into account the well-documented non-linear shift for the lower frequencies. In addition, the general focus has been only on vowel formant frequencies; the other factors critical to speech perception are usually ignored. In any case, the techniques used for unscrambling helium speech generally fall under two main subheadings: frequency domain processing and time domain processing. In frequency domain processing, the frequency of the incoming signal is manipulated in some manner—usually after being passed through a filter system. This method includes such techniques as frequency subtraction and vocoders. In time domain processing, the incoming time-varying signal itself is processed. Time domain processing would include such techniques as analytic signal rooting, tape recorder manipulation, convolution processing and digital coding schemes (see Fig. 7).

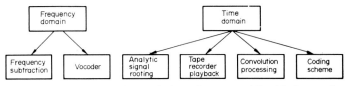

Fig. 7. Block diagram showing various techniques for unscrambling HeO$_2$ speech.

Frequency domain processing

The primary approaches to HeO$_2$ speech unscrambling used in frequency domain processing are frequency subtraction and vocoders.

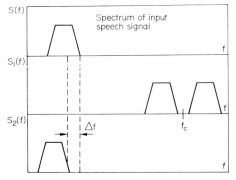

Fig. 8. Single band frequency subtraction.
  $S(f)$ = frequency spectrum of incoming signal.
  $S_1(f)$ = signal after first heterodyning operation.
  $S_2(f)$ = signal after second heterodyning operation.
  f_c = carrier frequency.

The frequency subtraction technique. When using the technique of frequency subtraction to lower formant frequency level, a fixed frequency is "subtracted" from the entire spectrum of the original speech signal. As may be seen in Fig. 8, the subtraction is generally accomplished by heterodyning a band-passed version of the incoming signal, $S(f)$, by a carrier frequency f_c, selecting one sideband and then heterodyning it down in frequency by $(f_c - \Delta f)$.

A modification of the technique is to split the incoming signal $S(f)$ into two sub-bands, $S_{11}(f)$ and $S_{22}(f)$, using bandpass filters. $S_{11}(f)$ and $S_{22}(f)$ are then heterodyned upward in frequency by a carrier frequency f_c. Subsequently, each of the sub-bands are then heterodyned down in frequency by separate tunable oscillators, $(f_c - \Delta f_1)$ and $(f_c - \Delta f_2)$, respectively (see Fig. 9).

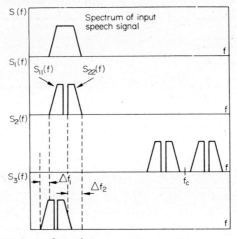

Fig. 9. Dual band frequency subtraction.

The frequency subtraction technique has been implemented by the Naval Applied Sciences Laboratory (NASL) and subsequently by Integrated Electronics, Inc.; this approach has been described by Copel (1966). Figure 10 is a block diagram of the NASL unscrambler; briefly it separates the helium speech input into sub-bands by bandpass filters a and b. The signals are then heterodyned up in frequency by a balanced modulator and the lower sideband is filtered off. At the next stage, signals are heterodyned down in frequency by two separately tunable oscillators (one for each sub-band), low-pass filtered and then mixed. By adjusting the two oscillators by means of external controls, the operator tunes for optimum intelligibility. The Integrated Electronics Corporation unscrambler (model 702A) works essentially on the same principle excepting that the lower sideband is used and the upper sideband is filtered off.

The vocoder technique. Since the HeO_2/P environment itself does not affect the fundamental frequency of diver's speech, it should be possible to use vocoder techniques to compress the speech spectral envelope.

This compression, presumably, would restore the vowel formant positions while preserving the harmonic structure of the speech signal generated. The formant restoring vocoder or FRV (see Fig. 11), developed by R. Golden (1966), compresses the spectral envelope of the helium speech by a constant scale factor. The helium speech signal is introduced into a contiguous bank of bandpass filters, each of bandwidth $k \times BW$. The outputs of each of these filters are full wave rectified and smoothed by low pass filters yielding a slow time-varying signal that is proportional to the energy in the bandpass of a given analyzing filter. These signals are used to balance modulate an excitation signal derived

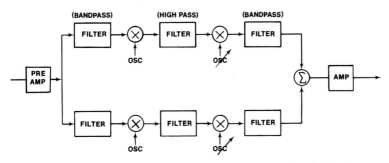

Fig. 10. Block diagram of NASL HeO$_2$ speech unscrambler (from Copel, 1966).

from the original helium speech and then resynthesized through a bank of bandpass filters, each of bandwidth BW; the centre frequencies of these filters are scaled downward by a factor of k. The vocoder technique has been implemented by a programme in an off-line mode (i.e., non-real time).

Time domain processing

Time domain processing involves direct manipulation of the time-varying speech signal. While this approach is different from frequency domain processing, the end result is the same, i.e., the lowering of formant frequencies. Some of the major methods of time domain processing, e.g., tape recorder manipulation, digital coding schemes, analytic signal rooting and convolution processing, are described below.

Tape Recorder Playback. One of the original and earliest techniques used to "unscramble" helium speech was to record it at one speed on a tape recorder and play it back at one-half that speed. By playing the HeO$_2$ speech at the slower speeds, Holywell and Harvey (1964) reported an increase in intelligibility. However, the time base of the signal is increased in proportion to the amount of speed reduction. Therefore, some off-line processing is required to restore the original time base.

A modified version of the tape recorder playback has been implemented through the use of rotating pickup heads which allow for processing to be executed in real time. Figure 12 illustrates this procedure. As shown in the figure, the loop of magnetic tape on which the helium speech sample was recorded rotates in a clockwise direction W at a speed of 15 ips, while the pickup heads rotate at $W/2$ in a counter-clockwise direction. By this means, a relative speed of 7.5 ips is maintained between the moving and the rotating heads. The results of this

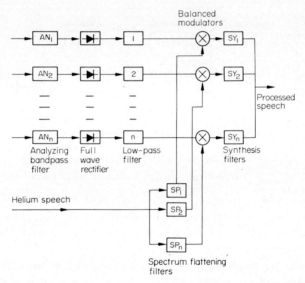

Fig. 11. Block diagram of a formant restoring vocoder—FRV (from **Golden**, 1966).

procedure is a speech output of one-half the frequency of the original speech signal.

Digital Coding. With the advent of microelectronics and the subsequent implementation of many digital logic functions in a small volume, digital realization of unscramblers has been accomplished. Recently, several companies (Raytheon, **IRPI**, Westinghouse, Singer/GP, **RELA**, **HELLE** and **DYKOR**) have designed and built unscramblers utilizing such digital circuitry. Generally, the digital coding technique discretely samples the helium speech in real time, stores the samples in a register, then reads the samples out at a slower rate while the signal is being band passed. In order to achieve proper frequency scaling and to be read out in real time, some of the information contained in the original signal is discarded.

Figure 13 is a block diagram of an unscrambler of this type—i.e., the Singer/General Precision (S/GP) unit. This system has a storage register composed of 160 condensers. As the information is being read into the storage condensers, at a rate determined by oscillator #1, it is simultaneously being read out at a slower rate, as determined by oscillator #2. This function is performed periodically in synchrony with the glottal period or at 16 milliseconds intervals when no glottal pulse is present. In the case of unvoiced sounds, the read-in and read-out are executed in the free-running mode; all storage elements are used and

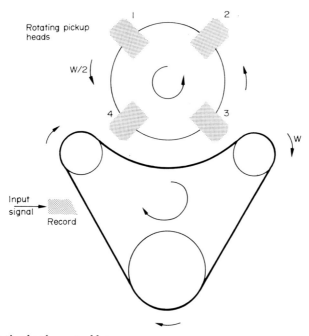

Fig. 12. Rotating head unscrambler

resetting is executed at the free-running sampling rate of 62.5 Hz (16 milliseconds). In the digital technique used here the last portion of each glottal period is ignored. This approach serves to shift the formant frequency in real time without affecting pitch repetition rate. In an earlier Singer unit, if pitch was not detected in the waveform, the unscrambler treated the signal as unvoiced. In that case, the signal completely bypassed the digital circuitry and was simply amplified to compensate for the apparent loss of consonantal energy as compared to vowel energy (see our earlier discussion on consonant-vowel ratio).

The Westinghouse Corporation unscrambler is similar to the Singer/GP unit. It is not discussed in detail because it is not currently in production and has not been available for some time. Finally, specific information is lacking on other available digital unscramblers (e.g., Raytheon, Helle, RELA and IRPI) but, from our observations, we presume that they are designed on substantially the same principles as is the Singer/GP device.

The most recent entry into the unscrambler field is DYKOR, developed at the Speech Transmission Laboratory, Royal Institute of Technology, Stockholm, Sweden. DYKOR is a time domain digital

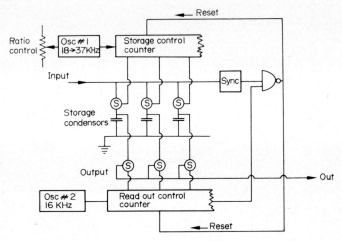

Fig. 13. Simplified block diagram of a digital helium unscrambler. Formant frequency reduction ratio is equal to the ratio of the oscillator frequencies (from Flower, 1969).

processor similar in basic nature to the other unscramblers of this type. DYKOR samples the helium speech input at a rate selected by the operator. The data stored is then read out at a 10 kHz rate. The ratio of the input sampling rate to the output reading rate determines the frequency division factor—i.e., the input frequencies are shifted down by this ratio. Further, DYKOR employs an input amplifier with variable pre-emphasis which is used to compensate for the disparity in consonant-vowel ratios or for deficiencies in microphone frequency response.

Convolution processing

The helium speech unscrambling techniques described above are in current use. We have evaluated many of them under various operational conditions; the results of these studies are reported below. However, it will probably be some time before advanced processing techniques become practical and available. One such device, the vocoder, has been briefly discussed above. Another future technique for helium speech processing which has been proposed for computer simulation is convolution processing (Moshier, 1969 and Quick, 1970).

When dealing with an helium environment, the shape of the vocal tract remains essentially as it is in air. Therefore, in the time domain the linear vocal tract transformation which is due to the helium atmosphere will differ from the normal transformation by a change in the time scale and by a constant multiplicative factor. Using these assumptions, the first step in convolution processing is to deconvolute the helium speech waveform. This is necessary in order to obtain the vocal tract impulse response function. Secondly, knowing the parameters of the transformation, one can construct its inverse. Appropriate time scaling is then performed on the constructed inverse transformation to compensate for the necessary change in the velocity of sound.

Analytic signal rooting technique

A second rather sophisticated method of helium speech unscrambling is analytic signal rooting. In this process, the speech signal is divided by filters into contiguous frequency bands spanning a suitable range. For use as an helium unscrambler, ten contiguous bands from 300 to 10 000 Hz has been proposed by Nelson (1970) who assumed that the output of each filter can be represented as a narrow band signal of the form:

$$S(t) = a(t) \cos [\phi(t)]$$

where $a(t)$ is the envelope of the signal and $\phi(t)$ is the phase. The Hilbert transform $s(t)$ is then found and from the transform one can calculate $a(t)$ and $\phi(t)$. A new function is then generated:

$$S\frac{1}{n}(t) = [a(t)]^{1/n} \cos \left[\frac{\phi(t)}{n}\right]$$

This function is equivalent to a downward scaling of the frequencies in the particular bands by $1/n$. The individual bands are then combined to produce the processed speech (for a complete description of the analytic signal rooting see Schroeder et al., 1967). This technique has not been put to use as of 1974.

Summary. Basically, helium speech unscramblers attempt to correct the upward shift of formants by bringing them back to "normal" levels. However, the correction factor used deals primarily with the linear frequency shift due to the change of the speed of sound in an HeO_2 mixture. It is our opinion that the non-linear shift of the lower frequencies has not been adequately handled in unscrambler design. Further, there have been virtually no attempts to compensate for the effect of HeO_2/P on formant transitions and consonants, two aspects of the acoustic speech signal known to be of critical importance for perception.

Although the primary approaches to helium speech unscrambling are either frequency or time domain processing, it seems probable that a hybrid approach utilizing both techniques will be used in the future. For example, with the digital time-domain technique, important transitional features of speech are probably lost due to the initial sampling. Thus, that approach to the problem is less than adequate since it does not include the processing of the consonant and transitional (from and to consonants) elements in a speech signal—nor does it improve the disparity between consonant and vowel energy. In any case, there are a number of HeO_2 speech unscramblers currently in use. Their ability to cope with the distortions produced by helium breathing mixtures and high ambient pressure will be discussed below.

C. On-Line Evaluation of HeO_2 Speech Unscramblers

The following section will present data from on-line evaluations of three unscramblers used at the Experimental Diving Unit in the U.S. Navy's Sealab programme and four unscramblers when used with five different microphones at the Westinghouse Ocean Research and Engineering Center. The first of these two on-line evaluations took place during 1969 during the Sealab-3 training dives. In order to make appropriate recordings of the divers' speech without excessive reverberation from the steel walls of the chamber, it was necessary to provide an area surrounded by acoustically absorptive materials. Since space considerations precluded the introduction of an acoustically isolating chamber into the already crowded habitat, environmental modifications were accomplished by using the fibreglass filled mattresses of a set of upper and lower bunks to form an enclosure. This configuration, with a fibreglass filled mattress placed in the rear and the talker acting as his own baffle at the front, served as a recording chamber (see again Fig. 6 for a photograph of this type of enclosure).

Special calibration procedures were carried out (for the microphone) as depicted in Fig. 14; i.e., the output from a Bruel and Kjaer beat-frequency oscillator, which was mechanically linked to a B and K graphic level recorder, was amplified to drive a 12 in. speaker in a simple baffle placed inside the chamber approximately one foot from the microphone. This position approximates that normally taken by the diver/talker; a blanket draped over the opening of the enclosure somewhat simulated his presence. The microphone output was coupled

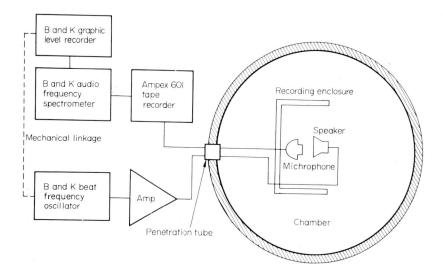

Fig. 14. Schematic of the configuration used for microphone calibration.

(via a penetration tube) through the tape recorder to a B and K audio-frequency spectrometer which was, in turn, linked to a graphic level recorder. The frequency response curves of the system—obtained (1) before the chamber was pressurized, (2) as soon as depth was reached and (3) at convenient intervals during depressurization—indicated the microphone to be responding to signals at frequencies considered adequate for studies of speech, i.e., less than 5 dB variations were obtained for frequencies through 10 kHz. Subjective evaluations of speech played through this system appeared to confirm that conclusion.

The HeO$_2$ unscramblers evaluated

The three speech processors used in this experiment were the Naval Applied Sciences Laboratory (NASL) unit and two prototype systems developed by the HRB Singer Company (HRBS) and the Westinghouse Corporation (W) respectively.

Test procedures

To be properly conducted, an assessment of the characteristics of systems such as those being evaluated, should follow a rational set of procedures; hence, we developed a rather exacting group of test protocols. They are as follows: (1) a number of talkers (three to five) were used for all systems and configurations because the use of too few

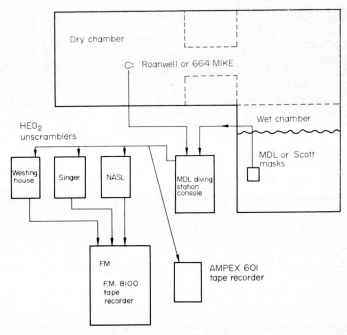

Fig. 15. Schematic of the recording array utilized. The outputs of the unscramblers were recorded on an Honeywell 8100 FM tape recorder and unprocessed speech was recorded simultaneously on an Ampex 601 tape recorder.

individuals could bias the results due to speaker variability, (2) standardized word lists (equated for difficulty) were utilized for much the same reasons, (3) all talkers read at least two word lists; they were recorded simultaneously via all equipment arrays if possible (see Fig. 15) and (4) at least 12 listeners were used to obtain the intelligibility levels.

The speech materials were recorded shortly after the diver/talkers reached saturation at a simulated depth of 600 ft. Four Navy divers read two PB_{25} word lists each (Campbell, 1965). All words were spoken in a "carrier phrase" context, i.e., "you will say . . .". ElectroVoice-

664 and Roanwell microphones were used in the chamber to transduce the speech; recordings were made on a Honeywell 8100 FM and Ampex 601 tape recorders which simultaneously recorded the unprocessed speech and the output of the three unscramblers situated on-line.

Evaluation of responses

Tape recordings of the diver/talkers' speech were spliced to allow three to 5 s intervals between words; all lists were randomized to eliminate order effects. These tapes were then played to groups of at least 12–15 semi-trained listeners; i.e., University of Florida students selected on the basis of (1) being native English speakers, (2) having normal hearing and (3) being capable of performing the required listening tasks. Before hearing the tapes, listeners were required to score at least 92% on a screening test which included 50 words from CID Auditory Word List 3-A (Hirsh recording) recorded in $+10$ dB S/N ratio of thermal noise, 25 words recorded in an HeO_2 environment, 25 words from diver communication system recordings and a 50 word CID Auditory Word List (4-A). The final 50 words constituted the screening test.

Each listener was asked to write down the words he heard as they were presented; the stimuli were the divers' word lists recorded via the output of the three unscramblers as well as in the unprocessed mode. Listener responses were scored for the number of words correct; the average percent (of correct words) for each unscrambler was then used as its overall intelligibility score.

The data for the second evaluation of HeO_2 unscramblers was collected at a depth of 600 ft during 1970 at the hyperbaric facility of the Westinghouse Ocean Research and Engineering Center, Annapolis, Maryland. The basic facility is similar to that at EDU; it consists of an entrance lock, a main chamber and a wet pot. The unscramblers tested were designed and fabricated by the (1) Industrial Research Products, Inc. (IRPI), (2) Raytheon Company (R), (3) Singer-General Precision (S/GP—a newer and different model than the previously evaluated HRB Singer) and (4) NASL. Included in this evaluation were five microphones. These are: (1) IRPI, (2) S/GP, (3) ElectroVoice 664, (4) U.S. Navy, Mark-8 and (5) U.S. Navy, Mark-11. The environmental controls, test procedures and data analysis established in our previous studies were also utilized in this evaluation.

Results of the first on-line evaluation

As stated, the units tested for this evaluation were the HRB Singer (HRBS), the Naval Applied Sciences Laboratory (NASL) and Westinghouse (W) unscramblers. Table 7 provides the mean intelligibility scores for processed and unprocessed speech as well as a comparison of the Roanwell and ElectroVoice-664 microphones when used in this environment. The Roanwell microphone was designed to be noise-cancelling, to operate under high ambient pressures and specifically to be the input transducer for the NASL unscrambler. The EV-664

TABLE 7. Comparison of mean intelligibility scores (in percent) for four diver/talkers in HeO_2 at 600 ft (182.9 m) utilizing the Roanwell and Electrovoice 664 microphones. Recordings of unprocessed speech were made simultaneously with those processed through three unscramblers situated on-line. Each diver/score is the mean of four PB_{25} lists. N = at least 20 listeners in each case.

Microphone	Unprocessed	Unscramblers		
		HRB Singer	Westinghouse	NASL
Roanwell				
Intelligibility Scores	20.0	22.7	38.5	52.2
% Improvement		13.5	92.5	161.0
EV-664				
Intelligibility Scores	7.5	10.7	27.0	27.6
% Improvement		42.7	260.0	268.0

is a laboratory quality dynamic microphone which had been especially calibrated for this research.

As shown in Table 7, mean intelligibility scores for the EV-664 range from 7.5% for unprocessed speech to 27.6% for the NASL unscrambler. Mean intelligibility scores for the Roanwell range from 20% for unprocessed speech to 52.2% for the NASL unscrambler. From the data it seems obvious that (under the operating conditions of this particular study) the limited frequency range but noise cancelling Roanwell microphone proved to be a superior input transducer to the broadband EV-664. However, a mean intelligibility level of 52% cannot be regarded as satisfactory although it begins to approach a level whereby at least some intelligible voice communication can be expected between aquanauts situated in a chamber or habitat and the support groups at the surface.

In addition to the evaluation of the above unscramblers as recorded in a dry chamber, it was judged necessary to investigate their effectiveness when coupled to various input configurations (masks) similar to those that would be used by a diver operating in the sea. Obviously, the addition of a restricted cavity to the vocal tract produces acoustic changes in the resulting utterances; so does the interface of the diver's head with the water, the regulator back pressure and so on. Accordingly, this sub-study was designed to add to the corpus of information already gathered on the unscramblers and to examine the relative effect of two available diver's masks: the Scott and the prototype MDL; both were used with the MDL microphone. Table 8 indicates that the unprocessed intelligibility for both masks is near the levels obtained in the

TABLE 8. Mean intelligibility scores in percent for the Scott and MDL Masks at 600 ft (182.9 m) recorded from the EDU wet pot. Each diver read two word lists on two different days.

Mask	Unprocessed	Unscramblers		
		HRB Singer	Westinghouse	NASL
Scott				
Intelligibility Scores	10.1	4.5	20.1	28.1
% Improvement		−55.4	99.0	178.2
MDL				
Intelligibility Scores	5.2	3.7	10.1	10.5
% Improvement		−28.8	94.2	101.9

chamber with the EV-664 microphone; the use of the Scott mask resulted in a significantly higher level of intelligibility than did the MDL mask with and without the aid of the unscramblers. Further, as stated above, the NASL unscrambler provided the greatest improvement; indeed, the performance levels among the unscramblers was significantly different with the HRB Singer prototype unit providing essentially no improvement (it is possible that there was an impedence mismatch between the microphones and the HRBS unit) and the Westinghouse unit operating nearly as efficiently as did the NASL device. However, as was the case in the dry chamber, these units did not seem to provide improvement sufficient enough to be useful in the actual diving situation.

Results of the second on-line evaluation

This particular (on-line) project was concerned with the evaluation of performance of four HeO_2 speech unscramblers when used with five different microphones. Three diver/talkers were recorded at a depth of 600 ft; it was unfortunate that all talkers did not read at least two word lists via all unscrambler/microphone combinations and each talker did not read via a specific microphone through all unscramblers and (unprocessed) simultaneously (see criterion 3 above). Therefore, the overall means presented in Table 9 are based on a varying number of scores—and ones that are sparse in some cases.

The main results of this project are apparent from examination of Table 9. First, none of the units improved speech intelligibility to levels

TABLE 9. Summary table of unscrambler evaluation. Mean scores of words correct for each HeO_2 unscrambler and microphone. Diver/talker depth was 650 ft (198.1 m) in HeO_2. There were N = 15 listeners for each PB_{25} Campbell word list.

Microphones	Unprocessed	IRPI	Unscramblers G–P Singer	Raytheon	NASL
IRPI	19.1 (6)[a]	43.4 (6)	16.8 (1)	53.4 (6)	31.8 (2)
Singer	20.2 (2)	32.4 (2)	22.3 (2)	—	—
EV-664-A	23.6 (6)[b]	50.0 (6)	27.3 (6)	—	—
EV-664-B	9.5 (6)	12.6 (6)	17.5 (6)	32.7 (3)	—
Mark-8	11.2 (4)	28.6 (4)	10.2 (4)	—	—
Mark-11	11.9 (6)	27.8 (6)	27.7 (6)	32.6 (1)	—
Category Means	15.9	32.5	20.3	39.6	31.8
Mean (all Scores)	15.7	32.5	21.5	45.1	31.8
Number talkers	30	30	25	10	2

[a] Values in parentheses are the number of word lists processed for each category. Unfortunately the engineers who made the recordings found it impossible to follow our specified protocols.
[b] The EV-664-A data were collected by the authors; the EV-664-B data by on-scene engineers.

that could be considered adequate for good communication; hence, it cannot be argued that any of them provided a realistic solution to the HeO_2/P communication problem. On the other hand, at least two of the units (Raytheon and IRPI) improved speech intelligibility by substantial amounts. Indeed, the Raytheon was the superior performer (providing nearly 200% improvement); the IRPI was second with about 100% improvement. With respect to the Singer/GP unit, it can be noted that it performed considerably better than did the attempts of the previous Singer devices but still not at the level of the other two units.

Microphone performance

The performance of five microphones when used with the four unscramblers, was evaluated. As may be seen from Table 9, the IRPI and Singer/GP transducers performed at about the same level (even though the Singer scores are based on only two talker/lists, a tentative comparison is considered to be possible). Secondly, the EV-664, the Mark-8 and the Mark-11 also performed comparably with each other, but were substantially poorer than the other two. It is suggested, therefore, that the Singer and IRPI microphones possibly will permit superior operation of HeO_2 speech unscramblers and should be considered for use in conjunction with such systems in future applications.*

Development of an off-line test

In the course of gathering basic data relative to the study of HeO_2 speech, we have collected an extremely large library of speech materials from a wide range of pressure and gas mixtures. Because of the availability of this extensive inventory of speech recorded in an HeO_2/P environment, it was possible to construct an off-line test for evaluating HeO_2 unscramblers. The advantages of an off-line test are obvious; further, such a test is particularly useful in the preliminary evaluation of an unscrambler during development—it obviates the necessity of actual chamber dives.

The basic criteria used for selecting materials for the off-line test were that they should be (1) rigorous and (2) representative of the varied conditions found in typical deep diving situations. The recordings comprising the test were selected from among "good" recordings, i.e., they were closely monitored to prevent any unrelated distortions from occurring. However, we have been careful to include as many of the actual constraints faced by the unscramblers as we were able. For example, noise is nearly always present in habitats, so "noisy" tapes were included. Moreover, a diver operates at various depths depending on the nature and reason for the dive; hence, he will be breathing various mixtures of HeO_2 through different mask and helmet configurations while using different microphones under varying conditions of ambient noise, and such conditions must be appropriately represented if a test of this nature is to be effective. In this regard, our off-line test is intentionally rigorous and favours no particular approach to the processing of HeO_2 speech. Finally, it must be remembered that the results of an unscrambler's performance become especially meaningful when they are directly compared to those from other unscramblers.

Specifically, the various word-lists chosen for the test represent as many and as varied conditions as possible and thereby permits a rigorous

---

* Several new microphones have recently become available and informal evaluations suggest that they will provide substantial improvement in HeO_2 voice communication. Included among these new units are the LTV (or Morrow, 1971) gradient microphone and one developed by NCSL.

evaluation of the performance of HeO_2 unscramblers. The criteria for the selection of the several categories were as follows:

(1) *High, medium and low intelligibility*—It is important that divers with different speaking characteristics are included in the test; one such parameter is intelligibility level. In this case individuals with high, medium and low intelligibility scores were included. (2) *Variation in depth* (i.e., HeO_2 *mixture and ambient pressure*)—Obviously, saturation diving will be carried out at a large variety of depths. Accordingly, lists read at 200 ft (61 m), 450 ft (137.2 m), 600 ft (182.9 m) and 825 ft (251.5 m) were included with the weighting of test items toward the greater depths. (3) *Noise*—A number of word lists recorded in noisy environments were selected; the rationale for including this material is that a noisy habitat is a reasonably typical situation. (4) *Last recordings before starting ascent (LBA)*—Lists included in this category reflect a diver's intelligibility after he has a chance to modify his speech during his stay in the habitat and has attempted to become more intelligible. (5) *Microphones*—Speech in the deep diving situation appears to vary as a function of the microphone used. The EV-664 and Roanwell microphones are the two which, at the time of this particular project, were most often used in saturation diving. (6) *Wet dives*—As all the above conditions occurred in a dry chamber, it was judged that an "in-the-sea" condition would add considerably to the evaluation of the performance of HeO_2 unscramblers since they will have to process divers' speech under all relevant conditions. In this case, diver/talkers wore either the Scott or MDL mask fitted with MDL microphones—and had been at depth for some time when the recordings were made. In summary, when all of the above conditions were met, it was found that a corpus of 57* separate word lists were needed in order to develop an adequate off-line unscrambler test.

D. Off-line Evaluation of HeO_2 Speech Unscramblers

Procedure

In order to conduct bench tests of HeO_2 speech unscramblers, the recorded speech materials are played back on an Ampex tape recorder whose line output fed to the input of the unscrambler being tested; the unscrambler's output is then fed into a Marantz amplifier coupled to a Marantz speaker. By maintaining this configuration for all unscramblers, we negate the effect of the response characteristics of the monitoring electronics as a variable. Further, prior to processing, an

---

* The original CSL off-line test was comprised of 49 word lists. The first off-line evaluation presented in Table 10, used only 49 lists.

attempt is made to "tune" the unscramblers according to the manufacturer's specifications. Specifically, three listeners perform a modified method of adjustment procedure in order to determine the system settings which produce the greatest intelligibility. This process is carried out by repeatedly playing a given signal while bracketing the area which gives best intelligibility. When agreement among the three listeners is reached, the unscrambler's output is recorded on a second Ampex. This bracketing technique is carried out for each unscrambler, talker and condition. Input and output levels are carefully monitored to prevent distortion of the signal.

Results of the first off-line evaluation of HeO_2 unscramblers

The CSL off-line test was first used to conduct an evaluation of the NASL, Westinghouse and the Industrial Research Products, Inc. unscramblers.† Since, in actual field use, unscramblers are often operated by less than qualified individuals, several semi-trained individuals were used to process the off-line test through the NASL and Westinghouse unscramblers; materials were obtained from the IRPI unscrambler by sending the recorded test directly to the manufacturer. The results of this effort may be seen in Table 10. The overall intelligibility of the unprocessed speech via an off-line procedure is 15.6%, the intelligibility scores for the unscramblers are: (1) Westinghouse: 13.5%, (2) NASL, 27.5%, and (3) IRPI, 33.3%. From these data it appears that the Westinghouse unscrambler, as used by less than qualified personnel, actually reduced speech intelligibility. During the previous on-line evaluation of the Westinghouse unscrambler at EDU, this unit was operated by the design and technical personnel responsible for its development—and in that case, the unit had demonstrated that it could enhance speech intelligibility. In any case, the other two units nearly doubled (NASL) or more than doubled (IRPI) the unprocessed intelligibility level; however, the improvement was not enough to insure good voice communication.

From the data presented above, it is obvious that the helium unscramblers evaluated did not provide enough speech improvement to allow for adequate diver-to-diver or diver-to-surface communication—at least, as evaluated by our own procedure. The off-line evaluation of unscramblers resulted in significantly better performance of the IRPI unscrambler than by the Westinghouse and NASL unscramblers. As indicated previously, the use of less than totally qualified personnel

† The HRB Singer experimental unit was not evaluated at this time as it was being reconstructed. On the other hand, by the time this first off-line evaluation was conducted, the Industrial Research Products, Inc. prototype unit had become available.

TABLE 10. Mean scores of words correct for seven categories (49 word lists) used to test the HeO_2 unscramblers off-line.

Depth (ft)	Depth (m)	Condition	Unprocessed Score	NASL	West	IRPI
200 ft (664 Mike)	61	High	73.5	31.6	32.4	49.1
		Med	48.4	62.4	28.0	74.3
		Low	26.6	39.7	15.8	62.7
		Noise	56.6	59.7	48.9	67.5
		LBA	67.8	55.2	42.3	63.0
Mean 200 ft			54.5	50.4	34.3	63.7
450 ft (664 Mike)	137.2	High	46.2	65.5	46.0	76.8
		Med	26.4	54.0	24.0	61.1
		Low	3.2	25.1	12.0	15.7
		Noise	17.5	38.4	17.3	34.3
		LBA	26.0	48.0	12.4	42.7
Mean 450 ft			20.0	42.0	20.1	40.1

600 ft (664 Mike)	182.9	High	8.7	23.6	8.0	46.3
		Med	6.8	21.2	10.0	30.6
		Low	1.8	11.9	3.4	18.7
		LBA	13.8	16.0	6.1	23.9
Scott Mask			9.2	16.7	10.1	21.4
MDL Mask			5.6	5.2	4.8	8.1
Roanwell			6.0	22.4	7.5	29.0
Mean 600 ft	182.9		7.1	15.2	7.1	21.3
825 ft Roanwell						
Mean 825 ft	251.5		2.7	28.4	7.5	32.9
Overall Mean			15.6	27.5	13.5	33.3

to operate the Westinghouse and NASL unscrambler could have accounted for (in part) the lower levels of improvement by these units. In any case, it is obvious that, in order to gain meaningful comparative data on unscrambler performance tested off-line, all units should be evaluated under similar and rigorously controlled conditions.

The second off-line evaluation of HeO_2 unscramblers

This section provides the results of an evaluation of a newer generation of unscramblers bench tested by personnel at the Communication Sciences Laboratory. In this study seven unscramblers were evaluated. They are the (1) Industrial Research Products, Inc. (IRPI), (2) Singer/General Precision (S/GP), (3) Raytheon (R), (4) Integrated Electronics, Corp. (IEC), (5) Helle (H), (6) RELA* Designs, Inc. (RELA), and (7) DYKOR* (D). The IEC unit is a commercial version of the NASL unscrambler.

Table 11 presents the mean intelligibility scores for the seven unscramblers tested as well as for unprocessed speech. The unprocessed score for the off-line test is 20.9%. The scores for the seven unscramblers were: 20.5% for the Raytheon; 24.2% for the IEC; 32.1% for the IRPI; 32.7% for the S/GP; 34.4% for the Helle; 38.7% for the RELA and 42% for DYKOR. From these data it is obvious that the RELA and DYKOR unscramblers performed significantly better than the rest with the Helle, S/GP and IRPI performing better than the IEC or Raytheon. However, as stated, the RELA and DYKOR devices were not tested at the Communication Sciences Laboratory under the same conditions as the other units, and this procedure may have affected the results. Further, the reversal in performance of the Raytheon unscrambler in the two evaluations was probably due to the fact that (1) the second evaluation was a different type of test and (2) the Raytheon unit tested was the same model previously evaluated whereas the other units with which it was compared were either new or improved models.

Previously, we indicated that the level of the intelligibility scores we obtained, while not satisfactory, occasionally approached a level where some intelligible voice communication could be expected. As can be noted from the table, the RELA and DYKOR especially, as well as possibly the S/GP, IRPI and Helle unscramblers (under certain conditions anyway) approach a processing level where some meaningful communication could occur. It must be stressed again that our off-line test (1) is intentionally rigorous, (2) is designed to favour no particular unit and (3) provides the most meaningful results when

---

* Test materials for the RELA and DYKOR unscramblers were obtained by sending the off-line test to the manufacturers where recordings were made by the technical personnel responsible for development of these units.

comparisons are made among units. Although a specific extrapolation is difficult and tenuous, we would predict even better performance from the DYKOR, RELA, S/GP, IRPI and Helle systems under favourable on-line situations.

E. Discussion

We have not been able to evaluate several European unscramblers. The reason for their omission is that they simply were not available or that the manufacturers refused to submit them to test. This is understandable when one considers that virtually every unscrambler manufacturer claims that his unit overcomes the problems of HeO_2/P speech.* An examination of our testing procedures and data prove otherwise. Further, discrepancies between the claims of manufacturers and the reports of in-the-field users are not due to either intentional distortions by the manufacturers or to incompetency on the part of persons attempting to use the systems. Rather, it appears that the testing of an HeO_2 unscrambler's capabilities during the design stage typically is carried out via a limited set of speech conditions and/or talkers. Rigorous and objective testing usually is not conducted under "real" conditions until a particular set of design parameters has been frozen and the unit fabricated. Moreover, the paucity of evaluations under "real" diving conditions may be understandable when one computes the cost and time involved in setting up a hyperbaric dive, especially when a large number of depths and conditions may be necessary to fully evaluate a system. Nevertheless, we strongly recommend that rigorous evaluations be carried out during all phases of system development; moreover, the use of our off-line test may prove helpful in this regard. In any case, before major improvements in unscramblers can be expected, further data is needed concerning the effects of helium and pressure on the various aspects of the acoustic speech signal to be processed. Only then will engineers and manufacturers have enough information to design devices that could completely solve the problem.

At this point it would seem profitable to delineate the weaknesses of current unscramblers and make suggestions concerning how they could be improved:

1. Primarily, all unscramblers evaluated to date exhibit somewhat lower than useful speech intelligibility scores.
2. Many units (especially the older models) operate differentially at different depths and for different talkers (even at the same depth).

---

* In fairness it must be remembered that several of the units we *have* tested are not commercially available; hence, no such claims have been made.

TABLE 11. Intelligibility levels (%) for the CSL off-line test obtained via seven HeO$_2$ unscramblers under controlled conditions. Scores have been corrected for unequal listener N's.

Depth (ft)	Depth (m)	Condition	Talker/List Combination	Unprocessed Combination Score	IEC	IRPI	Raytheon	Singer	RELA	Helle	DYKOR
200 ft (664 Mike)	61	High	1/1	73.5	35.7	64.9	55.2	60.0	68.0	41.7	64.7
		Med	1/1	48.4	62.8	72.6	62.0	74.5	75.7	88.0	89.4
		Low	2/2	26.6	38.2	44.7	34.0	51.7	59.4	64.9	67.6
		Noise	2/2	56.6	53.0	69.8	62.7	66.4	75.8	72.2	80.2
		LBA	2/2	67.8	42.1	68.3	43.4	46.9	58.5	69.8	67.7
Mean 200 ft	61			52.9	45.9	63.3	49.8	58.4	67.0	67.8	73.1
450 ft (664 Mike)	137.2	High	1/1	46.2	58.0	74.5	75.4	83.2	83.2	77.0	86.3
		Med	1/1	26.4	37.7	61.7	28.2	70.8	67.6	65.0	68.6
		Low	2/2	3.2	22.2	11.6	17.7	11.9	31.4	24.9	30.1
		Noise	2/2	17.5	28.2	31.7	16.3	41.1	41.8	36.6	52.2
		LBA	2/2	28.3	42.6	48.5	32.4	51.0	63.2	43.5	61.5
Mean 450 ft	137.2			21.3	34.9	40.2	29.2	45.8	52.4	43.3	54.8

600 ft (664 Mike)	182.9	High	1/1	8.7	23.7	34.3	14.3	27.4	32.0	29.4	39.1
		Med	1/1	6.8	21.9	39.2	11.0	29.5	44.8	34.4	40.7
		Low	2/2	1.8	9.0	21.6	9.9	17.2	22.0	15.7	18.7
		Noise	2/2	10.9	21.2	29.8	23.8	50.2	40.8	44.8	60.3
		LBA	2/2	13.8	12.4	15.4	9.4	14.0	20.6	17.5	27.2
Scot Mask			4/8	10.1	18.6	22.6	8.6	18.8	26.5	25.4	29.4
MDL Mask			3/7	5.4	9.0	11.8	6.9	7.4	12.1	11.6	12.6
Roanwell			4/6	6.0	25.4	27.2	16.8	28.5	37.4	30.4	40.7
Mean 600 ft	182.9			7.2	16.5	21.5	12.0	20.4	26.1	23.5	29.7
825 ft Roanwell Mean	251.5		4/12	2.6	22.8	29.4	14.4	34.8	43.1	33.2	41.7
Overall Mean			39/57	20.9	24.2	32.1	20.2	32.7	38.7	34.4	42.0
No. of Listeners				844	755	781	785	788	644	769	731

3. Some units require complex tuning for optimum adjustment; a time consuming and often laborious task which is a deterrent to use and may prove dangerous in an emergency.
4. Many unscramblers are fragile, bulky and/or unreliable over time.
5. Unscramblers, to date, do not take into account the differential effects of helium and pressure on the various aspects of the acoustic speech signal to be processed. This situation is not the fault of manufacturers; more research is necessary in the area of formant transitions and consonants.
6. Unscramblers do not, for the most part, attempt to compensate for the disparity in overall sound pressure level (SPL) between voiced and voiceless sounds.
7. Unscramblers usually do not take into account the relative loss of high frequency energy or the poor signal-to-noise ratio in the underwater environment.
8. Microphones and transducers currently used with unscramblers are not of sufficient quality to transmit the typical HeO_2 speech signal.

Fortunately, many of these problems are being investigated and some solutions are in sight. For example, the second generation unscramblers are easier to operate, are smaller, lighter, more rugged in their performance. Such improvements are important; we strongly recommend a continuation of this trend for third generation units. However, it is obvious that improved reliability of performance itself will not improve low speech intelligibility levels. In this regard, we present the following recommendations for the improvement of helium speech unscrambling:

1. Relevant scientific investigations must be carried out to determine the exact nature of the speech distortions caused by the HeO_2/P environment.
2. Unscramblers *must* deal, simultaneously, with both the non-linear and linear shifts in formant frequency caused by the HeO_2/P environment.
3. Unscramblers must be designed to compensate for the spectral distortions of both vowels and consonants. The acoustic nature of these two classes of speech sounds are vastly different and range from periodic signals of relatively long duration to broad band aperiodic signals of very short duration.
4. Unscramblers must be designed to compensate for the disparity in overall SPL, especially between vowels and consonants—and for the relative loss of high frequency energy found in the HeO_2/P environment.
5. Unscramblers must compensate for the poor signal-to-noise ratios common to most underwater and hyperbaric situations. Associated microphones should be noise cancelling and impervious to water or pressure effects.
6. Unscramblers must use broad band microphones and transducers which do not distort the speech signal (a minimum bandwidth of 250 Hz to 10 kHz is recommended).
7. Unscramblers should use solid state circuitry so as to improve reliability, reduce size and minimize the danger of fire in the hyperbaric chamber or habitat.

Our research in the area of speech in HeO_2/P environments continues. As indicated above, its major thrust will be concerned with the ability of the diver/talker to improve his speech intelligibility. In the near future, we plan to participate in a 1000 ft (304.8 m) dive (both wet and dry) at the Duke University Hyperbaric facility. This series of projects concentrates on (1) ways by which a diver can modify his speech in order to enhance communication and (2) the ability of a diver to utilize his feedback system (see section below on SCUBA speech) to improve the quality of his utterances. At the time of this dive, we also expect to conduct further on-line tests of the current generation of HeO_2 speech unscramblers.

4. SCUBA Speech

This portion of the chapter will focus on diver communication in shallow water (i.e., primarily compressed air) diving. Many of the research techniques and methodologies used here parallel those we have described relative to the study of diver's speech in the saturation diving milieu. Moreover, even in the SCUBA situation, knowledge concerning man's ability to use voice communication is in its early stages. Accordingly, let us reiterate a point made at the beginning of the chapter. In water, gestures and facial expressions are markedly reduced due to restricted vision; codes and writing are too limited, awkward and slow to be of much practical use. Indeed, the size of a functional lexicon for any of these approaches is too small. Therefore, if normally-paced communication is to take place underwater, speech must carry the main burden.

In order to answer fundamental questions about diver communication, it is necessary to acquire basic knowledge about the factors that limit or permit the use of speech when diving. Unfortunately, very little research has been carried out in this area; thus, in order to obtain appropriate data, we have conducted a number of projects. Our first area of investigation relates to the study of man's ability to produce intelligible speech under the constraints he encounters as a diver. That is, attempts at speech are inhibited underwater by factors such as ambient pressure, regulator back pressure, exhaust bubble noise, constriction of the articulators and the addition—and/or removal—of cavities to the vocal tract. A second area is concerned with the propagation of speech signals underwater and a third major area focuses on the analysis and appraisal of speech communication systems for the diver. Finally, a substantial number of tangential issues also have been investigated.

Research of this nature must be both basic and yet relevant to diving and divers. For example, consider the effects on speech intelligibility when an external "cavity" (such as a face mask) is added to the vocal tract; as you might expect, it will affect speech profoundly. Accordingly, we are conducting research in order to estimate the ideal size and configuration of such a cavity—both with respect to enhanced communication and life support considerations. Specifically, data abstracted from the laws governing the resonance characteristics of such cavities and from phonetic theories of speech production are being used to provide predictions of the size and configuration of the cavities optimal for intelligible speech. Later, actual prototype units can be built. These units will include the constraints imposed upon the oral cavities by the nature of respiratory physiology and the total configuration needed to provide a good seal over the facial structures. At that stage, empirical research can be utilized in order to discover which of the proposed muzzles most closely approaches the ideal. Hopefully some manufacturer would then decide to fabricate and market useable mouthcups based on our predictions. While such is not yet the case in fact, the above discussion does encapsulate one of our typical approaches to an issue in diver communication.

Before actually reviewing the state-of-the-art in this area, an additional problem should be discussed, viz., how does one go about doing underwater communication research with the same precision that a scientist enjoys when working in the usual laboratory (i.e. air) situation? Other things being equal, the greater the experimental control in research, the more rigorous and valid it is. In any case, in an attempt to achieve laboratory level precision in our research, we have developed specialized instrumentation and methodologies for our research. Some of these approaches will be described in the following section.

A. Research Environments

The nature of our underwater research often requires that most of our data be collected at various specialized facilities and in cooperation with other groups. For example, a large portion of our fresh water, near-field research in underwater speech and auditory functioning has been conducted at Bugg Springs, a field facility of the U.S. Naval Research Laboratory's Underwater Sound Reference Division; it is located near the middle of the state of Florida (see Fig. 16). The head of this spring is an elliptical cavity approximately 200 ft (61 m) by 100 ft (30.5 m) submerged in a nearly circular pool of about 400 ft (121.9 m) diameter. The side walls of the cavity drop almost vertically downward to a depth of about 175 ft (53.34 m) and the water temperature is constant throughout the year at 72°F. Although there is flow from the spring, there are no noticeable currents; ambient noise is approximately that of sea state zero and consists of wave slap, some hiss from the spring and fish sounds. Situated over the deepest point of the spring is a floating laboratory used primarily in the testing of Navy underwater sound equipment. This facility consists of a large floating barge with two laboratory rooms situated one on either side of a well through which specialized equipment can be lowered to the proper depth. The barge is kept in place by large cables extending to attachments on the surrounding shore.

A second facility we have found useful is the Naval Coastal System Laboratory's (NCSL) Stage 1 (see Fig. 17). This facility is located twelve miles from shore in the Gulf of Mexico and has been used for our research in sound propagation through water for the following reasons: (1) slope of the bottom is negligible in this area (a few feet in depth per

Fig. 16. Aerial view of Bugg Springs, a field facility of the U.S. Naval Research Laboratory's Underwater Sound Reference Division.

mile); therefore, effective angles are equal to incident angles, (2) no marked thermocline is present during most of the year, and (3) measures of sound velocity had been taken of both the water and sand bottom.

Moreover, we have used NCSL's hyperbaric chamber for studies on the effect of pressure alone on speech (reported on in a previous section) and for testing our divers for O_2 toxicity tolerance.

Other research sites that we have found useful include the Experimental Diving Unit (previously described) and such sites as the Florida Middle Ground (a region on the outer edge of the Northern Gulf Shelf), the Atlantic Undersea Test and Evaluation Center (AUTEC), Andros Island, Bahamas, and Lameshur and Buck Island Bays, U.S. Virgin Islands (during TEKTITE). These areas were chosen because they exhibited (1) good project support, (2) a compliment of experienced divers and (3) controlled or a generally protected (especially acoustically) underwater environment.

Fig. 17. U.S. Naval Coastal System Laboratory's (NCSL) Stage 1, located approximately twelve miles off-shore in the Gulf of Mexico.

B. Specialized Equipment

Development of a Diver Communication System (DICORS)

As we have stated, if rigorous underwater research is to be carried out successfully, the experimental procedures utilized must be carefully developed. For example, ordinary research on speech communication (in air) is based on a substantial variety of highly sophisticated techniques and methodologies. In an attempt to capitalize on such methodologies and minimize as many of the extraneous variables as possible, we have developed an underwater system which provides for experimental control of diver position, stimulus presentation, subject response and so on. In short, we have taken the laboratory into the water (Hollien and Thompson, 1967). The equipment configuration which we developed for this purpose and which has allowed us to carry out reasonably sophisticated communications research underwater has been named "Diver Communication Research System" (DICORS).

The overall configuration of DICORS (see Fig. 18) is that of an open framework in the shape of a truncated prism standing on one end. Its dimensions are as follows: height—80 in (203.2 cm); depth—34 in

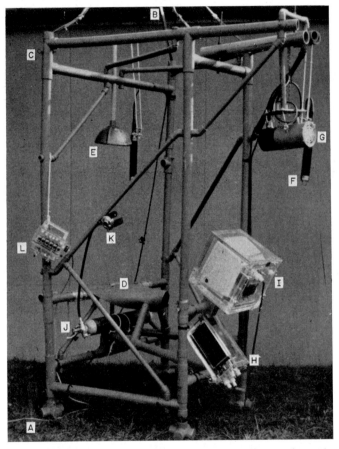

Fig. 18. Photograph of DICORS organized for studies of speech reception and speech discrimination. A—lead anchors, B—sling attached to suspension cable (not visible), C—eye bolts, D—seat, E—head positioner, F—hydrophone, G—transducer, H—TV camera, I—TV monitor, J—spare air tank and K—spare regulator.

(86.4 cm); width at the back (at diver's seat)—46 in (116.8 cm); and width at the front—22 in (55.9 cm). In addition, DICORS has a 22-in frontal extension which provides mounting for a number of items of equipment. The framework of the system is constructed of poly-vinyl chloride (PVC) tubing, a compound that is acoustically invisible underwater. The main frame consists of 1.5 in (3.8 cm) ID schedule 40 PVC tubing, the cross braces are of 3/4 in schedule 80 PVC tubing. The framework is free flooding with all potential cavities provided with air escape and water drain holes. Lead anchors (A), which have been attached to the bottom of each of the four main (vertical) structural members, provided adequate negative buoyancy to allow the unit to hang stably in the water from its sling (B) and suspension cable. The main support of DICORS, provided by nylon ropes attached to the sling, pass through the main vertical members and are secured to eye bolts within the lead anchors. In order to provide further stability and to prevent rotation of the unit on its axis, two guy wires are passed through eye bolts (C) attached to the top and bottom of the two rear vertical members. These guy wires also provide safety stops in case of the accidental release of the sling or suspension cable.

In order to control the diver's position with respect to research equipment he is situated on the seat (D) with his feet positioned either on the cross-members or hanging free. His head is placed in a positioner (E)—(three types of head positioners are used depending on the nature of the research being conducted)—and a weight belt placed across his lap, assists in holding him in the proper position. Under these conditions, not only is the diver positioned properly for the particular research procedures being employed, but he can also return readily to the same position for replications of the procedure or for other projects. Moreover, ingress and egress to DICORS are quick and simple.

DICORS was designed to allow maximum flexibility with respect to a large number of different types of studies in underwater communication and to provide the diver/subject with a relatively safe and comfortable environment. For example, the hydrophone (F) can pick up signals generated by the diver or, when placed by the head positioner we can calibrate the signal reaching the diver from the transducer (G). Further, the TV camera (H) is connected to a surface monitor (similar to the one presenting stimuli to the diver—(I) so the diver is under constant visual observation as he attends to the research tasks. In addition, an extra air tank (J) and regulator (K) are always attached to DICORS within easy reach of a subject. A final safety device is incorporated into our response systems (L)—which will be described below.

In addition to the basic DICORS configuration, we have developed several other related units—including a mini-DICORS. The mini-DICORS was designed for portability and safety when we conduct research in the open sea. It is easily transported, quickly assembled and, when surrounded on all sides by heavy-gauge welded wire, provides the diver with reasonable protection from marine predators.

Development of underwater switches

In order to record other than verbal responses from our diver/subjects, it is necessary to use some kind of response system. However, it

Fig. 19. Computer Keypunch system which records diver's responses.

Fig. 20. Two CSL data processors visually check diver's responses and maintain emergency watch. The large bulb and bell respond to a sixth position or emergency response.

became apparent very quickly that the state-of-the-art in underwater switch design was far behind our research needs. Therefore, we have had to design our own switches and, through a process of trial and error, we have developed three very reliable switching systems: (1) toggle,

(2) photocell and (3) reed. All of our units provide six possible responses; five of the responses are to stimuli choices. The sixth position is an emergency signal which, when activated, indicates that the diver/subject is experiencing some difficulty. All systems can be connected

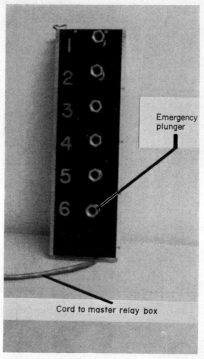

Fig. 21. Photocell response system. The sixth plunger is an emergency response.

to a visual indicator (lights) and computer card punch (Fig. 19) on the surface. When a response is made, a numbered light (1–6) indicates the subject's response. Simultaneously, a computer card is punched thereby recording the response for computer analysis. Surface observers check both responses to insure against error and as an emergency watch (see Fig. 20). Circuits are reset to an open position by pressing a reset button on the surface panel.

The earliest system developed consisted of mechanical toggle switches. These were encased in an oil-filled plexiglass box. The second system—the photocell units—also were enclosed in plexiglass (see Fig. 21). In this case, spring loaded stainless steel plungers, when depressed, interrupt a built-in beam of light, thereby causing deactivation of the sensor cell; in turn, this allows electrical activation of the relays which then trigger the desired response circuit. Our latest development in this area is a series of six independent reed switches enclosed in an hexagonal plexiglass container one-half inch thick and six inches in diameter. The circuit is closed by placing a magnetic "pen" in close proximity to a particular reed switch. Anyone wishing details concerning fabrication of these switches may have them simply by writing to us.

C. Diver/Subjects

Most of the subjects used in our underwater communications research are members of the Communication Sciences Laboratory's faculty and staff or are graduate students in our programme. Other members of our diving team are residents of the area who have developed extensive experience with our research. Therefore, we usually have available between ten and fifteen divers who can be considered trained phoneticians as well. These diver/subjects serve both as talkers and listeners for a wide variety of underwater speech studies. Occasionally, for contrastive purposes, we use either naive diver talkers and listeners or military divers.

D. Research Programmes

Propagation of sound and speech in water. 1: Filtered speech in quiet and noise

As would be expected, many aspects of speech communication, production, transmission and reception change appreciably when these processes are shifted from air to water. Essentially, the feasibility of underwater speech communication depends upon (1) the ability of man to perform in the new environment in some fashion parallel to his performance in air, (2) specialized equipment, and (3) the acoustic properties of water. Specifically, with regard to the final point, it may be said that the characteristics of the water medium suggest that the direct acoustic transmission of sound—and specifically speech—is practical. To confirm this position, it was necessary to study the transmission of underwater speech signals directly.

Our studies of the intelligibility of filtered underwater speech in quiet and noise had a primary goal of determining the minimal frequency band-pass necessary to transmit speech of adequately high intelligibility. The practical necessity for this type of information involved a number of factors including the size of underwater transducers that might be used in diver communication systems. If low frequency energy in the speech signal not necessary for adequate communication could be eliminated, the size, weight and cost of transducers could be significantly reduced. Three studies were conducted at Bugg Springs, Florida. The first study—methodological in nature, compared full range and high-pass filtered speech when (1) the signal was transmitted directly through the water and (2) when the water link was not part of the transmission line. The results (as determined by our listening sessions) showed consistency with results found for filtered and unfiltered speech in air. With our methodology established and found to be adequate, we carried out two further studies. These were (1) intelligibility of high-pass and low-pass filtered speech transduced through water and (2) effect of noise on underwater speech (both were done at Bugg Springs). In the first study the high-pass cut-off conditions were 900, 1200, 1800, 3600 and 4800 Hz; the low pass cut-off conditions were 600, 900, 1200, 1800, 2400 and 3600 Hz. The second study used the speech materials (Campbell PB_{25} lists) recorded during study 1 with the addition of two noise conditions; thermal noise (generated by a Grason-Stadler noise generator) and a composite of recorded underwater ambient noise (e.g., snapping shrimp, fish and dolphin, bubbles, sonar and propeller sounds) with a 3 dB octave slope.

The conclusions reached from both studies are that (1) speech material transduced through water behaves essentially as does similar material directly transduced and (2) in noise, speech intelligibility may be improved by reducing the signal bandwidth so that only those frequencies containing highly significant cues to word intelligibility are passed. Specifically, it is predicted that if a band-pass of from 700–3000 is used, no serious transmission degradation of the related speech intelligibility should be noticed.

Propagation of sound and speech in water. 2: Sound propagation at sea

Shallow water (the Continental Shelves) is usually considered to encompass those areas of the sea that are 600 ft (182.9 m) or less. It is in such areas (especially the very shallow water where most diving is carried out, viz., up to 70 ft (21.3 m) or 100 ft (30.5 m) that the physical characteristics of the milieu most radically depart from the basic assumption of most acoustic models, i.e., that of infinite space. Indeed, in the case of shallow water, the short distance between boundaries—surface (air) and bottom (sand, mud or rock)—results in an analysis of acoustic properties that is confounded by reflections and absorption of acoustic energy. Thus, as a result, accurate prediction of the acoustic transmission of underwater signals is not possible.

While the total problem is complex, a number of the acoustic properties of water, if studied individually, can be well understood and the characteristics of several of these may be advantageous to underwater speech communications. First, little energy is lost at any of the boundaries of water. At the interface between the water and air, for example, the magnitude of the difference in acoustic impedance prevents the transmission of sound energy from the water into the air and virtually all energy is reflected. At the lower boundary, somewhat larger percentages of the energy present in rays impinging upon the bottom actually are transmitted into the sand or rock. However, since the velocity of sound in the water is less than that in the bottom, at most incident angles, the energy is totally internally reflected. This phenomenon of sound energy being "trapped" in the water (as in a wave guide or reverberation chamber) effectively increases the transmission range of sound in this medium. In fact, the rate of attenuation due to divergence in shallow water is equal to that in air (i.e., inversely proportional to the square of the distance) only over short ranges. For longer distances, divergence is no longer spherical but is limited to horizontal spreading, and the rate of attenuation decreases to a proportion which is related simply to the inverse of the distance.

Accordingly, one of the first studies we carried out attempted to determine the magnitude of changes in sound pressure level (SPL) produced by interference patterns (phase effects) on sinusoidal* signals in shallow water. A single project was designed and carried out at Stage 1 (NCSL, Panama City, Florida) which is located 12 miles (19.3 km) off-shore in the Gulf of Mexico (Fig. 17). At this site, the bottom is hard sand and it varies only a few feet in depth per mile; measurements were taken only during relatively calm sea states. Our specific interest was to determine the feasibility of direct audio transmission of speech signals through water. Since transmission losses in shallow water for frequencies within the speech range are small, if the interference patterns caused by the phase shifts do not create large variations in SPL as a function of frequency, the transmission link would permit the use of a speech system for communication. The degree of compatibility of a speech system and the acoustic properties of shallow water can be estimated by measuring the response to sinusoidal signals covering the speech range.

Since theory would predict that signal transmission phenomena would be different near the surface or bottom than at greater differences from either boundary, three test depths were chosen; 1 m below the surface, mid-depth, and 1 m above the bottom (water depth is 32 m). The projector to hydrophone distances were 1, 16, 31, 100, 320 and 525 m.

---

* We started with sinusoids because speech waves were too complex for first approximations.

A frequency band from 100 to 5000 Hz is generally considered to more than encompass the speech range; this frequency band was swept by a tuneable oscillator and SPL was measured as a function of 18 discrete frequencies for each of nine projector-to-hydrophone configurations at each of six ranges.

The results of this project indicated that, indeed, interference patterns (represented by variations in SPL) were produced in shallow water. Additionally, the frequency regions where constructive or destructive interference patterns were located appeared somewhat random as a function of distance and transducer depths. Only two consistent patterns were found. First, when the projector was near the surface, destructive interference (reduction in SPL) was greatest at low frequencies (where low frequencies are most attenuated). Secondly, the response spectrum showed less variation when the transducers were at mid-depth or below.

Propagation of sound and speech in water. 3: speech propagation at sea

While the exact nature of the results of the previous study on underwater speech are not strikingly apparent, it is obvious that speech signals will undergo some distortion under conditions such as those studied. However, such distortion probably is not of a level that would prohibit the use of direct speech transmission as a primary approach to underwater communication. For example, if the spectrum of the pressure response to a speech signal can be varied as much as 15 dB by the transmission media without any severe loss of intelligibility, the empirical data suggest that direct speech communication is possible under a large number of practical conditions.

Accordingly, the above study was replicated for speech. Again, the project was conducted at Stage 1. The principal interest was the phase distortion effect of the ocean surface and bottom (acting as a wave guide) on the speech signal transmission. The depths and distances (projector to hydrophone) used were the same as those used previously. The speech materials transmitted were Campbell PB_{25} word lists. The results of this study indicated that, at a mid-depth range, the major degradation in speech intelligibility results from the masking effect of ambient underwater noises and *not* from phase distortion. The implication of this study suggest that direct speech transmission is a useful and feasible approach to underwater communication. However, the speech must be transduced at a high enough SPL to overcome the masking effects of the environment.

Development and standardization of speech materials for use in diver communication research

Now that we had demonstrated that speech is a viable means of communicating underwater, it became necessary to develop a corpus of speech materials that would provide a standardized and objective method for testing underwater systems and talkers. Monosyllabic word lists, equated for difficulty, have been developed for speech research in air. We felt it was necessary to validate these materials as being suitable for speech research underwater.

Two of our experiments in this area will be of interest. First we compared the underwater intelligibility scores of three standard word lists: two multiple choice discrimination tests (Black and Haagen, 1963 and Clarke, 1965) and the open-ended Campbell (1965) PB_{25} lists. Secondly, we wished to determine if alternate forms of the Campbell PB_{25} and Griffiths (1967) 50 Rhyming Minimal Contrast lists, which are equated for difficulty in air, maintain the same relationship when used in the underwater speaking situation. The advantage of utilizing alternate lists relates directly to reduction in the number of listeners required for a valid evaluation—a critical factor especially when using underwater listeners.

This particular research was conducted at Bugg Springs, Florida; the Aquasonics 420 diver communication system was utilized in transmission. In turn, an Aquasonics surface receiving unit picked up the speech;

TABLE 12. Summary of results of intelligibility tests using the Black-Haagen, Clarke and Campbell lists. Scores are expressed as a percentage of possible words correct.

List	Black–Haagen	Clarke	Campbell
Range	70.4–88.8	63.2–81.2	34.0–80.8
Mean	81.0	73.1	57.5
Median	82.9	73.2	60.4
Standard Deviation	6.8	6.3	15.0

it was recorded on an Ampex 602 recorder. Six males and four females, seated in DICORS, were diver/talkers in the experiments. Intelligibility levels were determined by means of standard listening techniques.

The results of the first study showed the Black and Haagen lists to have the highest intelligibility; they were followed by the Clarke and finally Campbell lists (see Table 12). As would be expected the closed-set lists produced scores considerably higher than did the Campbell lists. A second study was conducted. The results of this second study indicated that the Griffiths 50-word lists show intelligibility levels similar to the Black-Haagen lists and that the various forms of the Griffiths lists are equated for difficulty underwater (as they are in air). The results for the Campbell lists suggest that they are not truly equated and that talker variation may have a greater effect on open-ended listening tasks than on closed listening tasks. Our conclusions from these studies are that (1) the speech materials we use are valid in research of this nature and (2) from a preliminary analysis of phoneme type distortion, the communication/life support interfaces result in the greatest reduction in the intelligibility of fricatives (i.e.; the s, z, sh, f, v, h, th sounds). In short, by research of this type we are discovering (1) how to best test underwater systems, talkers and languages and (2) the words to avoid or change in order to obtain good intelligibility in voice communication.

Intelligibility of different types of speech modes

It has been observed that when a diver uses any underwater communication system, his attempts to speak seem to be characterized by a tendency to over-articulate, a reduction in speech rate and especially an increase in voice intensity (presumably he does these things in response to ambient pressure, constriction of the articulators and so

TABLE 13. Mean scores of words correct for each talker reading Campbell PB_{25} lists (1) twice in his usual "SCUBA" voice and (2) twice in a conversational voice. N = 10 listeners.[a]

Diver/Talker	Usual "Scuba" Speech	Conversational Type Speech
A	44.0	40.2
B	43.8	50.4
C	52.6	52.7
D	51.6	40.0
Mean	48.0	45.8

[a] All divers used an Aquasonics UO-42 system with a double-hose regulator and a Nautilus muzzle. Diver depth: 15 ft (4.6 m).

on). Since the exact nature of the intelligibility decrement in underwater speech has not yet been determined, it is possible that these diver-produced speech distortions result—in themselves—in a lowering of the speaker's intelligibility level. Thus, we were interested in discovering if different speech modes produced better or worse speech by the diver. For the first study we selected two types of speech: the usual "SCUBA" types of speech and a "conversational" speech mode.

In order to accomplish an appropriate comparison for diver talkers we used four male Experimental Phoneticians (from our corpus of divers) who were trained to be proficient with both types of speech. Word lists were read at a depth of 15 ft in Rainbow Springs (near Dunellon, Florida); the divers were situated in DICORS. Each diver was equipped with an Aquasonics 420 underwater communicator with a double-hose regulator and Bioengionics/Nautilus muzzle; responses were transduced by an Aquasonics surface unit coupled to an Ampex 601 tape recorder. Surprisingly, and as may be seen in Table 13, the usual loud, slow and over articulated "SCUBA" type speech resulted in intelligibility levels that were slightly better than those for the conversational mode. Apparently, the distortions resulting from the problems of speaking underwater in a "conversational" manner were just as severe as were those ordinarily encountered. In order to overcome the restricting and distorting effects of the various mask and muzzle straps to the articulators as well as the distorting and masking effects of regulator back pressure and ambient noise, it appears necessary to use an over-articulated, slower and louder form of speech to be intelligible. In any case, this research also led to the more sophisticated research described below.

Systematic variation of speech parameters

The results of the above study led us to investigate the effects of other types of speech modification on the speech of divers. Briefly, the purpose of this research was to advance our knowledge of what a diver could do to enhance his speech intelligibility and specifically, to discover under what conditions the highest intelligibility levels could be obtained. We also were interested in those speech sounds that are most intelligible in the underwater environments. Just as in our HeO_2 speech study of this type, in order to analyze the speech of divers producing the controlled utterances we specified, it was necessary to use only individuals who were either trained phoneticians or trained diver/talkers. Specifically, eight CSL divers read 50-word Griffiths list through an Aquasonics microphone mounted in a Bioengionics Nautilus muzzle which was hardlined to an Ampex 601 tape recorder on the surface. Each list was read in a different speech mode as follows: (1) normal articulation-most intelligible, i.e., anything the talker could do to make his speech more intelligible, (2) high and low fundamental frequency, (3) high and low vocal intensity and (4) slow and fast speaking rates.

During the recording sessions divers sat in DICORS—holding laminated word lists. Televised to a TV monitor in DICORS was the VU of a Sound Level Meter which enabled the divers to accurately control vocal intensity. Fundamental Frequency and rate were controlled by the experimenter (at the surface) by the use of flashcards televised to the TV monitor.

The results of this study confirmed some of our previous work—that is, the parameters of low f_0, slow rate of speech and high voice intensity appear to provide the greatest intelligibility. An extension of this study will contrast the above three parameters with those found to result in poorest intelligibility—high f_0, fast rate of speech and low vocal intensity; a control parameter—the diver's normal way of talking—also will be used. We hope to conduct this study shortly at the Puerto Rico International Undersea Laboratory (PRINUL). An additional project planned for PRINUL will examine the affect of stress on diver communication. This project is designed to prove a "rule of thumb" we have postulated. The rule is as follows: a diver's (externally non-redundant) message should be understood 90% of the time under ordinary working conditions—i.e., if he is swimming in the sea or communicating from an habitat. The message should be decodable 85% of the time under conditions of high stress and/or severe noise. With regard to correlation of word and message (not sentence) intelligibility, we propose that the relationship is such that if one can obtain 90% word intelligibility under experimental (i.e., optimum) conditions, about the same *message* intelligibility could be expected under operating conditions. Hence, our goal of 90% word intelligibility. In any case, planned studies will compare the intelligibility of monosyllabic word lists with 7–9 syllable messages from our diver lexicon under three diver/talker conditions: low stress, medium stress and high stress.

Speech feedback

A final study in the area of improving divers' speech deserves to be mentioned also; it is concerned with speech feedback. Again, our interest is primarily focused on developing methods that will enable a diver to enhance his speech intelligibility. Specifically, it is not likely that divers can be trained to produce effective speech patterns unless they are provided with some meaningful feedback on their performance. Obviously, the most effective feedback will be immediate and relevant, e.g., it will occur before the memory of the utterance fades and it will be a faithful reproduction of the produced phrase. In any case, we predict that an effective way in which to improve the quality of divers' speech is to present him with feedback—through the water—of his own speech production. We further suggest that, given immediate feedback of his own speech production, a diver can institute corrective compensations that will overcome the distortions inherent in an underwater speaking situation. The research was conducted at Bugg Springs and the subjects were our trained divers who had participated in many previous underwater speech studies.

In the experiment, the diver/subjects' speech production was transmitted through a hardline to the input of one Ampex tape recorder; the signal passed through the record head (by-passing the playback head) and then came into contact with the playback head of a second Ampex (by-passing its record head). The signal from the reproduce output of the second Ampex tape recorder was then sent through an amplifier to an underwater speaker. The distance used between record and playback heads produced a four-second delay between production and playback of a phrase when tape recorder speed was 7.5 i.p.s. This four-second delay was sufficient for bubble noise to fade away before the diver received his speech feedback. All divers wore Bioengionics/Nautilus muzzles equipped with Aquasonics microphones; speech materials consisted of 50-word Griffiths lists. After hearing his speech feedback each diver could, if satisfied with the intelligibility of the word just uttered, go on to the next phrase. If dissatisfied with his production, he repeated the phrase and word until he believed he could not make any further improvements.

The intelligibility of the first and last utterances of each phrase was compared. To our surprise, analysis of the data showed that no differences had occurred. However, when the experience and prior training of all the diver/talkers was taken into account we realized that they probably were performing so well at the beginning that they could not further improve their intelligibility. On the other hand, all subjects expressed the subjective impression that speech feedback helped them to become aware of the distortions in their speech output. We now plan to replicate this study using naive subjects; that is, divers who have had little or no experience communicating underwater.

The studies we have described above have clearly demonstrated that by one means or another divers can enhance the intelligibility of their underwater speech production. Additional research must be carried out in order to determine yet other ways of doing so. Further, programmes must be instituted that will train large numbers of commercial, military, scientific and recreational divers to utilize the procedures established by research in this area. Finally, there remains the need for the development of underwater communicators, muzzles, helmets and microphones that do not themselves degrade the intelligibility of the speech signal.

E. Equipment Evaluation

Microphones and mouthcups

As a correlary to the evaluation of Diver Communication Systems (discussed in next section), an extensive series of tests were carried out on the intelligibility of various underwater (diver) microphones and mouthcups. It was judged that intelligibility performance related to the microphones and mouthcups investigated would be an aid to the diving community. Moreover, no such evaluation—especially under systematically controlled conditions—had been conducted previously.

In the first phase of this research, ten microphones and mouthcup combinations were used by four diver/talkers who were submerged in a water-filled tank in a decompression chamber. Each talker read a Campbell PB_{25} word list at depths of 0, 12, 50 and 100 ft (0, 3.7, 15.2 and 30.5 m). The speech was transmitted to a tape recorder via hardline. Intelligibility scores were obtained from at least 10 listeners per word list read.

In a second project, these same units, as well as additional ones that had become available at a later time, were tested in fresh water at both shallow (12–15 ft, 3.7–4.6 m or 35 ft, 10.7 m) and moderate (100 ft, 30.5 m) depths. In both cases, 6–10 diver/talkers were utilized; they read either Campbell PB_{25} and/or Griffiths word lists. Recordings were made via hardline and intelligibility levels of all combinations of microphones and muzzles were obtained. The two masks/mouthcups that consistently showed the highest intelligibility were the Bioengionics/"Nautilus" muzzle and the Kirby-Morgan helmet; the performances of the Aquasonics and LTV microphones were best with respect to both intelligibility and reliability (see Tables 14 and 15 for detailed results of the various studies).

Intelligibility of diver communication systems

While impressive gains have been made in basic equipment for divers, systems for voice communication still remain at a relatively undeveloped level. Nevertheless, a number of underwater speech communication systems, both civilian and military, recently have become available. However, because so little is known about man's basic ability to communicate orally underwater, the design of these systems has been based almost solely upon electronic considerations: moreover, virtually no experimental evaluation (other than our past research) has been carried out on these units. The need remains, then, for an assessment of system efficiency in transmitting speech under conditions designed to duplicate actual diver-to-listener communication. Further,

TABLE 14a. Muzzles tested in the decompression chamber at MDL with various mikes at different depths. The word list read and the number of diver/talkers is shown at the right of the mean.

	Date	0 ft (0 m)	12 ft (3.7 m)	50 ft (15.2 m)	100 ft (30.5 m)
Bioengionics:					
w/Aquasonics Mike	12/68	79.1 C/4[a]	82.7 C/4	88.2 C/4	83.4 C/4
w/Aquaphone Mike	12/68	49.7 C/4	38.8 C/4	43.0 C/4	25.6 C/4
w/Bendix Mike	12/68	59.6 C/3	67.0 C/4	55.0 C/4	54.4 C/4
w/Dynamagnetic Mike	12/68	35.2 C/4	40.9 C/3	36.4 C/4	45.7 C/4
w/Yack-Yack Mike	12/68	74.1 C/4	80.0 C/4	85.0 C/4	76.2 C/4
Double Hose Bits (Standard Mouthpiece):					
w/Aquasonics Mike	12/68	53.2 C/4	61.0 C/4	67.2 C/4	53.9 C/3
MDL (Full Face Prototype):					
w/MDL Mike	9/68	35.1 C/6	18.1 C/4	39.0 C/4	52.1 C/3
Raytheon:					
w/Aquasonics Mike:	12/68	66.2 C/4	63.7 C/4	62.8 C/4	66.3 C/4
Scott Mask (Full Face):					
w/Aquasonics Mike	12/68	59.4 C/5	79.5 C/4	81.5 C/4	90.3 C/3

[a] Campbell PB_{25} word lists.

TABLE 14b. Muzzles tested in open water with various microphones at different depths. The word list read and the number of diver/talkers is shown at the right of the mean.

	Date	Location	10–15 ft (3–4.6 m)	35 ft (10.7 m)	100 ft (30.5 m)
Advanced Helmet:					
w/Advanced Mike	5/69	BS[a]	68.5 C/7	—	—
Bioengionics Muzzle:					
w/Aquas Mike	10.68	RBS[b]	66.4 C/8	—	—
w/Aquaphone	10/68	RBS	40.3 C/8	—	—
w/Bendix	10/68	RBS	48.7 C/8	—	—
w/Dynamagnetic	10/68	RBS	53.3 C/8	—	—
w/Morrow (LTV) Mike	7/71	BS	76.5 G/7[e]	75.5 G/7	80.1 G/6
w/Morrow (LTV) Mike	10/71	UFP[c]	79.1 G/8	—	—
w/Morrow (LTV) Mike	10/71	UFP	73.9 C/8	—	—
w/NSRDL	11/71	ZS[d]	68.7 G/5	71.8 G/5	65.2 G/5
w/Yack-Yack	10/68	RBS	69.3 C/8	—	—
Desco Mask:					
w/Aquas Mike	5/69	BS	25.3 C/4	—	—
Double Hose Bits:					
w/Aquas Mike	10/68	RBS	30.7 C/8	—	—
Kirby Morgan Helmet:					
w/Aquas Mike	5/69	BS	78.2 C/8	—	—
w/Aquas Mike	12/69	RBS	82.7 G/10	—	—
w/Aquas Mike	12/69	RBS	74.1 C/10	—	—
w/Aquas Mike	7/71	BS	86.2 G/6	86.6 G/6	82.2 G/5
w/Morrow (LTV) Mike	7/71	BS	85.1 G/6	83.2 G/6	—
w/Morrow (LTV) Mike	10/71	UFP	85.4 G/8	—	—
w/Morrow (LTV) Mike	10/71	UFP	70.1 C/8	—	—

MDL (Full Face Mask):			
w/MDL Mike	9/68	RBS	41.2 C/8	—
Raytheon Muzzle:			
w/Aquas Mike	10/68	RBS	42.1 C/8	—
Scott Mask (Full Face):			
w/Aquas Mike	10/68	RBS	67.8 C/8	—
U.S. Divers (Full Face Mask):			
w/Aquas Mike	5/69	BS	62.6 C/6	—

[a] BS—Bugg Springs. [b] RBS—Rainbow Springs. [c] UFP—University of Florida Pool. [d] ZS—Zuber Sink.
[e] Griffith 50 word list.

TABLE 15a. Microphones tested in the decompression chamber at MDL with various muzzles at different depths. The word list read and number of diver/talkers is shown at right of mean.

	Date	Location	0 ft (0 m)	12 ft (3.7 m)	50 ft (15.2 m)	100 ft (30.5 m)
Aquasonics Mike:						
w/Bioengionics Muzzle	12/68	MDL	79.1 C/4	82.7 C/4	88.2 C/4	83.4 C/4
w/Double Hose Bits (standard mouthpiece)	12/68	MDL	53.2 C/4	61.0 C/4	67.2 C/4	53.9 C/3
w/Raytheon Muzzle	12/68	MDL	66.2 C/4	63.7 C/4	62.8 C/4	66.3 C/4
w/Scott Mask (Full Face)	12/68	MDL	59.4 C/5	79.5 C/4	81.5 C/4	90.3 C/3
Aquaphone:						
w/Bioengionics Muzzle	12/68	MDL	49.7 C/4	38.8 C/4	43.0 C/4	25.6 C/4
Bendix:						
w/Bioengionics Muzzle	12/68	MDL	59.6 C/3	67.0 C/4	55.0 C/4	54.4 C/4
Dynamagnetic:						
w/Bioengionics Muzzle	12/68	MDL	35.2 C/4	40.9 C/3	36.4 C/4	45.7 C/4
MDL Microphone:						
w/MDL Full Face Prototype	1/70	MDL	35.1 C/6	18.1 C/4	39.0 C/4	52.1 C/3
Yack-Yack:						
w/Bioengionics Muzzle	12/68	MDL	74.1 C/4	80.0 C/4	85.0 C/4	76.2 C/4

TABLE 15b. Microphones tested in open water with various muzzles at different depths. The word list read and number of diver/talkers is shown at right of mean.

	Date	Location	10–15 ft (3–4.6 m)	35 ft (10.7 m)	100 ft (30.5 m)
Aquasonics Mike:					
w/Bioengionics Muzzle	10/68	RBS	66.4 C/8[a]	—	—
w/Desco Mask (used surface air supply)	5/69	BS	25.3 C/4	—	—
w/Double Hose Bits (standard mouthpiece)	10/68	RBS	30.7 C/8	—	—
w/Kirby Morgan Helmet	5/69	BS	78.2 C/8	—	—
w/Kirby Morgan Helmet	12/69	RBS	82.7 G/10	—	—
w/Kirby Morgan Helmet	12/69	RBS	74.1 C/10	—	—
w/Kirby Morgan Helmet	7/71	BS	86.2 G/6	86.0 G/6	—
w/Raytheon Muzzle	10/68	RBS	42.1 C/8	—	—
w/Scott Mask (full face)	10/68	RBS	67.8 C/8	—	—
U.S. Divers Full Face Mask	5/69	BS	62.6 C/6	—	—
Advanced Mike:					
w/Advanced Helmet	5/69	BS	68.5 C/7	—	—
Aquaphone:					
w/Bioengionics Muzzle	10/68	RBS	40.3 C/8	—	—
Bendix:					
w/Bioengionics Muzzle	10/68	RBS	48.7 C/8	—	—
Dynamagnetic:					
w/Bioengionics Muzzle	10/68	RBS	53.3 C/8	—	—

(Continued overleaf)

MDL Microphone:				
w/MDL Full Face	9/68	RBS	41.2 C/8	—
Morrow (LTV) Mike:				
w/Kirby Morgan Helmet	7/71	BS	85.1 G/6	83.2 G/6
w/Kirby Morgan Helmet	10/71	UFP	85.4 G/8	—
w/Kirby Morgan Helmet	10/71	UFP	70.1 C/8	—
w/Bioengionics Muzzle	7/71	BS	76.5 G/7	75.5 G/7
w/Bioengionics Muzzle	10/71	UFP	79.1 G/8	—
w/Bioengionics Muzzle	10/71	UFP	73.9 C/8	—
CSL Mike:				
w/Bioengionics Muzzle	11/71	ZS	68.7 G/5	71.8 G/5
Yack-Yack:				
w/Bioengionics Muzzle	10/68	RBS	69.3 C/8	—

[a] C = Campbell PB$_{25}$ lists; G = Griffiths lists.

since the available units utilize radically different means of transmitting speech, a comparison of the intelligibility levels of the selected systems would seem to be of particular interest.

In this regard, we have developed methodologies for evaluating underwater communication systems under diver-to-surface and diver-to-diver conditions. The results of these studies will allow individuals who need such systems to make selections based on at least some objective evidence and moreover, it will provide the firms designing and fabricating these units feedback of information in order that they may effect system improvements.

All approaches for the evaluation of underwater communication systems are designed to provide the most thorough and rigorously controlled *in situ* evaluation possible. They are as follows: (1) Near-field,* fresh water evaluations which include (a) a diver-to-surface procedure with the pickup transducers 3–10 ft (1–3 m) from the projector and (b) a diver-to-diver procedure at a distance of 30 ft (9.1 m); (2) Variable distance, off shore, saltwater evaluations which include (a) a diver-to-surface procedure with pickup transducers 50–4000 ft (15.2–1219 m) from the projector and (b) a shallow water diver-to-diver procedure at distances of 50–500 ft (15.2–152.4 m), and (3) Variable (medium) distance, harbour, saltwater, evaluations which include (a) a diver-to-surface procedure with pickup transducers 50–2000 ft (15.2–609.6 m) from the projector, and (b) a diver-to-diver procedure at distances of 50–500 ft (15.2–152.4 m).

To date we have completed investigations 1a, 1b and 2a; when all six investigations are complete, the obtained information should permit a simplified—and much shorter—set of evaluative procedures to be constructed. That is, only those techniques that are necessary to provide a good profile of the system would be retained, where possible they would be combined, and various combinations of the procedures could be used for evaluations conducted for special purposes. Thus, a standardized, yet viable and reasonably efficient, method of evaluating diver communication systems should result. Moreover, as these experiments are carried out, considerable information will accrue concerning the effectiveness of those systems that are available to divers at a given time. In any case, we can provide the diver with considerable information already about different communication systems when used under various conditions.

Because a large part of this programme involves standardized procedures which we have developed in order to permit the conduct of

---

* The term "near-field" as used in this chapter refers only to short ranges and *not* to the relation between the displacement and pressure phenomenon.

impartial but rigorous tests, the details of our approach are provided before we present the results of any of our completed studies.

F. Evaluation Procedure

Diver-to-surface: 1

For all research of this type, selected diver/subjects read standard intelligibility test materials through each of the devices. These samples are then presented to listeners in order to obtain intelligibility information for each of the communication systems being evaluated. Three basic types of communication systems have been evaluated; the first group consists of acoustic systems. An "acoustic" system (A type) includes a microphone, amplifier, power supply and transducer; it

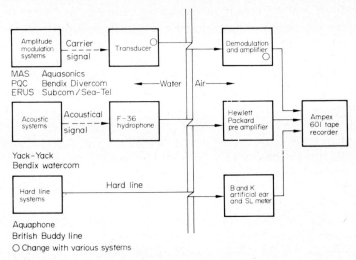

Fig. 22. Schematic drawing of the diver-to-surface protocols.

characteristically transduces speech directly into the water by means of the projector (underwater loudspeaker). The signal produced can be received by a hydrophone placed in the water, or by divers without any special receiving equipment. The second group of communicators consists of amplitude modulated (AM type) systems. In an AM type system, a carrier wave is utilized and is modulated by the speech signal. Such a system ordinarily consists of a microphone, power supply, amplifier, modulator and underwater transducer. Speech produced in this manner can be understood only by a diver or a surface observer having an appropriate receiver and demodulator. The third group of communicators, "hard line" (HL type) systems, employ a closed approach, comparable to a telephone, including (a) a microphone and

(b) a receiver. These systems require a physical connection (i.e., a cable) between the talker and the listener.

The stimuli and recording procedures have been standardized for our research programme. Diver/talkers are assigned appropriate word

Fig. 23. Photograph of a CSL diver, with the ERUS AM type system, settling into position in mini-DICORS for a diver-to-surface recording. The site is Buck Island, U.S. Virgin Islands.

lists equated for difficulty to be read for each communicator configuration. After familiarizing themselves with the task, they read their list to a trained listener who corrects any pronunciation errors. After approval of their reading, the divers descend to DICORS and activate the communication system undergoing evaluation. The stimulus words are usually televised via a TV camera at the surface to a TV monitor screen housed in DICORS. Divers read all stimuli, preceding each word with the phrase, "You will say . . .".

Recording procedures necessarily vary somewhat for the different systems. With the acoustic systems the signal is recorded by means of an F-36 hydrophone coupled through an amplifier to an Ampex 601 tape recorder. The modulated systems employ surface units whose output is coupled directly to the tape recorder (see Fig. 22 for a schematic drawing of the diver-to-surface protocols). All recordings are made at approximately the same VU level. It should be noted that each unit is evaluated in a manner based on the protocols specified by the manufacturer. In fact, many firms provide a company engineer to be on site during at least part of a project experiment.

Diver-to-surface: 2

This study differed from the previous one in that it was done in salt water. The purpose was to gain information on how various underwater communication systems would perform over distance in a typical off-shore environment with all the attendant problems arising from the ambient noise level, the corrosive effects of the salt water medium and so on.

The site chosen for this study was Buck Island, U.S. Virgin Islands. Not only did the bay provide protection from rough seas but it had a rubble bottom which, along with a slight surface chop, provided a typical offshore environment. The data gathering procedures were similar to those described above with two exceptions: (1) mini-DICORS was used and (2) the hydrophone was situated at distances of 50, 250, 500, 1000, 2000 and 4000 ft (15.2, 76.2, 152.4, 304.8, 609.6 and 1219.2 m) from the diver/talkers. Both the talker and pickup were always situated half-way between the bottom and surface (see Fig. 23).

Diver-to-diver

For this procedure, selected diver/talkers read standard intelligibility test materials through each of the devices; diver/listeners (in pairs) situated 30 to 500 ft (9.1 to 152.4 m) (depending on the study) from the talkers responded to multiple choice sets of words (projected on a TV monitor) by means of underwater switches. Responses were recorded on the surface both manually and on IBM cards. An overview of the procedures described in the following section may be seen in Fig. 24. Specifically, a pair of DICORS were placed back-to-back at varying distances from each other (see again Fig. 24); the distances varied between 30 and 500 ft. For the diver-to-diver evaluations the equipment mounted on DICORS-T (the talker unit) consisted of an underwater television monitor, a TV camera to monitor the diver, and a set of stimulus indicator lights; for DICORS-L (the listener unit), the equipment included a TV monitor and two response switches.

The stimulus/response procedure involved a diver/talker and two diver/listeners equipped with appropriate units descending to DICORS-T and DICORS-L respectively. When ready, the first set of words appeared on both TV monitors simultaneously. The talker was cued to say the proper word by a panel of lights wherein a numbered bulb

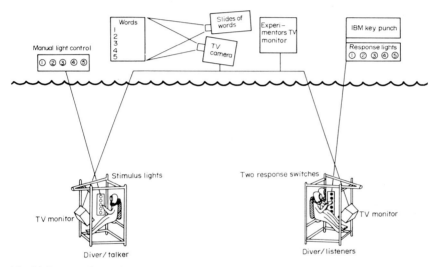

Fig. 24. Presentation and response apparatus for a diver-to-diver evaluation. Both talkers and listeners (in pairs) were situated in DICORS 30 ft apart in 30 ft of fresh water. Five multiple choice words were displayed by slide projector and transmitted via television camera to television monitors observable by experimenters, talkers and listeners. Talker was instructed to speak appropriate stimulus word by set of five manually controlled stimulus lights; listeners recorded their responses by activating switches coupled to IBM keypunch and sets of response lights.

corresponding to the word to be spoken was illuminated manually by the experimenter. Since all divers held their breath when they saw the set of five words appear, this stimulus light was turned on about two seconds after the word-set was transmitted to the TV monitor; such delay allowed a reasonable reduction of bubble noise. The diver/talker then spoke the carrier phrase, "You will say, . . .", followed by the stimulus word (the carrier phrase was used to capture the attention of the diver/listeners). The listeners then made their selection of the word they believed that they heard and activated the similarly numbered switch among the set of five available to them; their responses were recorded both manually and on IBM cards at the surface. Once the responses to a stimulus word were recorded, the next set of words

TABLE 16. Intelligibility levels, in rank order, of the diver communication systems evaluated. All scores are based on at least 10 listeners per talker.

Communication Systems	Number of Diver/ Talkers	Intelligibility Raw Scores	Corrected For Talkers Differences
BENDIX DIVERCOM (double hose regulator; Nautilus muzzle)	9	59.5	57.7
AQUASONICS 811 (double hose regulator; Nautilus muzzle)	7	56.5	53.8
AQUASONICS 420 (A) (double hose regulator; Nautilus muzzle)	12	52.3	52.3
AQUASONICS 420 (B) (single hose regulator; Nautilus muzzle)	12	46.4	46.4
PQC-3 (military) (double hose regulator; Nautilus muzzle)	9	42.5	41.0
BRITISH BUDDY LINE (double hose regulator; Nautilus muzzle; microphone placed on forehead)	4	40.2	35.9
RAYTHEON YACK-YACK (B) (double hose regulator; Nautilus muzzle)	5	37.2	33.2
RAYTHEON YACK-YACK (A) (single hose regulator; Raytheon muzzle)	12	22.9	22.9
SEA-TEL (prototype Canadian mfg.) (double hose regulator; Nautilus muzzle)	8	22.8	23.3
ERUS-2-3A (French mfg.) (double hose regulator; U.S. Diver's prototype full-face mask)	9	18.9	18.9
PQC-2 (military) (double hose regulator; Nautilus muzzle)	6	17.2	16.6
AQUAPHONE (A) (inexpensive model) (single hose regulator; Raytheon muzzle)	12	15.4	15.4
BENDIX WATERCOM (not in production) (double hose regulator; Bencom muzzle)	7	12.3	16.4
MAS (prototype) (not in production) (single hose regulator; modified Desco mask)	5	10.9	9.7
AQUAPHONE (B) (inexpensive model) (single hose regulator; standard mouthpiece)	12	6.5	6.5
PQC-1a (B) (military) (double hose regulator; Nautilus muzzle)	9	2.6	2.3
PQC-1a (A) (military) (double hose regulator; Scott mask)	9	1.0	0.9

were transmitted with information to the talker regarding the proper selection; the entire process was repeated until all 50 words were completed. For any particular run, the talker, communication system, list randomization and pair of listeners were randomly drawn—excepting that no listener heard any talker more than once on a particular unit.

G. Results of Diver Equipment Evaluation

Diver-to-surface: 1

Table 16 provides a compilation of the mean intelligibility scores for the underwater communication systems evaluated diver-to-surface in Bugg Springs. Under the conditions of the research, the most significant finding was that none of the systems evaluated were capable of transmitting speech at a high level of intelligibility. Although three different systems (i.e., A type, AM type and HL type) are represented in the seven configurations exhibiting highest intelligibility, it is the Bioengionics/Nautilus muzzle that is common to each. As we indicated above, the Bioengionics/Nautilus muzzle was superior to most other available mask/muzzle configurations. Another although secondary effect on intelligibility appears to be the use of a double hose as opposed to a single hose regulator. Note that the score of the Aquasonics 420 when used with the double hose regulator increased from 46.4% to 52.3%. The improvement in speech intelligibility may be due to less back pressure and/or more favourable microphone placement within the muzzle. The highest score—59.5% was achieved by the Bendix Divercom. This unit, which was evaluated at a somewhat later date than the other devices, is a prototype model which is now in production.

Diver-to-surface: 2

Table 17 presents intelligibility scores for the units tested over distance in salt water. Of course, the acoustic units were not expected to transmit over the entire range. All units were tested with a double hose regulator and Bioengionics muzzle with the exception of the Subcom device which incorporates the Kirby Morgan helmet. As can be seen, our modification of the Bendix Watercom (an A type—see Fig. 25) provides greatest intelligibility up to 250 ft (76.2 m), due perhaps to these design modifications and use of the B/N muzzle.

The Aquasonics 420 proved to provide greatest intelligibility over the entire range but is certainly not adequate for satisfactory communication. The relatively low and possibly adventitious score for the Aquasonic unit at 2000 ft (609.6 m) may have been due to the effects of the salt water environment; in any case, the mitigating factors are unknown. However, all units were constantly maintained as well as possible during the test period by CSL engineers familiar with their characteristics and operation. From Tables 16, 17 and 18 it is apparent that the PQC-2 (as well as the PQC-1a and PQC-3) is an erratic communicator requiring constant maintenance.

Fig. 25. Photograph of a CSL diver preparing for a diver-to-surface evaluation of the Bendix Watercom. Note the pressure compensation hose leading from the Watercom to the low pressure first stage port on the diver's regulator.

Diver-to-diver

Table 18 presents intelligibility scores for those units evaluated in a diver-to-diver mode at Bugg Springs. As can be seen from the table, scores are generally higher than are those found for the diver-to-surface evaluation. There are two explanations for this. These are: (1) Experienced talkers *and* listeners were used; the listeners were familiar

TABLE 17. Mean intelligibility scores for five diver/talkers at approximately 12 ft (3.7 m) depth over six ranges (distances) using nine underwater communication systems. Study was done at Buck Island (Virgin Islands) with diver/talkers reading Griffith word lists. All scores corrected for unequal listener N's.

Communication System	Range (Distance)					
	50 ft (15.2 m)	250 ft (76.2 m)	500 ft (152 m)	1000 ft (305 m)	2000 ft (610 m)	4000 ft (1219 m)
Bendix Watercom (modified)	65.4	48.8	47.7	—	—	—
Subcom/Inter. Rg. (300'–500')	50.0	40.1	28.5	—	—	—
Aquasonics 420	49.0	48.4	44.9	39.0	28.9	34.9
Erus	49.0	46.5	51.4	38.7	35.9	23.5
Yack-Yack	43.2	31.1	—	—	—	—
PQC-2[a]	43.1	44.4	34.5	40.7	39.3	31.9
Subcom/Short Range (50')	33.3	—	—	—	—	—
ScubaCom	28.9	—	—	—	—	—

[a] Both the Subcom and the Aquasonics 811 Surface Units were used.

TABLE 18. Intelligibility scores for eight[a] diver communication systems evaluated by a diver-to-diver technique. Five experienced talkers were utilized; each wore the communicator in conjunction with a double hose regulator and a Bioengionics (Nautilus) muzzle. Speech materials were the Clarke 50 work multiple-choice lists; the project was conducted in 30 ft (9.1 m) of fresh water at Bugg Springs, Florida.

Communication System	Number of Listeners	% Intelligibility
Aquasonics 811	34	72.4
Bendix Watercom	35	64.1
Aquasonics 420	36	62.9
Bendix Divercom[b]	16	58.7
Aquaphone	36	58.1
Raytheon Yack-Yack	36	57.4
PQC-2	34	54.3
PQC-3[b]	13	26.6

[a] Evaluation of the PQC-la was attempted also; unfortunately the available units could not be kept operational. The little data obtained for the PQC-la suggested an intelligibility level substantially below those of the other units.

[b] Partial test only.

with the talkers' speech modes; (2) a closed lexicon was used; the words were presented in a multiple-choice paradigm and were generally familiar to both the talkers and the listeners. In short, all possible procedures were employed that would operate to force the obtained intelligibility levels of these units as high as possible within the other constraints of this type of research. Under these conditions, it was judged reasonable to assign a level of 85% intelligibility as one which, if attained, would be indicative of adequate performance of a given system. An examination of Table 18 shows that no system reached the 85% target. Clearly, then, it must be assumed that satisfactory voice communication among divers is not yet possible. This is true even though the systems, for the most part, are reasonably well designed and the very best (for speech) available life support systems (i.e., double hose regulators and B/N muzzles) were used.

As a summary and conclusion to this section on equipment evaluation we can say with confidence that our methodologies for diver-to-surface and diver-to-diver communication evaluation are both objective and effective. No particular approach (i.e., A, AM, HL) to diver communication appears to be dominant with respect to intelligibility at this time. Further, none of the diver communication systems evaluated provide acceptable levels of speech intelligibility.

H. Diver Communication Performance during Work Tasks

Although we have shown in the above section that no diver communication system as yet provides an acceptable level of communication, some are beginning to approach levels where some reasonable (if minimal) amount of information can be transmitted. However, the need for more effective diver communication is growing more acute as man's involvement with inner space becomes increasingly complex in nature. Divers will no longer be able to function effectively as separate and isolated members of a team. Rather, their work will require a cooperative endeavour that will depend on their ability to communicate with each other. In order to study the abilities and problems of a diving team engaged in an underwater task requiring collaborative activities, we have initiated a series of "work" studies. These investigations attempt to objectively examine and quantify communication problems inherent in an underwater work situation and to suggest methods of training divers to effectively use underwater communication systems. Two of these studies has been completed; a short review follows:

a. Diver Performance With and Without Communication. This project, completed in Lameshur Bay, Virgin Islands using TEKTITE-2 and CSL divers, was conducted in an effort to study and quantify the communication problems inherent in underwater work situations. It is obvious that divers experience extreme difficulty communicating via voice in an underwater milieu and, from the results of this study, some of the reasons for such difficulty became apparent.

Specifically, the investigation was designed to determine whether a team of divers, aided by underwater communication systems, could

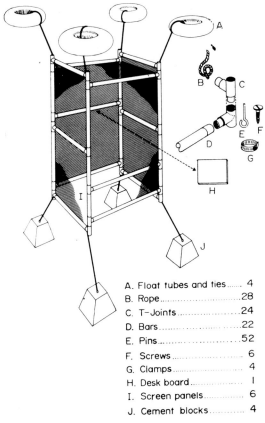

A. Float tubes and ties 4
B. Rope 28
C. T-Joints 24
D. Bars 22
E. Pins 52
F. Screws 6
G. Clamps 4
H. Desk board 1
I. Screen panels 6
J. Cement blocks 4

Fig. 26. Schematic drawing of the mini-DICORS used in an evaluation of diver work performance with and without communication. The various parts required for assembly are shown.

accomplish a complex construction task (requiring a high degree of cooperation among them) more effectively than could a matched team without communication. Two four-member teams equated in ability, but unused to working together or using communication systems, served as subjects. The work task required the assembly of a mini-DICORS (see Fig. 26) consisting of 132 parts (PVC tubing, nuts, eyebolts, cotter pins, ballast material, nylon rope and inner tubes) in 20 ft of water under conditions of low visibility; four separate units were employed for this purpose. The experiment proper was not initiated until a short general briefing was conducted by the experimenters and a short group meeting was held by the team. On the first trial, members of Team A wore Aquasonics 420 units; Team B had no communication gear. On the second trial (a day later), the communicators were worn by Team B; Team A had no communication aids. Time of assembly, parts not used or lost, and number and type of errors constituted the objective measures. Once these data had been obtained and weighted—and adjustments made for general team performance (one team proved to be more skilful than the other)—a performance index was calculated. In summary, not only was the index greater when the teams were not using the communication systems, but even when the paired trials were considered, the performance of the team *without* the communication gear was superior in both cases. Therefore, we concluded that to be aided by communication systems (1) divers must be extensively trained in their use, (2) they must be allowed to develop appropriate communication procedures and (3) systems with better intelligibility than those currently available should be used.

Our experience with this project led us to formulate a series of work-task studies, all similar in nature. The parameter differentiating these experiments is to be the amount of training given the divers in the use of a communication system. In the first experiment the divers were provided with a communication system; its basic nature was demonstrated and protocols for its use were reviewed (minimal training). A further study will present each team of divers with an extensive briefing on ways of using the particular system to best advantage. Following this briefing, a demonstration of the units in operation will be provided. In the final investigation the divers will use the system only after extensive training, i.e., training in the system's transmit and receive modes (in the water) will be conducted under two conditions: (1) with rigid transmission and reception protocols and (2) as circumstances dictate—until a predetermined level of communicative adequacy by the divers has been achieved.

All diving teams will be made up of three to four divers. Each team will be composed of individuals who have had no experience working underwater together (intra-group heterogeneity) but all teams will be matched for diving skill and ability (inter-team homogeneity).

The first of these studies was conducted in conjunction with Florida Aquanaut Research Expedition (**FLARE**) sponsored by the National Oceanic and Atmospheric Administration—Man Undersea Science and Technology (**NOAA-MUST**) programme. It was conducted in 20 ft

(6.1 m) of low visibility, relatively cold sea water. Unfortunately, the data from this study have not yet been processed and the results are not available at this time.

5. Summary

As may be seen from the discussions above, substantial advances are currently being made with respect to improved diver communication. Obviously, viable systems and approaches are not available at the moment but there is every reason to believe that they will be in the near future—at least if the present (or an increased) level of research can be maintained. Hence, the situation in this regard looks hopeful.

However, there is another perhaps equally important consideration. Many commercial and military divers still feel that underwater communications are unnecessary—at least for the jobs that they are required to do. Indeed, with respect to the demands being made on them, there is some justification for their position in this regard. For example, many diving operations are still organized in a manner that (1) either precludes the possibility of voice communication or (2) voice communication would not be needed even if it were available. In some cases the task is practiced to such an extent on land (or is such a simple one) that the diver or divers can do it in a virtually automatic way. In other cases the diver is expected to carry out only a small part of the task on a single dive.

It is becoming more apparent that dives such as these are being replaced by increasingly complex ones, e.g., the divers may be saturated, work for longer periods and at more complex tasks, or work out of a habitat. Moreover, as divers operate for longer periods and at greater depths, the environment develops an increasingly more dangerous situation. By necessity such operations require a great deal more co-operation and exchange of information among the divers and between them and surface support personnel than did those discussed above. In addition, as more significant and complex projects are being taken on, the cost effectiveness of the operations is becoming somewhat mor unfavourable. Improved diver communication will serve to markedly improve this situation. Indeed, in some instances it could shift a project from one that is unrealistic to one that is possible to conduct on the basis of reduced costs alone.

Accordingly, we strongly urge continued research in this area, research that will lead to the improvement of diver communication systems and procedures. Some of the necessary research is currently underway at our laboratory and elsewhere. For example, we are studying the design of mouthcups, muzzles and diving helmets in an attempt to optimize their design for both speech and life support considerations. Moreover, we are continuing our research efforts designed to determine ways a diver can enhance his own voice communication both in shallow water and under HeO_2 conditions. Finally, regardless of the variety or quality of available diver communication systems, they will be of little benefit if the diver is not trained to use them properly. Various approaches to this issue are currently underway. Moreover, our group, along with the U.S. Navy, NOAA's Manned Underseas Activities programme and the State University System of Florida, has initiated the Scientist-in-the-Sea (SITS) programme—a thrust designed to train young marine scientists of great potential to do research in the sea using the most modern equipment available. In short, the research currently being carried out here and at other laboratories will help the scientific, technical and military aquanauts of the future to realize the potential richness and utility of inner space.

Acknowledgements

This research was primarily supported by the Physiological Psychology and Engineering Psychology Programs of the Office of Naval Research and by the National Institutes of Health. Many people have contributed significantly to the program. In particular, we wish to thank the following: J. Bedingfield, R. F. Coleman, E. T. Doherty, T. Giordano, G. Fant, P. Hollien, J. Malone, T. Murray, P. B. Phillips, C. Slater, C. L. Thompson, G. Tolhurst, Cdr. J. M. Tomsky USN (ret.). The authors also wish to thank the U.S. Navy Experimental Diving Unit, Naval Coastal Systems Laboratory, the Underwater Sound Reference Division of the Naval Research Laboratory and the manufacturers of the many underwater communication systems and helium unscramblers for their kind cooperation.

References

Beil, R. C. (1962). Frequency Analysis of Vowels Produced in a Helium-Rich Atmosphere, *J. Acoust. Soc. Amer.* **34,** 347–349.
Black, J. W. and Haagen, C. H. (1963). Multiple-Choice Intelligibility Tests, Forms A and B, *J. Speech Hearing Disorders* **28,** 77–86.
Brubaker, R. S. and Wurst, J. W. (1968). Spectrographic Analysis of Diver's Speech During Decompression, *J. Acoust. Soc. Amer.* **43,** 798–802.
Campbell, R. A. (1965). Discrimination Test Word Difficulty, *J. Speech Hearing Res.* **8,** 13–22.
Clarke, F. R. (1965). Technique for Evaluation of Speech Systems, *Stanford Research Institute's Final Report* 5090, U.S. Army Electronics Laboratory, Contract DA 28–043 AMC-00227 (E), August, 1965, 1–65.
Copel, H. (1966). Helium Voice Unscrambling, *IEEE Transaction on Audio and Electro-acoustics*, AU-14, #3, 122–126.
Delattre, P. C., Liberman, A. M. and Cooper, F. S. (1965) Acoustic Loci and Transitional Cues for Consonants, *J. Acoust. Soc. Amer.* **27,** 769–773.
Fairbanks, G. and Grubb, P. (1961). A Psychophysical Investigation of Vowel Formants, *J. Speech Hearing Res.* **4,** 203–219.

Fant, G. and Sonesson, B. (1964). Speech at High Ambient Air Pressure, *Q. Progr. Rep. Speech Transmission Laboratory, Stockholm*, STL-QPSR, 9–21.

Fant, G. and Lindquist, J. (1968). Pressure and Gas Mixture Effects on Diver's Speech, *Quarterly Progress Report of the Speech Transmission Laboratory, Stockholm*, STL-QPSR 1–17.

Fant, G., Lindquist, J., Sonesson, B. and Hollien, H. (1971). Speech Distortion at High Pressure, *In:* Underwater Physiology. Proceedings of the Fourth International Congress of Diver Physiology, Academic Press, New York and London, 293–299.

Flower, R. A. (1969). Helium Speech Investigations, *Clearinghouse for Federal Scientific and Technical Information Report* # *AD* 702119.

Giordano, T., Rothman, H. B. and Hollien, H. (1973). Helium Speech Unscramblers—A Critical Review of the State of the Art, *Institute of Electrical and Electronic Engineers Transactions on Audioacoustics Technology*, AU-21, 436–444.

Golden, R. M. (1966). Improving Naturalness and Intelligibility of Helium-Oxygen Speech, Using Vocoder Techniques, *J. Acoust. Soc. Amer.*, **40**, 621–624.

Griffiths, J. D. (1967). Rhyming Minimal Contrasts: A Simplified Diagnostic Articulation Test, *J. Acoust. Soc. Amer.* **42**, 236–241.

Hollien, H. and Thompson, C. L. (1967). A Diver Communication Research System (DICORS), *CSL/ONR Progress Report No. 2*, Office of Naval Research, Physiological Psychology Branch, Grant Nonr 580 (20), January 15, 1967, 1–8 (AD 648–935).

Hollien, H. and Tolhurst, G. (1969). A Research Program in Diver Communication, *Naval Res. Rev.* **22**, 1–13.

Hollien, H., Coleman, R. F. and Rothman, H. B. (1970). Further Evaluation of Diver Communication Systems, *Proceedings of the Institute of Electrical and Electronic Engineers International Conference on Engineering in the Ocean Environment*, 1–IEEE Cat. No. 70 C 38–OCC, 34–36.

Hollien, H., Coleman, R. F., Thompson, C. L. and Hunter, K. (1970). Evaluation of Diver Communication Systems Under Controlled Conditions, "Undersea Technology Handbook", Compass Publications, Arlington, Virginia.

Hollien, H. and Rothman, H. B. (1971). *Chapter* 10: Studies of Diver Communication and Retrieval, The University of Florida COM-EX 2 Program, *Tektite* 11, *Scientist-in-the-Sea Report*, U.S. Government Printing Office, 2400–0682, Washington, D.C.

Hollien, H., Coleman, R. F. and Rothman, H. B. (1971). Evaluation of Diver Communication Systems by a Diver-to-Diver Technique, *Institute of Electrical and Electronic Engineers Transactions on Communication Technology*, **19**, 403–409.

Hollien, H. and Rothman, H. B. (1972). Evaluation of HeO_2 Unscramblers Under Controlled Conditions, *CSL/ONR Technical Report* #46, Office of Naval Research, Engineering Psychology Programs, Arlington, Va., July, 1972 (AD 748295).

Hollien, H., Thompson, C. L. and Cannon, B. (1973). Speech Intelligibility as a Function of Ambient Pressure and HeO_2 Atmosphere, *Aerospace Medicine* **44**, 249–253.

Hollien, H., Rothman, H. B., Feinstein, S. F. and Hollien, P. (1973). Speech Characteristics of Divers in HeO_2 Breathing Mixtures at High Pressures, *Proc. 5th Symp. Underwater Physiol.*, in press.

Hollien, P. and Hollien, H. (1972). Speech Disorders Created by Helium/Oxygen Breathing Mixtures in Deep Diving, *Proc. XV World Cong. Phoniatrics Logopedics*, Casa Ares, Buenos Aires, Argentina, 739–745.

Hollien, H. and Hollien, P. (1972). Speech Disorders Occurring While Diving in Shallow Water, *Proc. XV World Congr. Phoniatrics Logopedics*, Casa Ares, Buenos Aires, Argentina, 717–726.

Holywell, K. and Harvey, G. (1964). Helium Speech, *J. Acoust. Soc. Amer.* **36**, 210–211.

Hunter, E. K. (1968). Problems of Diver Communication, *Institute of Electrical and Electronic Engineers Transaction on Audioacoustics and Electronics*, AU-16, 118–120.

Lindblom, B. and Studdert-Kennedy, M. (1967). On the Role of Formant Transitions in Vowel Recognition, *J. Acoust. Soc. Amer.* **42**, 830–843.

MacLean, D. J. (1966). Analysis of Speech in a Helium-Oxygen Mixture Under Pressure, *J. Acoust. Soc. Amer.* **40**, 625–627.

Morrow, C. T. (1971). Speech in Deep Submergence Atmospheres, *J. Acoust. Soc. Amer.* **50**, 715–728.

Moshier, S. L. (1969). *Unsolicited Proposal to Investigate Time-Domain Convolution Processing of Helium Speech Signals*, Listening, Inc.

Peterson, G. E. and Barney, H. L. (1952). Control Methods Used in a Study of the Vowels, *J. Acoust. Soc. Amer.* **24**, 175–184.

Quick, R. F. (1970). Helium Speech Translation Using Homomorphic Deconvolution, *J. Acoust. Soc. Amer.* **48**, 130(A).

Rothman, H. B. and Hollien, H. (1974). Evaluation of HeO_2 Unscramblers Under Controlled Conditions, *Marine Technology Society Journal*, **8**, 35–44.

Rothman, H. B. and Hollien, H. (1972). Phonetic Distortion in the HeO_2 Environment, *In:* "Proceedings, VII International Congress of Phonetic Sciences", Mouton and Co., The Hague, 589–598.

Rothman, H. and Hollien, H. (1973). Evaluation of Helium-Precessing Systems, *Proceedings of BUMED-ONR Sponsored Workshop on Processing of Helium Speech*, in press.

Schroeder, M. R., Flanagan, J. L. and Lundy, E. A. (1967). Bandwidth Compression of Speech by Analytic-Signal Rooting, *Proceedings of the IEEE* **55**, 396–401.

Sergeant, R. L. (1963). Speech During Respiration of a Mixture of Helium and Oxygen, *Aerospace Medicine* **41**, 1963, 826–829.

Sergeant, R. L. (1967). Phonemic Analysis of Consonants in Helium Speech, *J. Acoust. Soc. Amer.* **41**, 66–69.

Sergeant, R. L. (1968). Limitations in Voice Communication During Deep Submergence Helium Dives, Paper presented to *National ISA Marine Sciences Instrumentation Symposium*, Cocoa Beach, Florida, January, 1968.

Speakman, J. D. (1968). Physical Analysis of Speech in Helium Environments, *Aerospace Medicine* **39**, 48–53.

Stover, W. R. (1967). Technique for Correcting Helium Speech Distortion, *J. Acoust. Soc. Amer.* **41**, 70–74.

Wathen-Dunn, W. (1967). Limitations of Speech at High Pressures in a Helium Environment, Proc. Third Symposium Underwater Physiol.

White, C. E. (1955). Report on Effect of Increased Atmospheric Pressure Upon Intelligibility of Spoken Words, *Memorandum Report Number* 55–8, U.S. Naval Medical Research Laboratory, New London, Connecticut.

Hearing in Divers

H. HOLLIEN
and
S. FEINSTEIN

Communication Sciences Laboratory, University of Florida, Gainesville 32601, U.S.A.

1. Introduction	81
2. Sound in Water versus Sound in Air	82
3. Anatomy of the Ear	83
4. Auditory Sensitivity in Air	86
5. Auditory Sensitivity in Water	87
A. The communication science programme on the hearing sensitivity of divers	88
6. Effects of Pressure on Auditory Sensitivity	99
7. Underwater Sound Localization	100
A. Sound localization in air	101
B. Arguments for and against man's ability to localize sound underwater	102
8. Programme on Underwater Sound Localization	104
A. The multiple sound source approach	105
B. Specific localization experiments	108
C. Acuity and precision of underwater sound localization	116
D. The joint programme on underwater sound localization	129
9. Some Tentative Theories Concerning Underwater Sound Localization	130
10. Diver Acoustic Navigation	132
Acknowledgements	135
References	135

1. Introduction

All divers are aware that the underwater world is noisy. It may seem quiet (at least at first) to the novice but the more advanced diver is aware of many underwater sounds. The pops, crackles and thumps of known and unknown origin, the easily identified roar of the breathing exhaust, the whine of an outboard motor, all form a continuous cacophony that is hard to ignore. However, with time, and change in adaptation level, the gross background noise becomes less conspicuous and the diver tends to notice only those sounds which are important to him or different from the background. It appears, then, that man hears in water much as he hears in air. Actually, this perception is a deceptive one because in reality man's auditory system is not anatomically adapted to the water milieu and as a result it does not function as well in water as it does in air.

It is obvious that the auditory system is important to the diver as a protective system and a communication channel but perhaps less obvious is the vital role acoustic stimuli can and should play in the spatial orientation of divers. In the normal air environment, man depends primarily on his visual system to provide him with spatial information, unless, of course, he is blind and is forced to depend on his auditory system to orient himself. Essentially this is the case underwater. In this instance, there is often such a dramatic reduction in visual cues (Luria and Kinney, 1970) that the diver is functionally blind (except for very short distances) or at least suffers a severe visual handicap. Hence, it should follow that the ability to use acoustic cues for spatial orientation underwater, in the same sense as the blind man uses them in air, would be a marvellous skill for the diver to develop.

2. Sound in Water versus Sound in Air

In order to understand the differences one will find between hearing in air and hearing in water, it first will be necessary to examine some of the basic principles of acoustics. We can begin with the fact that sound is due to small variations in the pressure of an elastic medium, such as air or water. The sound source causes the variations in pressure to travel outward at a known rate which is primarily determined by the density of the medium. For the purpose of the present discussion we will approximate the velocity of sound in air at 335.28 ms and in water at 1341.12 ms. Thus, the first obvious difference between sound in air and in water is that in the latter it is propagated about four times faster.*

The succession of rarefactions and compressions generated by a periodic sound source causes waves to flow through the medium, and if it were possible to take a "stop motion" picture of these waves, it could be seen that each wave essentially is a duplicate of the others and is repeated at regular intervals. The distance between similar points on adjacent waves is called the wavelength (λ). Further, the velocity (c) is the speed at which a particular point on the wave moves away from the source and the number of waves that pass a point in a given time reflect the frequency (f) of the sound. The unit of frequency is the Hertz (Hz), formerly known as cycles per second. These three characteristics of the wave are related by the equation: $f = c/\lambda$. For example, when comparing the wavelength ($\lambda = c/f$) of a 1000 Hz signal in air (1100/1000 or 1.1 ft) with its wavelength in water (4400/1000 or 4.4 ft), it will be noted that the wavelength at that frequency is about four times greater in water than in air.

* Our calculations are for fresh water; the speed of sound in sea water is somewhat greater.

The amount of resistance that must be overcome to propagate an acoustic wave in a medium is termed its acoustic impedance (Z). This quantity is the product of the density (ρ) of the medium and the sound velocity through it (c); hence, $Z = \rho c$. The Z of air is less than the Z for water and this mis-matching of impedances results in two consequences of concern. In the first place, sound transmission from one medium to the other is dramatically attenuated. Indeed, the intensity of a sound wave passing from air to water or water to air is reduced by a factor of 0.0010 or approximately 60 dB pressure transmission. Since the human head has an acoustic impedance very close to that of water (Ludwig, 1950; Goldman and Hueter, 1956), it becomes obvious that in air the head is an effective sound barrier while in water it is not. Thus, when immersed, the head loses some of its important properties for auditory perception.

The second effect of concern is that the high impedance of water causes it to be much more resistant to the propagation of sound than is air. To produce the same effect in water as in air, sound pressure level or SPL (expressed in dB reference 1 microbar) in the former must be much greater; or taking the converse, the acoustic energy in air is greater than in water when pressure is equated. Thus, it takes a much greater driving force to introduce sound into water than it does to do so in air. In this regard, it should be of interest to note some representative sound energies generated in our normal sonic environment (air) as well as some generated in the marine environment. Table 1 provides a sampling of such information.

3. Anatomy of the Ear

In Fig. 1 it may be seen that the human auditory system may be divided into five major parts: (1) the pinna, (2) the external auditory canal, (3) the tympanic membrane, (4) the middle ear (which includes the tympanic cavity and ossicles) and (5) the cochlea. The first division, the pinna (a), plays an important role in the localization of sound in air (Batteau, 1967; 1968; Fisher and Freedman, 1968). However, since the pinna consists of skin and cartilage it has the same Z underwater as does the surrounding medium and thus is acoustically transparent.

The external auditory canal (b) is a slightly curved tube, about 0.7 cm in diameter and 2.5 cm long, terminating at the tympanic membrane. If the canal is treated as a closed pipe, it is possible to calculate its resonant frequency—which turns out to be approximately 3 kHz. An empirical test of these measurements will reveal that a 10 dB increase in pressure (over that at the entrance) is found at the tympanum at 3 kHz and that significant amplification occurs from 2 kHz to 6 kHz. Because as much as one half of the canal is cartilage and muscle (Z approximately equal to water) its effective length is shortened by water and its resonant frequency increased to a level essentially outside the fundamental hearing range.

The tympanic membrane (c) is a flattened cone with an average area

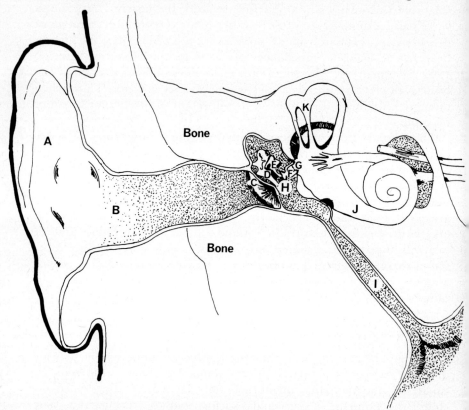

FIG. 1. The human ear including the pinna (a), external auditory canal (b), tympanic membrane (c), middle ear activity (h), ossicles (d, e, f), oval window (g), eustachian tube (i), cochlea (j), and semicirculuar canals (k).

of 66 mm² that lies across the auditory canal at a slightly oblique angle, with its apex facing inward. Behind the membrane is located a 2 cm³ cavity (h) that is contiguous with the throat via the Eustachian tube (i). A series of three small bones (ossicles) form a mechanical link between the tympanum and the fluid in the cochlea (j). The malleus (d) is attached to the tympanic membrane and vibrates with it while the other two bones (e, f) are linked to it in a way that reduces the magnitude of the excursions at the tympanum but preserves the energy so that the

TABLE 1. A comparison of acoustic pressure in air and water in dB re 1 μ bar as related to acoustic power in Watts/cm^2 (Tavolga, 1964).

Acoustic Pressure in Air (dB re: 1 μ bar)[a]	Acoustic Pressure in Water (dB re: 1 μ bar)		Acoustic Power (Watts/Cm^2)
		F-84 Jet Take-off at 80 ft	10^{-3}
50	90	Range of Dynamite Explosions	10^{-4}
	80		10^{-5}
40		Threshold of feeling at 1000 C.P.S.	
30	70	Single Motor Airplane at 15 ft	10^{-6}
20	60	Subway Train	10^{-7}
10	50	Symphony Orchestra at 20 ft	10^{-8}
		Nearby Outboard Motor	
0	40	Very noisy office	10^{-9}
		Foghorn Blast from Nearby Toadfish	
-10	30	Average Home Radio	10^{-10}
		Very Rough Sea (Shallow Location)	
-20	20	Average Office	10^{-11}
-30	10	Private Business Office	10^{-12}
		Shipping Noise (Cargo Vessels)	
-40	0	Quiet City Residence	10^{-13}
		Clicking of Many Snapping Shrimp	
-50	-10	Calm Sea	10^{-14}
-60	-20	Quiet Whisper	10^{-15}
		Threshold of Hearing for Squirrelfish at 800 C.P.S.	
-70	-30	Threshold of Human Hearing at 1000 C.P.S.	10^{-16}
-80	-40	Probable Threshold of Hearing for "Specialist" Fishes	10^{-17}

[a] To convert reference 1 μ bar to reference 2×10^{-4} μ bar or 20 μ N/M^2, add 74 dB.

stapes (f) pushes the fluid of the cochlea (via the oval window) with low amplitude and high force. In addition, if we consider the area of the tympanum in relation to the foot of the stapes (twenty times larger) and the mechanical advantage of the ossicular chain, we would expect to find an increase in pressure approximately thirty times greater than that which was transduced at the membrane.

Further, in recognizing that the cochlea is a hydraulic system, it becomes evident that the membrane and ossicles form not only a transmission link, but also an impedance-matching device between the air in the canal and the water in the cochlea. Obviously, if the canal is filled with water, a considerable impedance mismatch occurs. As a result the middle ear would be expected to be effectively non-functional underwater (Sivian, 1947).

4. Auditory Sensitivity in Air

The acoustic intensity that can be detected approximately 50% of the time at a particular frequency is called the threshold of hearing at that frequency. On the basis of thresholds obtained from a very large sample of listeners, a standard audiometric function can be obtained. This standard or mean threshold has a variability of approximately ± 10 dB and by convention it is expected that normally hearing individuals will fall within ± 10 dB of this standard. There are a number of

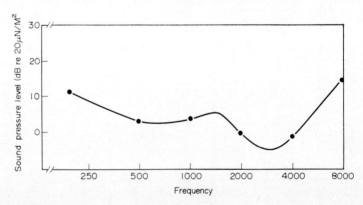

FIG. 2. Normal minimum audible field (MAF) thresholds as determined by an international standard (ISO R226-1967).

procedures utilized in obtaining hearing thresholds and each procedure produces a slightly different standard. Figure 2 presents a portion of the international standard (ISO R226-1967) which is one of several possible definitions of the "normal" threshold of hearing under free field listening conditions. To determine this standard, thresholds were obtained in an anechoic space and the effective pressures were measured at the point where the listeners ears would be during testing. This procedure is known as a minimum audible field (MAF) measurement. Another procedure involves the use of calibrated earphones to obtain thresholds and yields a minimum audible pressure (MAP). Yet, a different standard is used when hearing is tested for clinical reasons, viz., the reference point is the average threshold for the "normal" ear. This reference is called Hearing Level and the resulting graphs are called "audiograms".

5. Auditory Sensitivity in Water

In 1947 Sivian suggested that human hearing underwater would be attenuated by as much as 40 dB. He argued his position from a theoretical basis contending that the loss would result from the impedance mismatch between the tympanic membrane (and the middle ear) and the water in the external auditory meatus. Moreover, he expected additional attenuation to result from (a) unbalanced static pressure on the eardrum and (b) ambient noise in the diver's environment. Sivian's experiments, conducted in a swimming pool with a sound source suspended over the water were rather crude by today's standards; however, they seemed to support at least some of his assumptions. Later Ide (1944) tested sensitivity and found a difference of roughly 65 to 70 dB SPL between hearing underwater and that in air for a series of frequencies between 100 Hz and 6000 Hz.

Reysenback de Haan (1957) attempted to measure the underwater sensitivity of three men who had normal hearing. In one test, their external canals were filled with air and in a second test, they were filled with water. He found thresholds to be lower at 1 kHz but higher at 2, 4, 8, 12 and 16 kHz, in the latter condition. Unfortunately, no description was given of the experimental procedures he followed nor about experimental controls. A year later, Hamilton (1957) measured hearing thresholds using a modified Method of Limits. His subjects were tested two at a time as they sat in the water side by side; the frequencies investigated were 250, 500, 1000, 2000 and 4000 Hz. He reported that the thresholds in the water were on the order of 44 to 60 dB greater than in air. Further, he suggested that underwater hearing is by "bone conduction" because when his subjects occluded one ear with a finger, it did not seem to change sensitivity. In 1958, Wainwright used the Method of Limits (with two subjects) in order to determine hearing thresholds at seven frequencies between 250 Hz and 4000 Hz. His subjects did not wear rubber diving suits and closed circuit SCUBA was utilized. Minimum audible fields were compared for the same subjects both for air and water. Wainwright states the ". . . greatest loss in sensitivity in water occurred over the frequency range from 500 Hz to 2000 Hz and amounted to approximately 20 dB, while below 300 Hz the threshold intensity in water was lower than that in air". He reported that occluding the ears with the fingers had no effect on threshold.

Montague and Strickland (1961) apparently were concerned about the lack of agreement among the data that had previously been published. Accordingly, they redetermined underwater hearing thresholds in an attempt to resolve the discrepancies between the data obtained by the earlier investigators. Further, their subjects were tested with and without diving hoods in order to determine the amount of attenuation caused by the hood itself. Thresholds were obtained at 250, 500, 1000, 1500, 2000, 3000, 4000 and 6000 Hz using the Békésy technique. These authors reported that the SPL needed to reach threshold in water was about 40–70 dB higher than the MAP threshold in air; the greatest loss in acuity occurred in the region of greatest air sensitivity. Further, they reported that, at frequencies above 1000 Hz, the diver's hood yielded approximately 20 dB (or more) attenuation. While Montague and Strickland added significantly to the corpus of information concerning underwater auditory sensitivity, they did not totally resolve the problem—especially with respect to the *mechanism* by which the ear operates in this milieu.

A. The Communication Science Programme on the Hearing Sensitivity of Divers

It should be apparent that the reported research on underwater hearing sensitivity left many questions unanswered. First, the exact thresholds that could be expected when the head is immersed in water had not been established, nor had it been determined whether the ambient pressure and/or the water medium alone determines the sensitivity of the ear. Moreover, as has been stated, these early studies did not provide basic information on the *mechanism* of hearing underwater. Accordingly, early in 1966 we embarked on a programme of research designed to answer some of these questions. We now believe that we have reasonable information on both the "how" as well as the "how much". More importantly, we believe our model of underwater hearing will serve to explain why this human sensory system acts as it does in this milieu.

Experimental environment

The site of all of our experiments, except those so identified, was the Bugg Springs field facility of the Naval Research Laboratory's Underwater Sound Reference Division, Orlando, Florida (see Fig. 16, Chapter 1).

Research equipment

As was stated in the preceding chapter, if precise and rigorous research on diver hearing is to be carried out successfully the experimental procedures must be developed with care. Consequently, we developed an underwater system which provides for experimental control of diver positioning, stimulus presentation and subject response (Hollien and Thompson, 1967). Since this system (DICORS)

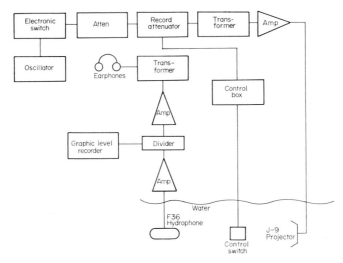

FIG. 3. A schematic drawing of the DICORS stimulus, response, and calibration system.

was described previously (see Fig. 18 in the preceding chapter) only specific features bearing on hearing research will be discussed here. First, a J-9 projector is mounted on the frontal projection of DICORS and an F-36 hydrophone on the rear frame at a distance of exactly one meter from the projector. The J-9 provides the sound source and the F-36 hydrophone provides the means by which calibration of the full system may be accomplished. That is, by placing the hydrophone in a position very near to where the diver's head would be during an experimental procedure, calibration of the entire system (including the diver) is possible. Of course, during an actual experiment, the calibration hydrophone is removed and the diver's head positioner is substituted. Figure 3 provides a schematic drawing of the stimulus, response and calibration system.

The large plexiglass box directly in front of the diver's knees (see again Fig. 18, Chapter 1), contains a TV monitor which can visually convey information to the diver/subject. For example, for a procedure involving speech reception thresholds, the diver is presented a particular stimulus word (auditorily); immediately thereafter he observes a group of words (among them the correct item) on the TV monitor. He then responds by activating that switch (among those in a set) which corresponds to the particular item chosen as the heard word. In Fig. 4,

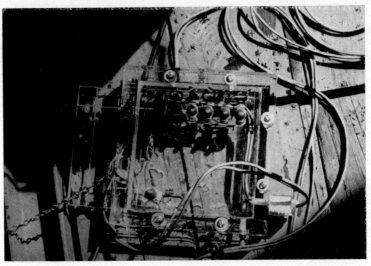

FIG. 4. A switching array used by the diver in DICORS to indicate his response to the experimenter on the surface.

one of several available switching arrays may be seen; in this case, six underwater switches are placed in a plexiglass box which in turn is held by the diver. When a given switch is activated, the response is observed and recorded topside by means of an IBM 010 key punch coupled to a bank of response lights. This overall system indicates immediately the diver/subject's particular response—as well as simultaneously punching the same information on an IBM data card for computer analysis at a later time.

When hearing thresholds were obtained (sinusoids), a single handheld switch was used to control attenuation of the Békésy audiometer. In these experiments, sinusoidal test stimuli generated by a beat-frequency oscillator (General Radio, Type 1304-B) were passed through an electronic switch (Grason-Stadler, Model 829D) and associated equipment to the J-9 transducer mounted on the frame of DICORS. Test frequencies of 125, 250, 1000, 4000 and 8000 Hz were utilized;

they were gated On and Off with a period of 500 ms, a 50% duty cycle, and 2.5 ms rise-and-decay times. The attenuation rate of the recording attenuator was 8 dB/second. The air conduction thresholds (for comparison) were obtained by standard audiometric procedures or by using a Rudmose (Model ARJ-4) automatic audiometer modified to allow presentation of the same frequencies as those used in the experiment.

Experimental procedure

All of our studies followed the same basic procedure. Only individuals who were competent divers and had experience in taking hearing tests in air and water were used as subjects. When the diver was in position and had equalized the air pressure in the middle ear against the

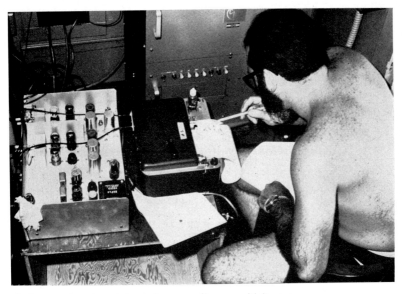

FIG. 5. An experimenter administering an auditory sensitivity test.

water pressure in the external auditory meatus, and was ready to begin the threshold test, he signalled the experimenter at the surface. The test stimulus was first presented at a high enough SPL to be clearly audible. The diver/listener adjusted the SPL of the stimulus so that it shifted back and forth across the audibility threshold level by activating a hand switch connected through a control box to a recording attenuator (Fig. 3). In all cases, the threshold measures were taken while the listener was holding his breath. This procedure was necessary because considerable noise is generated around the diver's ears when he exhales.

Figure 5 shows the experimenter administering the test and Fig. 6 provides an example of a typical response trace.

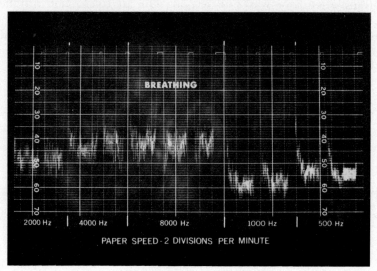

FIG. 6. A typical response trace from an auditory sensitivity test.

Experimental results: pure tone sensitivity

Our first series of projects were focused on the issue of defining exactly how much of a threshold shift occurred underwater. Accordingly, we carried out a series of experiments on a large population of divers with essentially normal hearing (Brandt and Hollien, 1967). Specifically we tested a large number of subjects (a) at 10.67 m only, (b) at 3.66, 10.67, 21.34 and 32 m, (c) with helium introduced into middle ear, (d) for speech reception thresholds and (e) for speech discrimination. Figure 7 provides a graphic representation of the results of the first or basic study. Moreover, the findings resulting from these several experiments may be summarized as follows:

1. Underwater thresholds are from 30 to 60 dB higher than for air conduction thresholds, the difference increasing with frequency. Specifically, underwater sensitivity varies with frequency from 67 to 80 dB SPL, a range of about 13 dB with a mean threshold of about 70 dB SPL. The pattern of the "loss" appears quite similar to that seen in patients with conductive hearing disorders.

2. Thresholds appear not to vary as a function of depth. Increases in ear depth from 3.66 m–32 m and the concomitant positive increases in water pressure (5.3, 15.6, 31.2 and 46.7 psi at 3.66, 10.67, 21.34 and

32 m, respectively) or corresponding increases in atmospheric pressure of 1.4, 2.1, 3.1 and 4.2 ATA, have no effect upon free-field underwater hearing thresholds in the frequency range between 125 and 8000 Hz. It should be noted also that these data confirm those obtained in the first experiment.

3. Helium in the middle ear—and the associated modification of middle ear impedance—appeared to have no effect, or at most a minimal effect, on underwater thresholds.

4. The effects on hearing appear to be conductive in nature rather than neurological. That is, when speech reception thresholds and speech discrimination scores were obtained and compared to the absolute

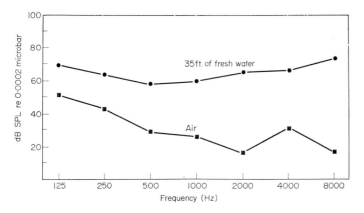

FIG. 7. The auditory sensitivity of divers in air as compared with their sensitivity in 35 ft of water.

thresholds for pure tones, it appeared that once the absolute threshold of hearing is reached for the underwater listener, normal speech reception and discrimination could be expected.

In summary, our early studies led to the conclusion that when a diver's head is submerged, (1) a stable, and calculable, shift in auditory sensitivity results and (2) this shift appears to result from a situation where the external and middle ear systems become essentially inoperable and hearing is accomplished by bone conduction.

Recently we have performed a number of experiments which have tested the bone conduction hypothesis. The experimental evidence is as follows:

1. There is no difference in auditory threshold whether the meatus is filled with water or contains a trapped air bubble. This experiment was first attempted by Reysenback de Haan (1957) but no details of his procedure were given. More recently, we obtained thresholds of human

hearing underwater for two conditions: (1) with the external auditory meatus completely water filled and (2) with a bubble of air trapped against the tympanic membrane. The first condition was accomplished by forcibly irrigating the external meatus underwater (Fig. 8); the second condition was obtained by placing ear plugs in the meatus (Fig. 9)

(a)

(b)

FIG. 8. Forcible irrigation of the external auditory canal to ensure that there are no bubbles trapped at the tympanum. 8(a) was taken during the actual research; (b) demonstrates the procedure in air.

and keeping them there until the head was underwater and the test (with the plugs removed) initiated. Subjects were seven divers who were tested in DICORS at a depth of 3.66 m; free-field thresholds were obtained for the frequencies 125, 250, 1000, 2000 and 8000 Hz. Threshold shifts (re: air) for both conditions of underwater hearing were consistent with those previously reported. On the other hand, the thresholds for the two experimental conditions were virtually identical for all frequencies except 250 Hz, where hearing was 6 dB better for the water-filled meatus condition. Apparently, the presence

Fig. 9. Ear plugs were used to maintain an air bubble at the tympanum.

or absence of air bubbles in the external meatus contributes little if anything to underwater hearing thresholds. Yet the impedance mismatch is so different for the two conditions that, if the external/middle ear system was even minimally functional, substantial differences in threshold would have been expected.

2. Prior to our second experiment in this series, Norman et al. (1971) reported an experiment in which they measured sensitivity at several frequencies for a diver with (a) a bare head; (b) with only the

ears covered with a set of neoprene ear patches fastened over both pinnae; (c) with most of the head covered, with the exception of the ears, by using a standard neoprene hood with holes cut at the ear-locations and with the pinnae pulled through the holes; and (d) with a full hood covering the ear openings. They found that "at high frequencies sound conduction through the ears does not appear to be important". The hood (even with open ear holes) attenuated 1 and 2 kHz stimuli by 30 to 37 dB.

FIG. 10. A diver wearing a neoprene hood with a ¼ in. rubber tube passing from the external auditory canal to the surrounding water.

We recently performed an experiment in which the thresholds of seven listeners were obtained under three conditions: (1) wearing a full wet suit with no hood, (2) wearing a full wet suit with a hood, and (3) wearing a full wet suit and hood with one quarter inch rubber tubes passing through the hood and into the meatuses. Figure 10 shows a diver outfitted for the third condition. We found no differences between conditions two and three (the hoods), but thresholds were significantly lower for the bare head condition and these thresholds were similar to those we previously reported. Quite obviously, if the external and middle ears are used to any great extent in underwater hearing, we should expect a substantial difference in thresholds between conditions two and three. Incidentally the data that may be seen in Fig. 11 demonstrate the frequency dependent acoustic attenuation of foam neoprene diving hoods (Montague and Strickland, 1961; Bogert, 1964; Smith, 1969 and Hollien and Feinstein, in press).

The approach taken by Smith (1965, 1969) to the question of bone

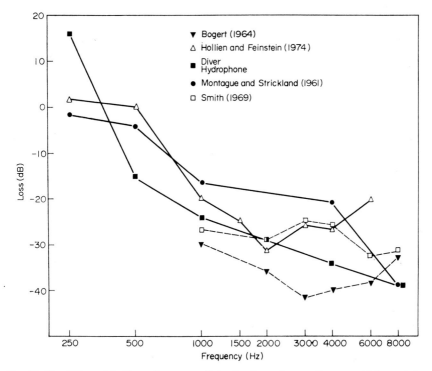

FIG. 11. The differential effects of neoprene diving hoods on diver auditory sensitivity and the sensitivity of reference hydrophones.

conduction has been quite different from those previously mentioned. Apparently, it is his feeling that it is necessary to know the bone conduction thresholds in air of the experimental subjects before it is possible to determine whether diver hearing is accomplished by this mechanism. This position is perhaps a bit stronger than is necessary but at the same time Smith's results and thinking provide further evidence supporting the bone conduction hypothesis. Indeed, he found that depressed air conduction hearing levels were not reflected as reduced underwater sensitivity except as the depressed air conduction hearing levels were accompanied by depressed bone conduction sensitivity. Further, he was able to demonstrate good correspondence between bone conduction thresholds (in air) and underwater thresholds.

The question of the mode of hearing underwater is still not completely settled, Bauer (1970) has suggested that the occluded bubble procedure is not a valid test of the function of the middle ear. His contention is based on his earlier work with underwater earphones (Bauer and Torick, 1966) and on a combination of an equivalent circuit of the tympanum, according to Zwislocki (1957) and of the external ear according to Bauer *et al.* (1967).

Bauer's arguments are based on the assumption that placing the sound source in front of the ears is equivalent to excitation of the middle ear. This may or may not be the case. For example, if it is assumed that an earphone in close proximity to the skull acts as a bone conductor then it might be expected that the middle ear is not necessary for the excitation of the cochlea. However, Bauer argues further that the earphone is not a bone conductor because ". . . if reception occurred through normal auditory channels, then directional perception via the earphones would be possible; whereas, if it took place through the bony portions of the skull, then earphones could not be used for directional perception." (He did, in fact, find directional perception with the earphones.) He then goes on to report that "the earphones were placed against the ears and against other portions of the head. A significantly louder sound was heard with the earphone held against the ear than for any other locations" and that ". . . when the earphone was held at arm's length, all sense of direction ceased".

In light of the now overwhelming evidence for human underwater sound localization (see data on this subject reported in a succeeding section of this chapter), one finds it hard to account for the inability of Bauer's subject to locate a sound source at arms length. Moreover, implicit in Bauer's own argument is the deduction that if an individual can localize a distant sound source (which is not in front of the ears) he must do so by bone conduction. Finally, since we do not know what methods he used to test the loudness of various placements of the earphones, it appears pointless to try to account for his results.

In any case, it is possible to hypothesize why underwater hearing is primarily bone conduction in nature. Basically, the external and middle ears appear to be removed from the acoustic pathway in a fluid medium because of impedance mismatches which can occur as a result of inappropriate force/amplitude and Z relationships. As we have pointed out, sound travels through air in a high amplitude, low force (Af) relationship; through water as high force, low amplitude (aF). The external and middle ear function to increase F in sound energy from its airborne level to a level that will interface properly with the fluid contained in the cochlea. Hence, for hearing in air: Af → aF. When man is underwater, however, the process is one of aF → Af → aF and the serial Z mismatches are so great that these conditions effectively negate the external auditory mechanism. Further, the acoustic impedances of the surrounding medium and the skull are so close that any mismatch can be considered to be essentially zero. Thus, sound energy should flow directly into the skull as if it were a continuation of the water medium.

6. Effects of Pressure on Auditory Sensitivity

There are at least two ways by which one can determine how pressure alone affects diver hearing. In the first procedure, hearing is tested at various depths in the water and in the second, hearing thresholds are obtained in the dry atmosphere of a hyperbaric chamber. We previously utilized the former procedure to test hearing up to 32 m and found no significant changes in sensitivity as a function of depth.

The first hyperbaric threshold measurements were obtained by Fluur and Adolfson (1966) at a simulated depth of 100.58 m of sea water. They observed reversible, depth related conductive hearing losses in 26 experienced divers; these losses approached 30 to 40 dB in the middle frequency range of hearing. They attributed the losses to impedance changes in the middle ear.

Hyperbaric observations of hearing thresholds also were made by Farmer *et al.* (1971). They obtained air conduction thresholds and sensory acuity levels (bone conduction) to a simulated depth of 182.88 m on six divers at 250, 500, 1000, 2000 and 4000 Hz. They tested at five depths during descent twice at 182.88 m and at six depths during ascent. During the two tests on the bottom, frequency difference limen (the ability to discriminate between frequencies) at 1.0 kHz were obtained. These investigators reported that "a proportional and variable elevation of the air conduction thresholds appeared during compression . . ." which, ". . . gradually decreased during decompression . . . .. No significant differences in these thresholds and surface thresholds at any frequency was observed at depths of 30.48 m and less". The maximum elevation in thresholds was reported to be 26 dB in the lower frequency ranges at 182.88 m. Finally, they found no significant changes in bone conduction or in ability to match frequencies. These findings support the conclusion that the loss in sensitivity is not sensorineural. This experiment represented the only attempt (to date) to properly control the effects of pressure and hyperbaric gas mixtures on auditory sensitivity. Unfortunately, however, the Farmer *et al.* experiments were not the primary mission of that dive so that the psychoacoustic experiments had to be run under less than optimum conditions.

We have recently conducted auditory threshold tests at 76.2, 137.16, 198.12 and 265.18 m under both dry and wet chamber conditions at the Duke hyperbaric facility. In this instance our experiments were a primary part of the mission and this situation allowed us to exert a level of experimental control not previously possible. As a result we were able, for example: (1) to choose divers whose hearing was normal by audiometric standards; (2) to test these divers when they were reasonably rested and had no conflicting duties to perform; (3) to allow sufficient time for the psychoacoustic tests so that subjects and experimenters were not hurried; (4) to provide sufficient time to run a signal detection experiment in order to determine the degree to which threshold shifts may be a function of psychological rather than sensory variables, and (5) compare the dry thresholds to the wet thresholds (cited above) at the various depths. We hope that the results of these experiments will allow us to advance a hypothesis explaining shifts in auditory sensitivity (if indeed any occur) in much the same manner as we have been able to provide a theory that explains man's auditory sensitivity underwater.

7. Underwater Sound Localization

Thus far, we have established that divers can hear underwater, however, with less sensitivity than in air. But having heard the sound, can divers localize the direction from which it is coming? As we know, in air, man depends on his directional sense of hearing to locate the source of sounds, to warn him of approaching objects, to help differentiate among competing signals and so on. Indeed, these processes are well-developed and provide important sensory information. On the other hand, it has been hypothesized that, in the underwater environment, humans have their sound localization capabilities so seriously impaired as to be essentially non-functional. For example, Bauer and Torick (1966) suggest that when an individual is submerged "Sounds appear to arrive from nowhere. The location of a friend or foe becomes a matter of dangerous conjecture and reverberant sounds mix with direct sound into an unintelligible jumble".

It might appear, then, that divers should not be able to localize underwater sounds. In order to understand this problem it is first necessary to review the nature of sound localization in air.

A. Sound Localization in Air

Directional perception of sound in air is based on the utilization of time-of-arrival (phase) and/or intensity information provided by the arriving signal to the auditory mechanism. For the low frequencies, time differences appear to be the most important and the arrival of sound at one ear (versus its arrival at the opposite ear) can vary up to 0.6–0.7 ms. At higher frequencies, the head creates a shadow effect which in turn produces a marked difference in intensity between the two ears. Mills (1958) suggests that the division of the frequency range (for sinusoids anyway) occurs at 1400 Hz with temporal cues dominant below that level and intensity cues dominant above it. Quite obviously, time or intensity data of these magnitudes are adequate for effective sound localization in air.

More specifically, the differences cited above appear to depend on the fact that the head acts as an acoustic baffle which deflects, diffracts, and reflects sound energy. For example, when a sound source consists of short pulses of thermal noise (a noise consisting of random frequencies produced at random amplitudes) and is gradually moved in a circle whose centre is between the ears, the sound is detected at one ear before the other. This is known as an interaural time difference (ITD). The ITD, empirically, is approximately 180 μs at 23° azimuth, 369 μs at 45° azimuth, 486 μs at 60° azimuth and 600 μs at 90° azmuth (Jeffress, 1957). It has also been noted that the head acts as an increasingly effective acoustic baffle with increasing frequency so that the high frequency components of the stimulus ($>$1400 Hz) are attenuated at the far ear. These interaural intensity differences (IID) are complex functions of frequency and angle. When broadband thermal noise is employed as the stimulus and an artificial head is rotated in the sound field, the IID reach maxima at approximately 60° and 120°. As the source moves behind the head into an arc extending from approximately 120° to 240° of azimuth there is a reduction in intensity at both ears due to the pinnae (Nordlund and Liden, 1963).

Moreover, the manner in which sound localization experiments are conducted tend to influence the data—at least to some degree. Of course, the choice of experimental procedure depends to an extent on the kinds of information the experimeter is trying to extract from the investigation. Thus, certain investigators have been interested in determining the precision with which a sound source can be located as related to specific stimulus parameters and they have utilized procedures which require an egocentric reference (a reference which emanates from the perceived spatial orientation of the listener), e.g. the median plane of the head. Typically these procedures require the listener to point to the source either naming its azimuth in degrees or indicating its location diagrammatically (or on some prearranged scale).

Other investigators have examined the way various stimulus parameters effect the ability of an individual to match the location of some external reference. These experiments utilize procedures which require the listener to match the location of the reference source with a probe in the same, or a different, modality.

A third approach requires the examination of the effect of various stimulus parameters on the acuity of the localization response. This method is analogous to examination of the ability of the visual mechanism to distinguish small spatial separations between portions of the visual field. In this case, the procedure is to require the listener to discriminate between two sources which are gradually brought closer together until they are no longer discriminable as separate entities. The discrimination threshold is described as a Minimum Audible Angle (MAA), (Mills, 1958). A final procedure is one that attempts to determine the way in which individual cues determine the perceived location of a sound. Obviously, intensity and time are confounded to some degree when a distant source is utilized, hence time cannot be changed without a related change in intensity or vice versa. In order to overcome this problem it is only necessary to present the stimuli via earphones in order to manipulate IID and ITD independently. This paradigm yields a response which is described as laterization because the sound image is most often perceived as being inside the head. Typically the listener is required to adjust the sound so that its position is located at the median plane of the head by nullifying one cue with another or by indicating diagrammatically the location of the sound.

In any case, it is possible to summarize relevant data concerning sound localization in air. That is, certain relationships are generally agreed upon; they are that: (1) localization is most precise for complex sounds such as clicks or thermal noise; (2) the most precise localization occurs with respect to the median plane (for any stimulus); (3) the ITD threshold varies (depending on author, procedure utilized and/or stimulus) between 7 and 20 μs; (4) just noticeable differences (jnd) for phase vary with signal duration to a maximum of six microseconds at 700 ms duration; and (5) the IID threshold is probably between 0.5–1.0 dB.

B. Arguments For and Against Man's Ability to Localize Sound Underwater

Earlier in this chapter it was noted that: (1) the speed of sound is much greater in water than it is in air; (2) in water there is no longer an appreciable impedance mismatch between the head and the surrounding medium (as there is in air) and (3) underwater hearing is accomplished by bone conduction rather than via the middle ear.

The result of these changes is that interaural time differences are reduced by a factor of 4.5 or more; further, interaural intensity differences are drastically reduced because the head is no longer an acoustic baffle. Indeed, since one hears by bone conduction underwater, many divers and scientists believe that the ears simply are not isolated one from the other.

Other than the work by Bauer and Torick (1966), there are only two experiments which argue against human underwater sound localization. First, Kitagawa and Shintaku (1957) were unable to demonstrate such localization by using "high percussion sounds", these sounds were produced by a buzzer or by hitting two small stones or two small bottles together. These sound sources were moved back and forth in front of the subject who reported their apparent location and the investigators reported that images of the percussive sounds (except for "certain individual variations" and changes caused by head movement) were fixed near the occipital region. When the buzzer was used as the sound source, the image was fixed in front of the forehead (at least for the distance of 30 cm); if the subject occluded his ears with his fingers, the sound image shifted to the occipital region.

Secondly, Reysenback de Haan (1957) reported an experiment in which divers were found unable to localize sounds underwater. In this experiment, the subject floated on the surface of the water 3 m above and 25 m distant from the sound source; in order to hear the stimulus, he ducked his head below the surface of the water. However, it is clear from the geometry of this experiment that there was little likelihood that the subjects could have been expected to localize the sound. Specifically, surface noise and reflections would have severely reduced the signal to noise ratio and it is doubtful that the listener could have maintained his orientation to the sound source. Thus, it may be seen that there is, at least, some evidence that underwater sound localization is not possible.

Opposed to these negative findings and the subjective reports of some divers, there are reports to the contrary by still other divers plus a very powerful early study by Ide (1944). He found that the sound produced by an ammonia jet could be localized immediately by some divers and by others after a few hours practice with an "anti-masking helmet" (a 10.2 m strip of foam rubber 1.3 cm thick running from forehead to base of skull):

> Several of the men felt, after practicing binaural listening for a short time each day for several days, that they no longer needed the helmets in order to get good

bearings on the jet... The effectiveness of the underwater binaural sense was demonstrated by several men who "homed" to the sound from a distance of about 300 yd, with their face plates blacked out so that they could not see. This was accomplished, although with a constant bearing error, even by a man with unequal ear sensitivities, one ear having been damaged. An excellent demonstration was made by an underwater swimmer wearing the Lambertson diving apparatus. This man, swimming entirely underwater and guiding himself by binaural perception of the jet, followed a 300 yd curved course through a strong cross-current and came up right beside the jet. He stated that the sound gave more satisfactory directioning than the wrist compass, owing to the difficulty of making compass corrections to allow for the current.

Ide (in a personal communication, 1966) stated that "Since we had a number of swimmers, all healthy young men, available for experimentation, we had no difficulty in selecting men who were clearly successful in locating point sources of audible sound in water. It was also clear that this type of capability improved up to a point with increased experience... The variability from one man to another finally led us to abandon the use of binaural localization for our problem".

8. Programme on Underwater Sound Localization

Our programme on underwater sound localization developed in the middle 1960s. However, initially we asked somewhat different questions about the problem and conducted our early research at different institutions. Accordingly, this section of the chapter is organized in a manner that presents first one set of studies, then the other. Following these discussions, our joint programme will be outlined and some preliminary attempts at theory construction will be made.

One of us (SF) became interested in the area because he had observed that some marine animals (such as harbour seals and sea lions) did not exhibit any apparent anatomical adaptation for underwater localization yet were able to orient to sounds underwater and possibly use a form of passive sonar. His early experiments were run in both a reverberant and an anechoic tank. A boom suspended a projector in front of the diver who indicated whether the sound source was to his right or left. Some divers experienced difficulty in localizing the sound at first but were able to do so eventually; others were able to localize immediately. He concluded that divers are able to localize sounds at least within a given quadrant. Moreover, he found that these divers could discriminate between good and poor reflecting materials—a discrimination which sea lions had found impossible (Feinstein, 1966). As a result of this early experiment a series of further studies was carried out to determine the acuity and precision of the underwater sound localization response (Feinstein, 1971).

At a slightly later point in time, (HH) became intrigued by what appeared to be a rather serious conflict between the two sets of available data; also between the situation theory would predict and some subjective evidence to the contrary. Hence, he undertook first a pilot study (Hollien, 1969) and then, when the evidence seemed to suggest that underwater sound localization was possible, a series of experiments investigating a number of appropriate relationships (Hollien et al., 1970; Hollien, 1973; Hollien et al., 1973). The data from these studies are reported first.

A. The Multiple Sound Source Approach

The first study in this series was pivotal in nature and provided the basic methodologies for the other five. Hence, it will be described in some detail. A necessary first step was to develop a rigorous methodology that would permit valid and appropriate research of the desired nature to be carried out. It was imperative, (1) that subjects could be placed in a reasonably anechoic space with no reflective surfaces within 12 m of them; (2) that all stimuli, responses and subject positions could be controlled with near laboratory precision; (3) that none of the sound sources (projectors) would be placed close to the subjects; (4) that the entire experimental milieu permit calibration; (5) that the experimental stimuli consist of a number of different acoustic signals; (6) that the method permit a large number of subjects to be studied; and (7) that the subjects used should be experienced divers with average to good auditory acuity. Criteria 1, 2 and 4 were met by the utilization of USRD's research facility at Bugg Springs (see again Fig. 16 of Chapter 1).

In order to meet the other criteria, a Diver Auditory Localization System (DALS) was designed and constructed; its design was based on that of DICORS. In general, DALS is an open framework diving cage, constructed of polyvinyl-chloride tubing (PVC tubing is acoustically invisible underwater); the primary modification consisted of coupling a series of five 3 m arms to the top of the system. These five arms were located to allow J-9 projectors to be placed at ear level at a reasonable distance from the centre of the subject's head and at angles to the diver/subject of 0°, 45°, 90°, 270° and 315°. A rough schematic drawing of DALS may be seen in the lower half of Fig. 12; a partial photographic view in Fig. 13. The photograph does not show the DALS system in its entirety because it was so large it could be assembled only underwater—and even there, it was too large to photograph. However, the general pattern of the system may be seen from the figures.

As stated, five J-9 projectors were used to provide the sound sources for the project. The J-9's were selected because of their omnidirectional characteristics and because it is possible to calibrate them in a very precise manner. In order to do so, an F-36 hydrophone was fixed to DALS at a position corresponding to the centre of the diver's head. The signals from the J-9 projectors were received by the hydrophone and transmitted by cable to an amplifier (Ithaca model 250) and a divider network on the surface. The signal was then fed to a graphic level recorder (General Radio type 1521-B) coupled mechanically to the beat-frequency oscillator. The signal voltage and frequency were

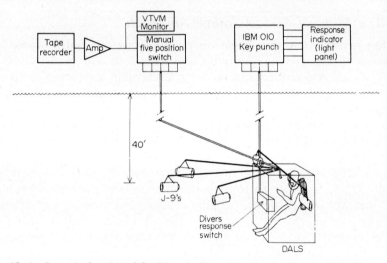

Fig. 12. A schematic drawing of the Diver Auditory Localization System (DALS).

monitored by a voltmeter (Ballantine model 302C), a frequency counter (Hewlett-Packard model 512A), and an oscilloscope. All of the surface equipment was located in a large, air-conditioned laboratory room on the test facility platform. Each of the five J-9 projectors was calibrated to produce the same SPL reading at the F-36 hydrophone (for all experimental signals) in order to assure that subjects would not receive cues from intensity differences.

The experimental stimuli selected for the basic experiment were 250, 1000, 6000 Hz sinusoids and thermal noise. The stimulus presentations consisted of five pulses of the particular experimental frequency set up as 500 ms bursts at 110 dB SPL (re: 0.0002 μ bar) or 40 dB re: average underwater hearing threshold for the diver/subjects. Each of the stimuli within the set of five were gated ON and OFF with the duty cycle of 1 s and 25 ms rise-fall time. Subjects were 17 adults (10 males and 7 females) recruited from the diving teams at the Communication Sciences

Laboratory, University of Florida, and the U.S. Navy Mine Defense Laboratory (now NCSL), Panama City, Florida. The mean ages of the males and females were 29 (range: 18–48) and 27 (range: 20–35) years, respectively. All subjects were competent divers with experience in

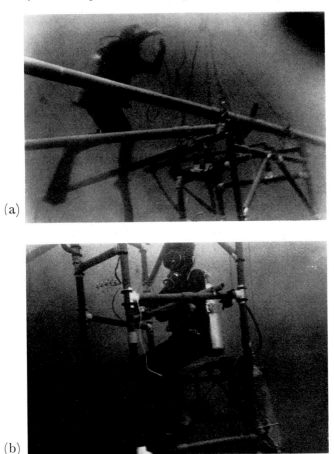

Fig. 13. Views of the DALS system: (a) shows it being constructed underwater; (b) shows a diver/subject in position and ready for a test.

taking hearing tests in air (all had essentially normal hearing) and had participated in underwater projects of this general nature many times.

A general understanding of the experimental procedure may be obtained by re-examination of Fig. 12. DALS was lowered by winch to an ear depth of 40 ft. The diver/subject, wearing open-circuit SCUBA equipment and a wet suit jacket, descended to the cage, seated himself, locked his arms over a bar provided for subject positioning (see again Fig. 13), and placed a lead weighted belt over his legs to keep him firmly on the seat. During the experiment, subjects were free to move their heads but not their bodies. Stimulus presentations were patterned so that they were received by subjects only when they were holding their breath and after the bubbles of their previous exhalation had reached the surface.

The experimental signals were presented to diver/listeners five times from each of the five transducers, for a total of 25 presentations of each stimulus. As stated, they responded by means of the five-position underwater switch coupled to an IBM key punch at the surface. Moreover, these responses were individually verified (by a research assistant who checked a set of lights coupled to the key punch) before subsequent stimuli were presented. In this manner, errors in recording data were avoided and subjects were given ample time to respond to each stimulus presentation. After the subject's response was recorded, a new stimulus was presented and the procedure was continued until all 25 presentations of each frequency were completed.

The results of the investigation demonstrated a much higher underwater sound localization ability on the part of the subjects than had been expected. These results are detailed in Table 2 which lists: (1) the ranked individual scores obtained for each subject, (2) their mean overall scores, (3) the mean for each stimulus and (4) the standard deviations associated with each of the four stimuli. Inspection of the table reveals that the overall performance of the 17 subjects was clearly above chance level as, on the basis of chance alone, it would be expected that the scores would cluster around 20%. Only 10 of the 68 scores were close to chance, only one individual (subject 17) had an overall score of less than 30%, and the overall mean of 43.8% (based on 1700 S-R presentations) is more than double chance. Thus, it was concluded that humans show better ability to localize sounds underwater than theory would predict—and, incidentally, that localization is best for low frequency or broad-band signals.

B. Specific Localization Experiments

Once the validity of the underwater localization response had been established, it appeared appropriate to begin to investigate specific parameters related to this auditory ability in order to better understand its nature and extent. Accordingly, four additional studies were completed at Bugg Springs over the next two years. The purposes of these experiments were to investigate: (1) the contribution of head movements to the localization process; (2) the relative efficiency of different types of sound stimuli; (3) the effects of different signal amplitudes and (4) the effects of training on the ability of divers to improve their localization skills.

TABLE 2. Percent of correct localization responses to each of four stimuli by seventeen subjects. All stimulus presentations were at 110 dB SPL; the diver was located in DALS at an ear depth of 12 m.

Subject	250 Hz	1000 Hz	Stimulus 6000 Hz	Noise	Mean
1	40	52	76	68	59
2	72	52	32	68	56
3	64	36	64	44	52
4	56	52	52	48	52
5	56	48	40	56	50
6	72	48	28	36	46
7	44	52	24	56	44
8	60	20	12	84	44
9	40	52	32	48	43
10	64	36	28	40	42
11	44	40	28	52	41
12	68	16	16	64	41
13	40	24	40	56	40
14	44	28	20	56	37
15	36	40	24	44	36
16	36	36	28	40	35
17	24	28	24	32	27
Mean	50.6	38.8	33.4	52.5	43.8
Standard Deviation	14.0	11.9	16.3	12.9	

Head movement versus fixed head

In this experiment, the basic protocols were followed with one exception; the subject was immobilized by having him replace his dive mask with one that was directly attached to DALS. This approach proved effective since the diver could not move his head even slightly without flooding the mask.

The results of this experiment may be seen in Table 3. From the data, it may be observed that subjects appear to be able to localize sound underwater equally well with their heads immobilized as they could when they were free to scan by head movements. The exception to

TABLE 3. Percent of correct localization responses to each of four stimuli by seven subjects. All stimulus presentations were at 110 dB SPL; the diver was located in DALS at an ear depth of 12 m. Data compare the effects of scanning of the stimulus source by head movements versus responses with the head immobilized.

| | Stimulus | | | |
Condition	250 Hz	1000 Hz	6000 Hz	Noise
Fixed Head:				
Mean	37.1	37.7	28.6	58.3
Movable Head:				
Mean	61.1	37.7	34.8	54.8

this generalization is found for the 250 Hz condition. Here the differences between the two scores are statistically significant with the fixed head responses the poorer. However, it should be remembered that the 250 Hz condition was always presented first and that divers are notoriously uneasy (and for good cause) when their freedom of movement is curtailed. Hence, it was fairly obvious (especially from reports by the subjects) that they were attending as much to life support considerations in this case as they were to the experimental stimuli. They further reported that they became more comfortable with the procedure about the time the second set of stimuli (1 KHz) were presented; the data, then, appear to reflect these conditions. Another interesting result of the experiment was a dramatically reduced response latency for this experimental procedure when compared with any of the others. For whatever the reason, in this case the diver/subjects responded to the stimuli at rates that were about four times quicker than for the other procedures. In any event, it can be tentatively concluded that divers do not have to avail themselves of head scanning in order to localize sound underwater.

Different classes of sound stimuli

In the second experiment, two major changes in protocol were made: (1) the stimuli were presented at 95 dB (SPL) and (2) only 15 presentations of each stimulus were provided. The purpose of the study was to investigate different classes of signals in order to ascertain whether or not different stimuli might be more effective for underwater sound localization than those that had been used previously. Pulses, glides and dolphin whistles were included among these new stimuli.

The results of this experiment may be seen in Table 4. No statistically significant differences were found among the stimuli or between these

TABLE 4. Percent of correct responses to 15 presentations of each stimulus type by six diver/subjects. Stimulus presentations were at 95 dB (SPL); ear depth was 12 m.

Subject	Pulse Trains (PPS)					Glides Hz				Dolphin Whistles
	1	25	50	100-400	400-100	500-2000	2000-500	1500-6000	6000-1500	
1	27	13	47	40	47	40	67	47	40	53
2	13	20	60	53	20	53	40	27	40	20
3	87	93	100	53	93	67	100	53	47	60
4	40	40	40	53	40	60	53	60	67	60
5	33	53	13	73	73	53	60	33	33	33
6	40	20	53	67	40	27	40	27	20	13
Mean	40	40	52	57	52	50	60	41	41	40
S.D.	23	27	26	11	24	13	20	13	14	19

stimuli and those used in the previous (basic) experiment. However, some differences are apparent among the resultant scores; ones which require comment. From the previous studies, it will be remembered that broad-band noise and low frequency sinusoids proved to be more powerful stimuli for underwater sound localization. In these data, the pulse train of 50 pps and the low and middle frequency glides exhibit the higher scores. Hence, this study, along with the earlier ones, provides some indication of the signal classes that may have potential for enhanced acoustic localization underwater. Such data are particularly meaningful when studies of diver acoustic navigation are undertaken.

Intensity effects

The third experiment focused on signal strength with the procedure replicated at 80, 95 and 110 dB. The stimuli used were: 50 pps, 250 Hz, 6000 Hz plus 2000–500 Hz and 1500–6000 Hz glides. Six subjects were presented the stimuli in random order but always with the 95 dB condition first and the 110 dB condition last (due to calibration considerations).

The results of this research may be seen in Table 5. Although not

TABLE 5. Percent of correct localization responses to each of five stimuli at three intensity levels. Six diver subjects were utilized; ear depth was 12 m.

Condition	50 PPS	250 Hz	Stimulus 6000 Hz	2000–500 Hz	1500–6000 Hz
80 dB:					
Mean	29	52	47	48	43
S.D.	18	17	21	17	15
95 dB:					
Mean	36	38	41	33	46
S.D.	7	13	17	9	15
110 dB:					
Mean	52	62	52	48	50
S.D.	19	16	21	17	26

statistically significant, the highest intensity condition exhibits the greatest overall localization scores and the middle intensity the lowest (80 dB: 43.8%; 95 dB: 38.8% and 110 dB: 52.8%). It is suggested that this result could be due to two variables—acting singly or interacting with each other. As was pointed out earlier, the 95 dB condition was always presented first because of the difficulty in calibrating the system;

the 110 dB condition last. Hence, training, or increased experience, may be an important variable in these studies (we have already noted that diver performance tends to improve with increased participation in underwater hearing experiments). The second variable emanates from the possibility that divers utilized tactile cues for the high intensity condition (110 dB). Both of these possibilities appeared feasible; thus they became the subject of subsequent experiments. In summary, however, it appears that divers can localize soft and loud underwater sound sources albeit with slightly different levels of success.

Effects of training

Determination of the effects of training on seven divers who had little or no experience with underwater sound localization constituted the thrust of the fourth experiment. In this case, diver/subjects were given a pre-training test consisting of 15 presentations each of two stimuli (25 pps and 1000 Hz signals) at 95 dB. Following this test, they received two sets of 30 training trials (for each stimulus) in which the correct location of the sound source was provided immediately after a response to the stimulus and while the stimulus was still being presented. Once this process had been completed, subjects received a post-training test similar to the first test. The 1000 Hz stimulus was always presented first.

The results of this experiment may be found in Table 6; improvement

TABLE 6. Pre- and post-training percent of correct responses by seven diver/subjects. Stimuli were 25 pps and 1000 Hz presented at 95 dB; ear depth was 12 m.

Condition	Stimulus	
	25 pps	1000 Hz
Pre-training		
Mean	43	34
S.D.	20	21
Post-training		
Mean	51	51
S.D.	15	18

was noted for both stimulus conditions. Statistical tests indicated that while improvement for the 1000 Hz condition was significant at the 5% level of confidence, the 25 pps condition just barely failed to reach that level of significance. It is possible that (1) a greater number of training trials would have provided a more marked improvement and (2) ease of training is a function of type of signal. In any case, it can be tentatively concluded that training appears to improve an individual's ability to localize sounds underwater.

It is of some interest to note that recently Norman et al. (1971) report an experiment in some ways similar to the four just discussed. Specifically, they had divers indicate which of seven sound sources was energized on a given trial. They found that their subjects could localize minimally when their heads were bare or when they wore a hood with ear holes but that neoprene patches over the ears caused localization performance to become much worse. This experiment was conducted in a swimming pool and in spite of the confidence expressed by these investigators that their research was not contaminated by the effects of multiray paths, it is necessary to be cautious in accepting their results—which they themselves point out to be "paradoxical". In any case, the results of their research are generally consistent with the data described above.

Localization over distance at sea

It is of some importance to investigate the underwater sound localization abilities of divers at distances other than the relatively close range (i.e. three metre) situation employed in the experiments cited above. Even more important is the need to obtain at least preliminary data about diver's abilities to localize sounds in the open sea. Accordingly, a set of two related experiments were carried out during TEKTITE-2. However, since Greater Lameshur Bay was found to be extremely noisy due to steady small boat traffic, the research actually was conducted in the bay of a small island just off Charlotte Amalie, St. Thomas, US Virgin Islands. This site, Buck Island, was made available to the project by the college of the Virgin Islands. A view of the bay may be seen in Fig. 14.

The procedures basic to the previous experiments were followed with certain exceptions. Transducer distance was 12.19 m and subjects were placed in a mini-DICORS (Figs. 23 and 26 of the preceding chapter) in 9 m of sea water. The J-9 transducers were positioned at ear depth (4.5 m) on taut lines tied at one end to cement clumps and at the other to partially filled inner tubes; they were rigged to face the diver. This arrangement prevented the equipment from varying in distance or position due to the effects of currents, etc. The diver's bodies were fixed in relation to the transducers but they were free to move their heads. The first study duplicated the four stimuli (25, 1000, 6000 Hz and thermal noise) utilized in the original experiment; the second employed two pulse trains, four glides and dolphin whistles; signal level was 100 dB SPL re: 0.0002 μbar.

The results of the first study were somewhat disappointing as the scores for the five diver/subjects were lower than expected (250 Hz: 38.4%; 1 kHz: 32.8%; 6 kHz, 20.8%, and thermal noise: 40.0%);

FIG. 14. Views of the Buck Island research site: in (a) the observer is looking East from the research area toward the laboratory building; in (b), West toward the site of the experiments (between the first marker and the boat).

incomplete data was obtained on a sixth diver and his scores are not included. However, in retrospect, these depressed scores appear to be due to several extraneous conditions. First, three of the five subjects had not participated in underwater sound localization experiments of any type. Secondly, the experimental site was considerably more noisy than Bugg Springs and the signals sometimes were difficult to hear. Finally, a number of barracuda were feeding on schools of small fish in the vicinity of the project during the experimental runs. This activity was somewhat distracting to the diver/subjects. In any case, except for the 6000 Hz condition, the mean responses were clearly above chance.

The results of the second study appeared to confirm the notion that divers possess some ability to localize underwater sounds in the open sea; these results may be seen in Table 7. Three of the four subjects in this experiment were those who exhibited the poorest scores on the first study; the fourth was a diver who had not participated previously in any investigations of this type (except for an abortive attempt at the first procedure). Nevertheless, the resulting scores are of the same magnitude as those obtained from similar experiments at Bugg Springs. Indeed, they are quite high—especially those for the pulse trains and the dolphin whistles. In short, it appears that, once divers are in command of the task, it can be expected that they will localize sounds at sea at the same levels of competence that they can in the relatively benign environment of fresh water springs. The development of appropriate skills in this regard is of special importance to diver navigation by means of acoustic beacons.

C. Acuity and Precision of Underwater Sound Localization

The research programme developed by the other of us (SF) focused on different issues and techniques than those described above. The objectives of the following experiments were (1) to determine the acuity (or resolution) of the underwater localization process and (2) to determine its functionality (i.e.: the precision by which localization is specified).

Minimum audible angle (MAA) experiments

Mills (1958) describes a procedure which provides an index of the acuity of the auditory localization system, i.e., the Minimum Audible Angle (MAA). In this procedure, one sound source is located in the intersection of the coronal and midsagittal planes of the head and the second source is to the right or left of centre. The experimental stimuli are presented in succession as pairs of pulses; the first always coming from the central position and the second from either side. The subject reports whether the sound came from the right or left speaker and the difference in degrees between the standard and comparison (which can be detected 75% of the time) becomes the Minimum Audible Angle (MAA). Investigators who have worked with marine mammals have changed this procedure slightly. In these cases, the standard source is omitted

TABLE 7. Percent correct responses of four diver/subjects for seven stimuli in open water sound localization experiment. Projectors were 12 m from subjects; signal strength at listener was 100 dB SPL re: 0.0002 μbar and ear depth was 4.5 m.

Subject	Stimulus						
	15 pps	50 pps	500–1000 Hz	1000–500 Hz	4000–6000 Hz	6000–4000 Hz	Dolphin Whistles
1	66.6	60.0	66.6	33.3	26.6	33.3	46.6
2	53.3	26.6	33.3	66.6	40.0	20.0	53.3
3	40.0	53.3	33.3	46.6	40.0	33.3	33.3
4	40.0	46.6	26.6	6.6	26.6	26.6	66.6
Mean	50.0	46.6	40.0	38.3	33.3	28.3	50.0

and the animal indicates whether it hears a signal to its right or left. Thus, the MAA can be defined as the difference in degrees between the place of the sound source and an imaginary line extending from a point between the eyes.

In the series of MAA experiments reported here the procedure was as follows: Diver/subjects were situated in the underwater equipment configuration that can be seen schematized in Fig. 15. As may be noted, the projectors were set at a predetermined angle, and the diver was

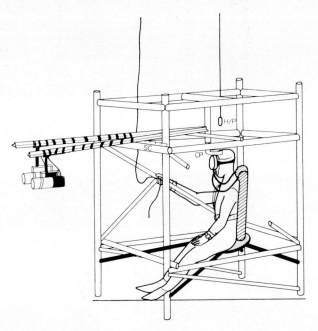

Fig. 15. MAA aparatus.

seated in the stage with his head against a positioning ring. Subjects were instructed to hold their breath and signal when ready. They then listened for the stimulus and responded either "right" or a "left" on each trial. The experimenter energized one of the projectors according to a pre-determined random sequence of 15 "right" and "left" responses; the number of angles tested on a dive (in blocks of 30 trials) depended on the temperature of the water. Divers were instructed to surface after the first series if they were chilled or to continue with the experiment if they were able; this procedure was necessary because of the low water temperature (30°F to 50°F). Responses to the right or left source were expected within 15 s of the stimulus onset and with the exception of the first few trials and the smallest angles, nearly all responses were initiated well within that limit. Although the diver/subjects wore arctic wet suits and hoods, they were limited to approximately one hour of work per six hour day in order to avoid the effects of hypothermia. Three stimuli (3500 Hz, 6500 Hz and white noise) constituted the experimental stimuli; they were pulsed (100 ms duration at 1/pps) and the intensity level was 40 dB re 1 μbar. Eight divers served as subjects. However, two could not perform the localization task so they were not included in the analysis.

Figure 16 summarizes performance at each angle tested for each stimulus and indicates the MAA's were: (1) 21.5° at 3500 Hz, (2) 14.5° at 6500 Hz, (3) 9.75° at white noise.

Since, divers do not normally depend on auditory information to orient themselves underwater, it was hypothesized that their performance would be less than asymptotic in the MAA task until they learned to utilize the available auditory cues effectively. In order to test that hypothesis, a second experiment was designed to give them a large number of trials (300) with feedback at an azimuth greater than the point at which correct responses occurred only 50% of the time (3°) but smaller than the MAA (9.75°). In order to monitor the immediate effects of feedback on performance, a block of feedback trials was always followed by a block of no-feedback trials. Therefore, in the first half of this experiment, two blocks of 30 trials were run at seven degrees on each dive. Pulsed thermal noise was utilized as the stimulus because it yielded the greatest acuity in the first experiment and was thus expected to provide more easily detectable cues for this procedure. In part two of this experiment, the MAA was redetermined using the same procedure as in the first experiment.

Four divers who had participated in the preceding experiment served as subjects. The mean MAA's of these four subjects, obtained from that experiment, were compared with their scores after training. Figure 17 summarizes their performance and demonstrates that training improves the acuity of sound localization underwater.

Anderson and Christensen (1969) also reported an experiment based on the MAA procedure. They tested seven divers in a free field environment both "off shore" and in an enclosed harbour. Their subjects sat in a wooden chair and transducers were suspended 6 m away in a half circle around them at angles of 10°, 15°, 20°, 30°, 45° and 90°. Subjects held a hand switch by which they indicated whether the sound came from the right or left of the median plane. These authors investigated localization acuity at 1, 2, 4, 8 and 16 kHz and concluded that, although there was considerable variability among subjects, it appeared that, "Directional hearing underwater seems to work on the

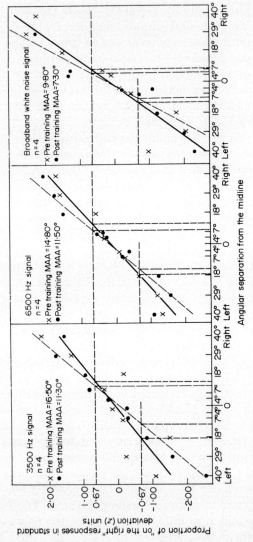

Fig. 16. MAA's obtained from six divers prior to and following training.

same parameters as in the air, with allowances for the longer wavelength in water". They did not determine the actual MAA's. However, from the minimum angle used and the relatively poor performance at frequencies comparable to those used in our experiments, one would expect them to be quite large.

Table 8 presents a comparison of the MAA's obtained from three

TABLE 8. Comparison of sound localization acuity underwater in several marine mammals and man.

Animal	Investigator	Stimulus	Minimum Audible Angle
Porpoise	Dudok van Heel (1959)	6.0 kHz	7.9°
		3.5 kHz	11.0°
		2.0 kHz	12.6°
Sea lion	Gentry (1967)	6.0 kHz	10.0°
		3.5 kHz	15.0°
Seal	Mohl (1968)	2.0 kHz	3.1°
Man	Feinstein (1973)	6.5 kHz	11.3°
		3.5 kHz	11.5°
		White noise (20–2000 Hz)	7.3°

marine mammals and from man. When evaluating these data it must be remembered that, while the test procedures followed the same basic paradigm, different statistical procedures were used to determine the MAA; also, there were differences in test environments and signal characteristics. Therefore, any comparisons that are attempted must be made with some caution. With these limitations in mind, it is still possible to note how closely the performance of the trained human listener approximates that of the trained marine mammal. This level of performance would be totally unexpected in the context of earlier hypotheses that sound localization under water requires special anatomical adaptations. Further, in view of the data obtained for the porpoise and the sea lion, it could be predicted that human localization performance would improve as a function of increasing frequency. This was not found to be the case even though in the first experiment the MAA was bigger for the 3.5 kHz signal than it was for the 6.5 kHz signal. After training this difference disappeared, a result which may indicate that the earlier difference in MAA at these frequencies was a function of the audibility of the signal. A further implication may be that human sound localization underwater is not wavelength dependent at these frequencies. This conclusion seems to be a reasonable one in light of the compelling evidence that human hearing under water is a bone conduction phenomenon (Brandt and Hollien, 1967; Hollien and Brandt, 1969; Smith, 1965).

Finally, the comparability of the sound localization acuity of humans to that of marine mammals was an unexpected result. Assuming that binaural difference cues are used in underwater localization, it might have been predicted that acuity would have been severely reduced. However, such a prediction would have been based on a false assumption—namely, that the magnitude of the binaural cues in air is necessary for

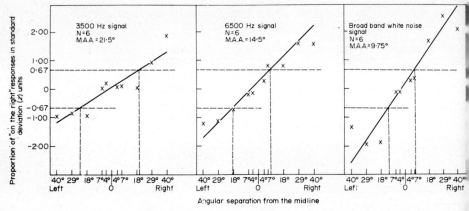

FIG. 17. MAA's obtained from four divers after training.

localization; in reality, it is more than sufficient. For example, Fedderson et al. (1957) found that the interaural time differences for noise were 180 μs at 23° azimuth, 369 μs at 45° azimuth, 486 μs at 60° azimuth, and 600 μs at 90° azimuth. These results may be contrasted with those of Klumpp and Eady (1956) who found a threshold of correct lateralization at 11 μs for repeated clicks, 28 μs for single clicks; or with Tobias and Zerlin (1959) who demonstrated that just noticeable differences for phase vary with signal duration, reaching an asymptotic value of 6 μs at 700 ms duration. Clearly, man ought to be able to utilize less than the maximum stimulus magnitude available to achieve sound localization. On the basis of these data it may be said that the development of human underwater sound navigation procedures appears to be a reasonable prediction.

A third MAA experiment was required because the first two experiments of this series were carried out in the cold and relatively noisy waters of a Canadian harbour. The divers had to wear hoods while being tested and the combination of these hoods and the ambient noise level made it necessary to use a very intense sound source (40 dB re: 1 μbar). Thus, it was possible that the obtained MAA's could be the result of both auditory and nonauditory (tactile) cues. Therefore, a replication of the earlier experiments was carried out at Bugg Springs. In this experiment, the seven divers who served as subjects wore full wet suits (except for the hood) and stimulus intensity was 26 dB re: 1 μbar. Inspection of Fig. 18 makes it clear that no remarkable differences were found between the performance of these subjects and that of divers

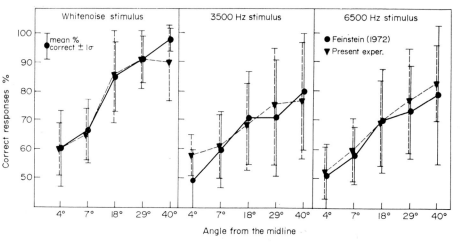

FIG. 18. Comparison of % correct responses in cold water MAA and warm water MAA experiments.

from the earlier experiments. In terms of acuity, the mean MAA at 3500 Hz was 22°, at 6500 it was 19° and with white noise it was 10°. This experiment provides what we believe is strong evidence that our subjects do not use non-auditory (tactile) cues in the sound localization procedures. This is not meant to indicate that were such cues available they would not be used. Rather, they are simply not necessary for the levels of discrimination that we have found in all our research. The data also indicate that MAA's are relatively insensitive to a wide range of conditions, e.g., level of background noise, water temperature, water salinity and absolute loudness of the stimulus. Finally, it appears as if the presence or absence of a neoprene diving hood is not likely to effect the capacity of divers to localize sounds, provided that the signals are loud enough to be heard.

Precision of underwater sound localization

In order to make a meaningful comparison between sound localization in the air and underwater, it is necessary to determine the precision with which a sound can be located in the horizontal plane as well to determine localization acuity (MAA). The precision of the localization response in this case, was defined as the average difference between the objective and subjective azimuths of a single sound.

A pointing procedure was selected for this series of experiments

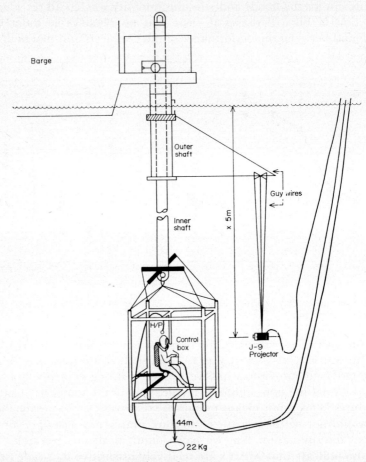

FIG. 19. Localization precision experimental apparatus.

because it provided the listener with a response that was: (1) easily made under water; (2) comprised of well-defined motor behaviours which could be standardized for all subjects and (3) capable of reflecting the impression of location. The pointing response as defined here also provided the experimenter with a numerical readout that was accurate to $\pm 1°$ of the indicated position.

In this experiment, the diver/subject sat in the same PVC framework that was used in the MAA experiments; the configuration of the system may be seen in Fig. 19. The system utilized was suspended at an ear depth of 6 m by ropes that went from the corners to the centre shaft of a unit designed to position accurately for calibrating sonar domes and other specialized heavy listening equipment. The framework was tied to prevent twisting and a 22 kg anchor was attached to the bottom centre of the framework to prevent sway. The sound source (a J-9 sound projector) was suspended from a boom attached to the outer shaft of the positioning device. The horizontal boom placed the projector face 1.83 m from subjects' ears and the vertical shaft dropped 3.66 m from the boom to place the projector at ear level. The sound source and framework could be rotated independently by means of a

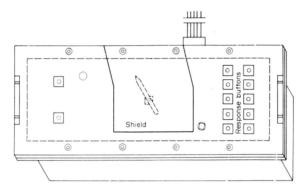

FIG. 20. Diver pointing system.

servomotor system on the deck. In turn, the servomotors were keyed to a polar recorder which could be read to the nearest degree.

The control system used by the divers to indicate the azimuth of the source is shown in Fig. 20. The system consisted of a plexiglas watertight box; on the top centre of this box a flat metal pointer, sharp at one end and rounded at the other, was located beneath a metal shield. The shield was adjusted to allow the subject to move the pointer without seeing either his hand or the pointer.

During the actual test period, the diver set the azimuth indicator at zero. He then removed his dive mask and put on the blacked out mask attached to the head positioning ring. When ready, he signalled the experimenter. At that point, the experimenter set the azimuth and turned on the signal. The diver responded by directing the pointed end of the azimuth indicator at the sound source. Once the azimuth setting was recorded, the next trial was begun. Subjects required approximately 30 minutes to complete a 24 position session.

In order to establish baseline performance against which sound localization could be compared, an experiment was run in which the diver could see the J-9 but not his hand or the indicator. Fifteen random permutations of the 13 positions were completed by each of four subjects. The average error in this experiment was 5.98°, which is an indication of the maximum sensitivity of the pointing response.

Following this visual localization procedure, a series of auditory localization experiments were carried out. In these experiments the sound source was moved to any one of 24 predetermined positions between 15° and 360°. Evaluation of the data obtained from this

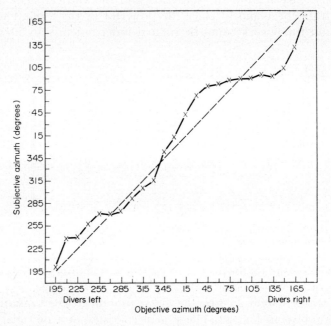

FIG. 21. Precision of the localization response when the source is located in the horizontal plane.

experiment will reveal an average error (AE) within the full circle of 19.20°. However, by reference to Fig. 21, it can be seen that there were obvious asymmetries; ones which appear as greater precision (1) in the front semi-circle (19.35° AE) than in the rear (23.44° AE) and (2) on the left side (12.75° AE) than on the right (27.23°). Moreover, response reversals were observed 10% of the time—a situation that approximates the classic localization data (in air) obtained by Stevens and Newman (1934; 1936).

The reason for the noted asymmetry of the curves seen in Fig. 21 becomes obvious when it is remembered that all subjects moved the pointer with their right hand. To illustrate this effect, point your right index finger while rotating your hand on a pivot point. When doing so, you will note that pointing to a location in front of you and to your left is accomplished with ease, whereas points to your right and right-rear require an unnatural posture and are made with some difficulty. Thus an AE of approximately 13° probably is a more valid indication of the precision attainable underwater and is directly comparable to the Stevens and Newman data.

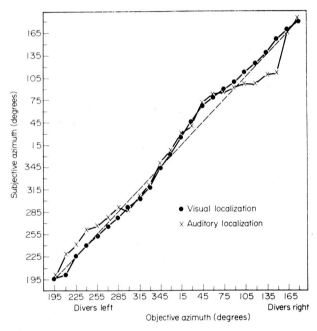

Fig. 22. A comparison of the precision of a visual localization response with an auditory localization response after training.

In a second experiment an attempt was made to determine whether or not some level of training would improve performance in localization studies such as these. In this case, it will be remembered that the diver/subjects were required to point an indicator at the sound source but were given no information concerning the precision of their judgements. Since they had not previously attempted a task such as this one, it was reasoned that they were not likely to be operating at their asymptotic level of performance. Hence, in the second investigation, the divers were provided with a substantial number of training trials, and they were given an opportunity to correlate the auditory location of the sound source with its visual location. It was expected that this procedure would constitute an effective way to bring performance to asymptote quickly.

The method roughly paralleled that of the first study except that blocks of two dives were run each day for 10 days. On the first dive, the diver/subject wore a clear mask. He was instructed to keep his eyes closed between trials and to open them only after he had set the indicator. When the diver did so, he was allowed to make any correction in the azimuth indication that he judged necessary. On the second dive an opaque mask was worn and responses to source location were made by sound alone.

Figure 22 compares the visually corrected localization response (AE-6.98°) to the auditory localization response (AE = 12.20°). Once again reversals occurred but in this case they were substantially reduced.

In any case, the reduction in the average error, over the full circle, is striking when these (auditory alone) data are compared to those from the first precision experiment (see again Fig. 21). However, it is not possible to determine at this time whether the improvement in performance resulted from an increased sensitivity to available binaural information or to an increase in coordination between the ears and hand. In either case some advantage appears to accrue from training.

All of the signals used in this set of experiments were presented at an acoustic pressure of 40 dB re: 1 μbar. This value was chosen because it constituted a level the diver could hear reliably when wearing a wet suit and hood. As discussed previously, there is a possibility that this intensity level was high enough to provide the diver/subject with tactile sensations; in turn, these cues may have formed the basis of the localization response. One way to test this hypothesis is to replicate the experiment with reduced signal intensity and examine the data for deterioration of the localization response. Such a project was carried out and it was found that, as the signal became less intense (and therefore more difficult to hear), it was localized with less precision, e.g., at 35 dB (re: 1 μbar) the average error was 17.9°, at 30 dB it was 20.9 dB, at 25 dB, 27.7° and at 20 dB, 28.4°. These results support the hypothesis that underwater sound localization is more a function of audibility than of tactile cues. However, the data are not in total agreement with two of the previous studies. First, degradation in response noted here was not found in the last MAA experiment where signal strength was relatively low. Secondly, in the intensity experiment utilizing multiple sound sources, no systematic reduction in sound localization ability was noted. It must be remembered that in the present experiment the signal to noise ratio was not as great as it was in the other experiments, i.e. even at reduced intensities the signal was always clearly audible in the later MAA experiment as well as the multiple sources experiments, while in the present experiment relatively small reductions in intensity yielded marked reductions in audibility.

D. The Joint Programme on Underwater Sound Localization

Recently the two programmes of research described above have been merged into a single one and a rather extensive and vigorous research programme has resulted. For example, the MAA experiment in warm, quiet water (described above) was conducted under the auspices of this joint programme. Other related subjects, too numerous to list, have become the focus of our research thrust in this area. However, to date, data are available only for one set of experiments—which were conducted by the authors in collaboration with Dr. Jo Ann Kinney.

It will be noted that almost all of the MAA and precision experiments were carried out with signal levels of 40 dB re: 1 μ bar. The only exceptions were the two studies in which signal strength itself was studied. Moreover, many of the experimental conditions utilized in the multiple source experiments employed signal strengths of 100–110 dB SPL. In these cases, the diver/subjects (who wore no wet suits or hoods; or only a jacket) often reported that they were aware of tactile sensations—especially on the legs—and they wondered if these cues were not being used as an aid to the auditory localization process. Accordingly we judged it necessary to conduct experiments that would directly assess the presence or absence of such tactile cues.

Two experiments were carried out. Both utilized the multiple sound source approach (see again Figs. 12 and 13) and were conducted at USRD's Bugg Springs' facilities. Stimuli were 250, 1000, 6000 Hz and thermal noise. Signals were presented randomly among the five transducer positions in sets of five pulses; each pulse had a duty cycle of 1 s, a duration of 500 ms and a 25 ms rise and decay time. Experienced diver/subjects were utilized in the experiments; they wore nothing but bathing suits, SCUBA gear and a face mask.

It was apparent that, if we were to evaluate the role tactile cues play in underwater sound localization, we would have to develop procedures by which these (tactile) cues were maximized and the auditory cues minimized. In order to do so, we felt that at least two procedures would be necessary. In the first procedure, we presented all stimuli at 55 dB SPL. Of course, these signals were below the auditory threshold (see Fig. 7) but we reasoned that, while not audible to the diver, they might be strong enough to be felt. In the second, experiment we wanted to utilize a signal at an intensity level that we knew from experience could be felt by the submerged diver. Hence, in this study, stimuli were presented at 110 dB SPL. Naturally, a signal of this strength could be heard clearly. Hence, in this experiment, all diver/subjects wore either one or two hoods (depending on preference) and received (via underwater phones) a sawtooth noise emitted at a high enough intensity level that it effectively masked the experimental stimuli.

The results of the first study yielded only chance levels of localization response* (250 Hz: 22.4%, 1000 Hz: 14.4%, 6000 Hz: 16.8% and thermal noise: 18.4%). In this case, it could be concluded that the stimuli were not presented at levels intense enough for either tactile or auditory localization. The scores for the second experiment were slightly higher than for the first (250 Hz: 30.0%, 1000 Hz: 16.5%, 6000 Hz: 22.8% and thermal noise: 16.0%); however, only one of these scores is significantly above chance. Moreover, the diver/subjects reported (1) that while they often could "feel" the signal, they could not tell (from these tactile cues) where the sound was coming from and (2) that sometimes they seemed to be able to hear the stimulus over the masking signal; in this case they went ahead and localized by means of these sensations. In summary, it appears that the diver wishing to locate underwater sound sources or navigate by means of acoustic signals may be able to do so via the auditory modality but probably cannot depend on any supplemental information processed tactually.

9. Some Tentative Theories Concerning Underwater Sound Localization

Enough research now has been completed to conclusively demonstrate that man is indeed capable of underwater sound localization. It is conceded, of course that this capability is somewhat poorer in water than it is in air but, in sum, the data suggest a reasonably similar type of processing for both environments. In any case, it would seem appropriate at this point to attempt to explain (at least tentatively) why localization is possible in water and suggest some of the mechanisms that account for this human ability. Three such explanations are offered; the one we believe to have the greatest basic merit is presented first.

It has been demonstrated that the mechanism for underwater hearing is primarily one of bone conduction; that is, the two "external" ears do not operate to provide differential auditory information to the cochlea via the normal mode. Moreover, the speed of sound in water is so great—relative to its speed in air—that this factor also operates in a manner detrimental to the usual localization process. Indeed, based on these relationships and the inter-aural distances found in man, it can be

* The value for 250 Hz actually is above chance; however, it was the signal that was "heard" by the subjects (see below).

expected that sound arrives at one cochlea only microseconds before it arrives at the other. However, if very small arrival-time differences (in this case, on the order of 0 to about 67 microseconds at angles of 0° and 90°, respectively) can be processed to provide appropriate information, at least some underwater sound localization should be possible. Evidence that such processing capabilities are present in humans has been reported, viz., that the ear is able to resolve time differences on the order of only 7 μs (Tobias and Zerlin, 1959; United Research, Inc., 1962; Zerlin, personal communication, 1969). Thus, if the central nervous system can resolve time differences across the head of 10 μs and less, the type of processing described above would permit considerable underwater sound localization to be accomplished. That time of arrival differences (or inter-aural time delays, ITD) are utilized in underwater sound localization receives further support from MAA experiments. They suggest that the acuity of the human system is approximately 10°; theoretically, an ITD of 10° would be 11 μs or greater underwater.

Tonndorf (1968) has taken the position that the bone conduction mechanism is most likely to transduce ITD. He bases this hypothesis on a series of experiments directed at the determination of the mechanism of bone conduction (Tonndorf *et al.*, 1966). In this regard, he indicates that there are three independent "components" contributing to bone conduction: an external ear component, a middle ear component and an inner ear component. The first of these depends on acoustic radiation from cartilagenous walls into an air filled external ear canal; presumably, this component would be ineffectual in an underwater environment. However, the middle ear component remains active underwater. This component is a function of ossicular inertia. Tonndorf hypothesizes that ossicular inertia allows the introduction of phase differences between the two ears (except at 0° azimuth). These phase differences would be somewhat, but not completely obscured by the inner ear component (Tonndorf, 1968); however, they should still provide considerable information to the central nervous system. The fact that we have found low frequencies to be localized best is consistent with this notion.

At this time, it is impossible to describe the effect of the head on the sound field underwater; however, we are presently studying this issue. If the head were to act as it does in air, then it would be expected that, at frequencies at least four times higher than 1500 Hz, intensity cues would form the basis of underwater localization. However, little information is available in this regard since the highest frequency used in any of our experiments has been 6500 Hz and localization scores at that frequency were not markedly different from those observed for the lower frequencies.

Brandt (1970) in a personal communication has proposed that inter-aural intensity difference cues will be available to the diver/subject as a result of the differential sensitivity of the skull to bone conducted sound—and that intensity cues rather than time cues determine underwater localization. He cites experiments reported by Isele et al. (1968) with respect to differential sensitivity of the auditory system for bone conduction (in air) as a function of five different placements of a bone conduction oscillator. The results of the cited study suggest that a subject may be able to localize this type of stimulus on the basis of intensity differences alone. On the other hand, experiments in which the sound source was driven at high energy levels at a distance from the head have indicated no differential sensitivity of the skull at either 0° or 180° (Nixon and von Gierke, 1959; Zwislocki, 1957). Moreover, Brandt's proposal appears limited in its scope since it ignores relevant ITD relationships.

We have suggested on a number of occasions that the areas of the skull immediately adjacent to the external auditory meatuses are probably the most sensitive regions for underwater bone conduction. We now are encouraged in this hypothesis by the results of McCormick et al. (1970) who were able to conduct some sophisticated "body-mapping" experiments on two species of dolphin. From their research, they were able to demonstrate that sounds are transmitted to the dolphin's ear (primarily) through the tissues in its immediate vicinity. We plan to carry out similar mapping experiments on humans; the results of this effort should provide information concerning both ITD and IID relationships in the underwater environment.

In the final analysis, and until more data are obtained, we must take the position that both temporal and intensive cues probably are utilized to provide water sound localization. Admittedly, available data better support ITD as a factor in this regard but the potential utilization of IID cues cannot be ignored. Moreover, it is very probable that these cues interact in the localization process and they do so in ways that are predictable from measurements made in air.

10. Diver Acoustic Navigation

There are a number of reasons for studying diver sound localization, for example, such research: (1) adds to our basic understanding of binaural hearing (2) may allow a diver to determine the location of objects or individuals emitting acoustic signals of one kind or another and (3) is directly relevant to the implementation of acoustic diver navigation systems. It has been apparent almost from the inception of SCUBA that one of the major limitations to diver effectiveness is the absence of adequate and efficient methods of directing him from one specific location to another. The traditional solution to such navigation problems has been the use of an underwater compass to permit him to use a "dead reckoning" procedure. However, Anderson (1968) has pointed out, that his tests with well-trained subjects showed:

> ". . . the average performance accuracy . . . was plus or minus 53 feet from the centerline of the measurement array or 3.98 degrees in compass error . . . In an operational situation when a diver might be engaged in an underwater search task or in accurate placement of underwater equipment, this level of performance would be marginal."

Indeed, such performance would be less than marginal under the more difficult conditions that often are encountered by divers attempting to navigate underwater.

Diver "retrieval" devices are sometimes utilized as navigation aids. Such retrieval has been and still is accomplished by means of explosive devices such as the M-80 explosives used by the U.S. Navy or the Thunderflash device of the Canadian Forces. In this regard, it has been demonstrated by Hollien that explosives are only marginally effective even for signalling recall (they are especially inadequate at long ranges) and are totally unsuitable for acoustic navigation due to the very brief signal they produce.

Two relatively new approaches to the navigation-retrieval problem are presently evolving; one approach is to develop electronic devices for sound navigation and ranging (SONAR); the other is to study and possibly develop human biological SONAR capabilities. The former has resulted in a proliferation of devices (they sometimes consist of rather large and cumbersome units) which the diver carries in his hands or on his person—and upon which he must depend for navigation. The second approach is the approach to which we subscribe; it is based on the evidence that it is possible for man to localize sound underwater with considerable precision—and without the aid of electronic listening devices. Specifically, our assumption is that in most cases the human auditory system can be used underwater to provide the necessary information for effective navigation.

Earlier in this chapter we described the results of Ide's (1944) attempt to determine whether underwater sound navigation could be used by commando swimmers to locate and return to small boats at night. Briefly, he found that this acoustic navigation task could be accomplished by some of his swimmers but he felt it was too unreliable to be practical for military operations. The matter then rested for some 25 years.

The issue of diver sound navigation came under scrutiny again when one of us (SF) attempted to replicate Ide's experiment in deep open water using a signal which had proved most effective in related sound localization experiments (Feinstein, 1969). In this experiment the sound source (a J-11 transducer) was placed one meter above a thermocline (average depth: 6 m) at the centre of one end of a large barge anchored in 45 m of water. The diver/subject entered the water at a predetermined point approximately 270 m from the barge and descended until he could hear the signal. In all cases he had a 15 m safety line attached to him at one end and to a marker float at the other. On hearing the signal, the diver/subject attempted to swim toward it. A second (safety) diver in a tender maintained contact with the float via a long safety line.

Four divers completed a total of 22 acoustic navigation runs; of that total, 69% succeeded in reaching the barge. Moreover, all of the misses were in the general vicinity of the target and the diver never swam a course away from it. Thus, these results were much the same as those reported by Ide.

Subsequently we have run an additional 23 subjects for a total of 70 trials, in various other marine and fresh water environments, using different signals and different sources. Of that total number of trials, the subjects reached the sound source nearly 93% of the time. These results would appear to indicate that the variability found in diver sound navigation is determined to a great extent by extra-aural perceptual parameters. For example, Leggiere et al. (1970) report an experiment which yielded evidence that non-aural perceptual factors contribute to the inter-diver variability in sound navigation. Their navigation course was 12.19 m long and they allowed a maximum of five minutes to complete the run. Four subjects completed 20 trials in this experiment and 60% of the trials were successful. The signal used was an 800 Hz tone on for 1 s and off for a quarter of a second. They suggest that success in underwater sound navigation is "dependent on being oriented to the bottom plane, making the search essentially a two dimensional exercise". They also suggested that diving skill and confidence in one's ability as a diver perhaps were more important than the opportunity to learn to utilize sounds for navigational purpose.

To date, all of the acoustic navigation experiments that have been reported have utilized a beacon consisting of a single sound source. It is possible, however, that an array of two or more sound emitters in the proper temporal and spatial relationship could prove to be a more effective target for the diver than will the single source. One of us (HH) has proposed that a line array of six J-9 transducers energized in sequence to produce an apparent auditory movement (the *Phi* phenomenon) is likely to be more effective than a single pulsed source. In theory, the array will provide the same kind of phenomenal motion cues as are utilized in landing light systems for aircraft runways. A diver listening to the beacon would be able to orient himself so that, for example, the sound would move first from right-to-left and then from left-to-right. The impression of spatial location then should be so strong that deviations from a true course would be recognized immediately.

Experiments which examine the utility of the *Phi* approach to underwater acoustic navigation are presently in progress. In addition to these experiments, we are continuing to work with and improve the single source concept. In any case, we are confident that within a few years a form of acoustic navigation which is completely dependent on the binaural capacity of the diver will be a standard diving procedure.

Acknowledgements

The authors thank their colleagues, H. Rothman, D. Brown, J. McNulty, P. Hollien and J. Feinstein for their contributions to both the research reported and the present chapter. Much of the work reported by the authors was supported by the Office of Naval Research contract number N00014-68-4-0173-0008; the Defense Research Board of Canada, contract number 9401-49, and a National Institutes of Health grant, number NS-10121.

Special thanks are due to the Swimmer–Diver branch of the Naval Coastal Systems Laboratory in Panama City, Florida, the Underwater Sound Reference Division of the Naval Research Laboratory at Orlando, Florida and the Defense Research Establishment Atlantic at Dartmouth, Nova Scotia. Without their outstanding support much of our research would have been impossible.

References

Anderson, B. (1968). Diver performance measurement, *Tech. Rept. No. 16*, U-417-68-030, Electric Boat Division, General Dynamics, Groton, Conn.

Anderson, S. and Christensen, O. O. (1969). Underwater sound localization in man, *J. Aud. Res.* **9**, 358–364.

Batteau, D. W. (1967). The role of the pinna in human localization, *Proc. Roy. Soc. (Biol.)* **168**, No. 1011, Series B, 158–180.

Batteau, D. W. (1968). Listening with the naked ear, *In:* "Neurophysiology of Spatially Oriented Behavior" (Ed. S. J. Freedman). The Dorsey Press, Homewood, Illinois.

Bauer, B. B. (1970). Comments on effect of air bubbles in the external auditory meatus on underwater hearing thresholds, *J. Acoust. Soc. Amer.* **47**, 1465–1467.

Bauer, B. B., Rosenheck, A. H. and Abbagnaro, L. A. (1967). External ear replica for acoustical testing, *J. Acoust. Soc. Amer.* **42**, 204–207.

Bauer, B. B. and Torick, E. L. (1966). Analysis of underwater earphones, *J. Acoust. Soc. Amer.* **39**, 35–39.

Bogert, R. J. (1964). Attenuation of sound by a neoprene diving helmet. *USN/USL Calibration Memo*, 2568–2610.

Brandt, J. F. and Hollien, H. (1969). Underwater hearing thresholds in man as a function of water depth, *J. Acoust. Soc. Amer.* **46**, 893–894.

Brandt, J. F. and Hollien, H. (1967). Underwater hearing thresholds in man, *J. Acoust. Soc. Amer.* **42**, 966–971.

Dudok van Heel, W. H. (1959). Audio-direction finding in the porpoise (*Phoncaena phocaena*), *Nature* (Lond.), **183**, 1063.

Farmer, J. C., Jr., Thomas, W. G. and Preslar, M. (1971). Human auditory responses during hyperbaric helium oxygen exposures, *Surgical Forum* **22**, 456–458.

Fedderson, W. E., Sandal, T. T., Teas, D. C. and Jeffress, L. A. (1957). Localization of high frequency tones, *J. Acoust. Soc. Amer.* **29**, 988–991.

Feinstein, S. H. (1966). Human hearing underwater: are things as bad as they seem?, *J. Acoust. Soc. Amer.*, **40**, 1561–1562.

Feinstein, S. H. (1969). Underwater sound navigation, Unpublished paper given at Men's Performance in the Sea, Seminar, Communication Sciences Laboratory. University of Florida.

Feinstein, S. H. (1971). The acuity and precision of underwater sound localization, Doctoral Dissertation. Dalhousie University, Halifax, Nova Scotia.

Feinstein, S. H. (1973). Acuity of the human sound localization response underwater, *J. Acoust. Soc. Amer.* **53**, 393–399.

Feinstein, S. H., Hollien, H. and Hollien, P. (1972). Diver auditory sensitivity: another look at bone conduction, **52**, 170 (Abstract).

Fisher, H. J. and Freedman, S. J. (1968). The role of the pinna in auditory localization, *J. Aud. Res.* **8**, 15–26.

Fluur, E. and Adolfson, J. (1966). Hearing in hyperbaric air, *Aerospace Medicine* Vol. 37, August, 783–785.

Gentry, R. L. (1967). Underwater auditory localization in the California sea lion (*Zalophus californiaunus*), *J. Aud. Res.* **7**, 187–193.

Goldman, D. E. and Hueter, T. F. (1956). Tabular data on the velocity and absorption of high frequency sound in mammalian tissue, *J. Acoust. Soc. Amer.* **28**, 35–37.

Hamilton, P. M. (1957). Underwater Hearing Thresholds, *J. Acoust. Soc. Amer.* **29**, 792–794.

Hollien, H. (1973). Underwater sound localization in humans, *J. Acoust. Soc. Amer.* **53**, 1288–1295.

Hollien, H. (1969). Underwater sound localization: Preliminary information, *J. Acoust. Soc. Amer.* **46**, 124–125 (Abstract).

Hollien, H. and Brandt, J. F. (1969). The effect of air bubbles in the external auditory meatus on underwater hearing thresholds, *J. Acoust. Soc. Amer.* **46**, 384–387.

Hollien, H. and Feinstein, S. H. Contribution of the external auditory meatus to auditory sensitivity underwater. (in press).

Hollien, H., Feinstein, S. H., Rothman, H. B. and Hollien, P. (1973). The auditory sensitivity of divers at high pressures. *Proc. Fifth Symp. Underwater Physiol.* (in press).

Hollien, H., Kinney, J., Feinstein, S. and Hollien, P. (1973). Subliminal underwater sound localization, *J. Acoust. Soc. Amer.* **53**, 336 (Abstract).

Hollien, H., Lauer, J. L. and Paul, P. (1970). Additional data on underwater sound localization, *J. Acoust. Soc. Amer.* **27**, 127–128, (Abstract).

Hollien, H. and Thompson, C. L. (1967). A diver communication research system (DICORS), *Progress Report CSL/ONR No. 2*, ONR Grant Nonr 580(20), 1–5, AD-648, 935.

Ide, J. M. (1944). Signalling and homing by underwater sound for small craft and commando swimmers, NRL Sound Report No. 19.

Isele, R. W., Berger, K. W., Lippy, W. H. and Rotolo, A. L. (1968). A comparison of bone conduction thresholds as measured from several cranial locations, *J. Aud. Res.* **8**, 415–419.

Jeffress, L. A. (1957). Note on the interesting effect produced by two loud speakers under free space conditions by R. L. Hanson and W. E. Kock, *J. Acoust Soc. Amer.* **29**, 145.

Kitagawa, S. and Shintaku, Y. (1957). Direction perception and subject auditory direction line, *lcta Oto-Laryng.* **47**, 431–443.

Klumpp, R. G. and Eady, H. R. (1956). Some measurements of interaural time difference thresholds, *J. Acoust. Soc. Amer.* **28**, 859–860.

Ludwig, G. D. (1950). The velocity of sound through tissue and the acoustic impedance of tissues, *J. Acoust. Soc. Amer.* **22**, 862–866.

Leggiere, T., McAniff, J., Schenck, H. and van Ryzin, J. (1970). Sound localization and homing in divers, *Mar. Tech. Soc. J.* **4**, 27–34.

Luria, S. M. and Kinney, J. A. (1970). Underwater vision, *Sci.* **167**, 1454–1464.

McCormick, J. G., Wever, E. G., Palin, J. and Ridgway, S. H. (1970). Sound conduction in the dolphin ear, *J. Acoust. Soc. Amer.* **48**, 1418–1428.

Mills, A. W. (1958). On the minimum audible angle, *J. Acoust. Soc. Amer.* **30**, 237–246.

Mohl, B. (1968). Auditory sensitivity of the common seal in air and water, *J. Aud. Res.* **8**, 27–28.

Montague, W. E. and Strickland, J. F. (1961). Sensitivity of the water-immersed ear to high- and low-level tones, *J. Acoust. Soc. Amer.* **33**, 1376–1381.

Nixon, C. W. and von Gierke, H. E. (1959). Experiments on the bone-conduction threshold in a free sound field, *J. Acoust. Soc. Am.* **31**, 1121–1125.

Nordlund, B. and Liden, G. (1963). An artificial head, *Acta Oto-laryng.* **56**, 1–7.

Norman, D. A., Phelps, R. and Wightman, F. (1971). Some observations on underwater hearing, *J. Acoust. Soc. Amer.* **50**, 2, 544–548.

Reysenback de Haan, F. W. (1957). Hearing in whales, *Acta Oto-laryng. Suppl.* **134**, 1–114.

Sivian, L. J. (1947). On hearing in water *vs.* hearing in air, *J. Acoust. Soc. Amer.* **19**, 461–463.

Smith, P. F. (1969). Underwater hearing in man: 1-Sensitivity, U.S. Naval Submarine Medical Center, Groton, Connecticut, *Report No.* 569, 1–23.

Smith, P. S. (1965). Bond Conduction, air conduction, and underwater hearing, U.S. Naval Submarine Medical Center, Groton, Connecticut, *Memorandum Report* No. 65-12, 1–7.

Stevens, S. S. and Newman, E. B. (1934). The localization of pure tones, *Proc. Nat. Acad. Sci., Wash.* **20**, 593–596.

Stevens, S. S. and Newman, E. B. (1936). The localization of actual sources of sound, *Am. J. Psychol.* **48**, 297–306.

Tavolga, W. N. (1964). Psychophysics and learning in fishes, *Natur. Hist.* **73**, 34–41.

Tobias, J. V. and Zerlin, S. (1959). Lateralization threshold as a function of stimulus duration. *J. Acoust. Soc. Amer.* **31**, 1591–1594.

Tonndorf, J. *et al.* (1966). Bone conduction studies in experimental animals: A collection of seven papers, *Acta Oto-laryng. Suppl.* **213**, 1–132.

Tonndorf, J. (1968). A new concept in bone conduction, *Archives of Otolaryng.* **87**, 595–600.

United Research, Inc. (1962). Localization of sound, U.S. Naval Ordinance Test Station, China Lake, Calif., NOTS TP 3103, No. 1.

Wainwright, W. N. (1958). Comparison of hearing thresholds in air and water, *J. Acoust. Soc. Amer.* **30,** 1025–1029.

Zwislocki, J. (1957). In Search of the Bone-Conduction Threshold in Free Field, *J. Acoust. Soc. Am.* **29,** 795–804.

Zwislocki, J. (1957). Some impedance measurements on normal and pathological ears, *J. Acoust. Soc. Amer.* **9,** 1319–1317.

Improving Underwater Viewing*

S. J. COCKING

Marine Technology Support Unit, Atomic Energy Research Establishment, Harwell, Oxfordshire

1. Introduction	140
2. Short and Long Range Viewing	141
3. The Nature of the Problems	142
4. The Nature of the Solutions	143
5. Light and its Interactions in the Sea	144
A. Absorption	144
B. Refraction	146
C. Scattering	146
D. The total attenuation coefficient	148
6. Natural Light in the Sea	149
7. Artificial Light Sources	150
A. Incandescent filament lamps	150
B. Discharge lamps	151
C. Lasers	152
8. Imaging Receivers	152
A. Image intensifiers	153
B. Television tubes	153
C. Spectral response	154
D. Contrast reproduction	156
E. Resolution	157
F. Sensitivity	158
9. Transmission of a Light Beam through Water	160
10. Transmission of an Image through Water	162
A. Resolution and the modulation transfer function	163
11. Back Scatter and Light Source/Receiver Geometry	164
A. Contrast loss by back scatter	164
12. Signal to Noise Ratio, Detectability and Limiting Resolution	165
A. Inspection, recognition and detection	167
B. Extensions for through-water viewing	168
13. System Performance in Terms of its Components	168
A. Performance of receivers	169
14. Calculation of Back Scatter	171
15. Geometrical Isolation	174
16. Target Scanning Methods	177
17. Time Dependent Isolation: Range Gating	179
18. Continuous Time Dependent Isolation	181
19. Application of Polarizing Filters	182

* For a list of the abbreviations used in this chapter see Appendix.

20. Image Processing	183
21. Concluding Remarks	185
References	186
Appendix	190

1. Introduction

Underwater swimmers around the United Kingdom are familiar with the surprisingly rapid disappearance of fellow swimmers as they become separated. At a range of less than 1 m a companion can be recognized and the colour and details of his clothing and equipment distinguished. At about 3 m, colour and geometrical detail are no longer discernible.

FIG. 1. The decrease of discernible detail with increasing distance even in relatively clear water is shown. A diver disappeared from view at a range of 10 m, the most distant diver is at 6 m. (*Photo:* Author).

Further separation leads to a blurred outline of the companion, barely visible against the grey-green of the surrounding water, and finally to his disappearance apparently through this green veil (Fig. 1).

This difficulty of adequate range in underwater viewing is not only the concern of amateur divers. The problem of visually locating and of closely inspecting objects under water is of increasing practical importance in the exploration and use of the seas. Man's eyes have been supplemented by the use of cameras, both photographic and closed circuit television for underwater viewing, and the depths now being visually explored far exceed the capabilities of the free swimmer; submersible devices have carried men and cameras to the deepest waters of the oceans.

In the clear waters of the Mediterranean and Caribbean and in much of the deep oceans the optical quality of the water offers no major problem in direct viewing or camera work to ranges of perhaps 30 m. Nevertheless through-water viewing in most commercial operations, which are usually in less favoured waters, is restricted by water clarity.

Because of the increasing interest arising from both military and commercial needs, studies of the fundamental processes which limit the underwater viewing range have been made, most extensively in the United States of America. This chapter aims to summarize the principal physical processes in image transmission in real waters and the means which have been developed towards overcoming its limitations.

A review adequately covering viewing by eye, by photographic camera and by electro-optical systems would be too extensive for the present volume. The emphasis in this review is therefore on discussion of electro-optical receivers, although many of the features discussed apply generally. Further, discussion of natural light in the sea is minimal here, since artificial illumination can be adapted to the special requirements of the viewing systems discussed. Valuable summaries of underwater viewing by eye with natural illumination are given by Lythgoe (1971, 1972). Photography in the sea is extensively covered in the excellent book by Mertens (1970) while up-to-date summaries of the status of natural light in the sea may be found in books by Jerlov (1968) and Tyler and Smith (1970).

The review has four parts; Sections 1–4 outline its scope and Sections 5–8 give the basic information needed to analyze through-water viewing. This information is applied in Sections 9–13 to quantify the problems as far as possible, while in Sections 14–21 methods aimed to improve underwater viewing are outlined, together with the results of practical demonstrations where these have been made. Since the present review was completed an extensive lecture series, has appeared. These lectures carry much of the present material further, and the interested reader is referred to this series for further study. (AGARD lecture series No. 61, 'Optics of the Sea', 1973.)

2. Short and Long Range Viewing

It is useful to distinguish between short range, high resolution viewing of objects of known position and the viewing needs in finding an object in turbid water.

Given the position of an object under water, it is usually possible to approach it closely for viewing. The short range viewing requirements are then to be able to see all parts of the object with which one is immediately concerned, for example, in visual inspection of engineering work or in mechanical assembly. Minimizing the through-water path is then advantageous and photographic techniques such as the use of very wide angle lenses and mosaic strip photography (Rebikoff, 1966) have been developed for very short focusing ranges. In the most turbid waters, such as ports and estuaries, even short range viewing is seriously limited. A variety of techniques to replace the turbid water with clear water are possible (see Section 4) and may well be the only solution to adequate viewing in these circumstances.

A different problem arises in visual search for objects or in underwater navigation among obstructions. At sufficient range this is the province of acoustic methods, since visual range is, even in the clearest water, shorter than acoustic range. Nevertheless an object *detected* by acoustic means is at present usually *identified* by visual means. Identification of most objects lying on the sea bed, particularly if they lie among rock, requires the more detailed information possible with visual observation. This has been demonstrated in search for lost aircraft and submarines and in the famous search for the hydrogen bomb off Palomares (Patterson, 1966).

The present article is mainly concerned with longer range viewing, in particular in underwater search and identification. The importance of maximum possible identification range in making a planned search of an area is obvious, the reduction in cost of a search operation often justifying some sophistication in the apparatus used.

3. The Nature of the Problems

A visual target is detected by a difference of light intensity or colour received from it compared with that from its surroundings, that is, the contrast must be greater than some minimum which varies between different optical detectors. Further, the received intensity must be in the range where the detector can operate satisfactorily, that is, give a response dependent on the received light intensity and above the inherent noise of the detector. To make a detailed visual examination of the target we also require to be able to resolve small, closely spaced elements of the target, both in terms of intensity and of colour.

To summarize, our ability to gain information about a target can be limited by either (a) image contrast, (b) image intensity or (c) image resolution.

If the sea were a totally homogeneous liquid, the leading problems in underwater viewing would be low intensity and loss of colour information due to the selective absorption of light in water. The problem is more severe in real waters which contain small suspended particles. In this case, light scattered by these suspended particles causes a loss of image contrast and of image resolution as well as increasing intensity loss. The loss of contrast arises since extraneous light scattered by the suspended particles is added to the target image, while loss of resolution is due largely to forward scattering of the target image. These basic facts have been noted in several discussions of underwater viewing, e.g. Wall and Coleman (1968) and of atmospheric viewing through haze or fog. In Sections 9–17 these effects are examined in some detail in order to quantify these problems.

4. The Nature of the Solutions

Since the viewing problem is much aggravated by the presence of suspended matter in natural waters, an obvious solution in principle is to replace this unsatisfactory optical medium with a more acceptable one. Thus perspex cones and clear water-filled bags have been used to displace the turbid water between the observer's eyes or camera and the target. On the US Deep Submergence Rescue Vehicle a jet of clean water provides a clear sight path to assist in controlling the locking-on operation. These simple and effective techniques are limited in range only by considerations of practical convenience. At more extended ranges the existing conditions must be accepted and more subtle means adopted to minimize the limitations.

Image intensity using illuminating lamps can be maximized by attention to the optical control while careful choice of the wavelength of emitted light can improve penetration. Far higher increase of intensity can be achieved with the use of electro-optical devices which offer intensification of received image by factors up to 10^5. These can be used either for direct viewing or placed in front of a camera, while electro-optic imaging tubes now available allow clear pictures on a television monitor of scenes illuminated by starlight only.

To restore image contrast lost by the presence of extraneous scattered light, a variety of means have been proposed. These methods all seek to exploit some difference between wanted and unwanted light. Methods discussed in Sections 15–17, known as geometrical isolation, laser scanning and range gating, all exploit the different geometrical positions of origin of wanted and unwanted scattering. The different responses of target and particle scattering to polarized light have been examined (Section 19), while image processing techniques aim to emphasize features of the target image which differ from the particle scattered light (Section 20); for example, sharp edge gradients can be emphasized over featureless scattered light. A difference which has apparently not been exploited is the motion of suspended scatterers in a current relative to a fixed target.

While image intensity and contrast can undoubtedly be enhanced, the loss of resolution over large ranges is a more intractable problem. Examination of the potential practical applications of extended range viewing suggests, however, that high resolution at long range is rarely a leading requirement, since given enough image information to identify a target, one can normally approach more closely in order to make more detailed examination. Attention to the former problems can thus provide a sufficient solution to the viewing problem in many cases.

5. Light and its Interactions in the Sea

The leading physical parameters of water of importance in this review are here noted. The present discussion is greatly extended by Mertens (1970) and Williams (1970).

Light, visible to human beings, is a relatively narrow band of wavelengths from 350 to 750 nm (1 nanometre = 1 nm = 10^{-9} m) in the spectrum of electromagnetic radiation. Electro-magnetic radiation exhibits an oscillatory electric field in the plane perpendicular to the direction of propagation. Unpolarized light has components with the electric fields distributed evenly in all directions in this plane. However, most interactions of light with matter, such as refraction and scattering, result in an uneven distribution of the directions of the electric fields; the light is then said to be partially plane polarized. The degree of polarization relative to a plane (e.g. the plane of scattering) is defined as

$$\frac{I_1 - I_2}{I_1 + I_2}$$

where I_1 and I_2 are the intensities with polarization components parallel and perpendicular to the plane. When light has two linear polarized components which are out of phase by 90° the light is said to be elliptically polarized, or, when the two components are equal, circularly polarized.

Light travels in air at a speed of 0.30 m per ns, the velocity of light in water is appreciably lower (0.225 m per ns).

A. Absorption

Many familiar materials including glass and water transmit light with little apparent loss. However, with increasing path length, the absorption of light in these media becomes apparent and the wavelength dependence of the absorption can be marked. Figure 2 shows the fraction of light of different wavelengths transmitted by pure water, almost all of the light lost being absorbed by the water. Sea water contains approximately 3.5% by weight of dissolved mineral salts, but these have a negligible effect on the absorption in the visible region. This is not so of other constituents of sea water such as the so-called "yellow substance" (Kalle, 1966), which is a complex mixture of products of the breakdown of organic materials often of terrestrial

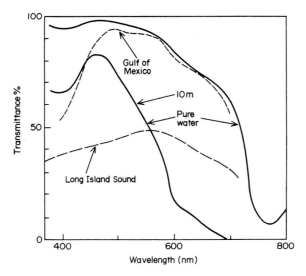

FIG. 2. The transmission of light by several waters (after Luria and Kinney, 1970) as a function of light wavelength. The transmission of 1 m of water is shown except the curve for pure water showing a 10 m path. This curve shows the marked emphasis of a narrow wavelength band near 480 nm due to the selective absorption of pure water.

origin, or of the various plant and animal constituents of biologically productive waters. Absorption thus varies very greatly both geographically and at different times of the year. A typical transmission curve for coastal water is indicated in Fig. 2. Keith-Walker (1970) and Jerlov (1968) present data on the marked variation of absorption with depth at a variety of locations.

Absorption thus has the effect of reducing light intensity and, further, acts as a spectral filter. For transmission through more than a few metres of water the red end of the spectrum is effectively lost and the light received tends with increasing range to be only that at the minimum absorbed wavelength (blue-green for clear oceanic waters and green for typical coastal waters). Light sources aimed to transmit light through appreciable distances should have their output concentrated in this maximum transmission region, while receivers should be chosen with high sensitivity in this region (see Sections 6 and 7).

B. Refraction

Light travelling through an air-water interface is deflected from its normally straight path; a measure of the bending of the light path is given by the refractive index (μ). The important consequences in design of lenses for through-water viewing have been discussed by Mertens (1970) and the resulting visual distortions by Ross (1971). For standard sea water at 20°C and light of wavelength 501.7 nm, μ = 1.343, decreasing by 0.36% over the wavelength range 501.7 nm–632.8 nm (Stanley, 1971), μ also varies with the water temperature (10^{-2}% per °C) and with concentration of dissolved minerals, that is with salinity in the case of sea water.

The sea contains regions of differing salinity and temperature. Light travelling through such variations can be bent from its normally straight path. Yura (1971) states that the observed spreading of light beams through angles less than 1° is due to such salinity and temperature macrostructure. In regions of extreme mixing, the multiple refraction effects can be seen by eye in a similar way to the refraction effects in the atmosphere over a hot surface.

C. Scattering

Pure water scatters light due to the microscopic statistical fluctuations of its density. The probability of scatter is, however, seven times lower than the absorption probability even at the minimum absorption wavelength. Scattering by suspended particles in the water is usually of more importance.

The probability of scattering for a particle is usually given as an area, being the effective area which that particle presents to the incident light. Thus if a volume of water contains a distribution of $p_i(A)$ particles per m³ of effective scattering area A_i m², the total scattering coefficient for this laden water, b, (in m^{-1}) is

$$b = \sum_i p_i(A) A_i \tag{1}$$

A general theory of light scattering by spherical particles, the Mie theory, covering a wide range of particle sizes has been extensively reviewed by Kerker (1969). The sea contains a broad distribution of particle sizes from submicron to several millimetres. Since in general, the shapes are variable and non-spherical, scattering theory can be used in only a semi-quantitative way, but is extremely useful in understanding the scattering phenomena observed in the sea (Cocking, 1974).

Summarizing the available evidence for oceanic waters Jerlov (1968) concludes that the predominant light scattering arises from particles of diameter greater than 2 μm (2.10^{-6} m). The average particle size is larger in coastal waters according to Jerlov (1968) and most under-

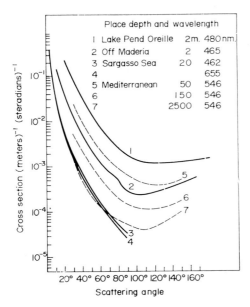

FIG. 3. The angular distribution of scattering in various waters (data of several authors collected by Jerlov, 1968).

water swimmers can testify to the considerable quantity of material which can be resolved by eye. A considerable quantity of silt, sand and products of plant decay can be disturbed from the bottom by wind, (Lee and Folkard, 1969; Visser, 1970) current or by the observer. This material is largely in the size range >50 μm as evidenced by its relatively rapid re-settling.

The angular distribution of scattering is described by the volume scattering coefficient, $\beta(\theta)$ in m^{-1} steradians^{-1}, where θ is the scattering angle and,

$$b = 2\pi \int_0^\pi \beta(\theta) \sin \theta d\theta.$$

Figure 3 shows the observed angular distribution of scattering by several waters. The dominant forward peak is characteristic of particles with dimensions greater than the light wavelength, that is greater than, say, 1 μm (Ashley and Cobb, 1958; Hodkinson, 1966). Duntley (1963) has observed that such angular distributions obtained for a wide variety of waters have similar shape and has suggested that a universal angular distribution with different normalizing factors could be used to describe a wide variety of waters. Although particle sizes in different waters are not the same, the result of averaging over the various particle size distribution could lead to a broadly similar distribution (Deirmendjian, 1969).

D. The Total Attenuation Coefficient

The intensity of a parallel, monochromatic light beam falls from I_0 to I over the distance r according to the equation,

$$I = I_0 e^{-\alpha r} \qquad (3)$$

where $\alpha = a + b$.

α is the total attenuation coefficient, a the absorption coefficient and b the scattering coefficient; all are normally given in units m^{-1}. $\frac{1}{\alpha}$ is known as the attenuation length.

Instruments designed to measure these parameters have been detailed by Jerlov (1968). Williams (1970) has made pertinent criticism of some existing designs. Table 1 gives data from recently published measurements.

TABLE 1.

Location	α m^{-1}	b m^{-1}	λ nm	Reference
North Sea	0.39–0.05	0.19–0.03	460	Keith-Walker (1970)
Lizard	0.30–0.1	0.16–0.075	460	Keith-Walker (1970)
English Channel	0.70–0.1		460	Keith-Walker (1970)
Various	1.0 –0.05	0.3 –0.04	440	Jerlov (1968)
Pacific		1.0 –0.4	546	Beardsley et al. (1970)
Bermuda	1.1 –0.8		546	Morrison (1970)
Long Island Sound	3.0 –1		530	Morrison (1970)

The above discussion and the measurements of Table 1 relate to monochromatic light. Usually light sources are polychromatic so that an appropriate average over the spectrum must be made to calculate the intensity. In general, then, the exponential fall-off of intensity with distance will not apply until the selective absorption of the water has filtered the light so that only a narrow wavelength band remains.

6. Natural Light in the Sea

Natural light offers a variable illumination which decreases markedly with depth. Table 2 gives the order of magnitude of illuminance at sea level.

TABLE 2.

Bright day	$10^4 - 10^5$ lux
Very dark day	10^2
Full moon	10^{-1}
Quarter moon	10^{-2}
Starlight	10^{-8}

Sunlight penetrating into deep water is attenuated by absorption, the spectral dependence of this absorption resulting in an increasing preponderance of the least absorbed colour, that is blue light near 480 nm for clear oceanic water and green light near 540 nm for coastal waters (Kampa, 1970a, b). To the directly penetrating light will be added scattered light. The angular distribution of this component tends with increasing depth towards an asymptotic distribution with a dominant vertical component and smaller contributions from all other directions including upwards (Tyler and Smith, 1970).

Fall-off of light intensity with depth D, measured with an irradiance meter, is often represented

$$I = I_0 e^{-KD} \tag{4}$$

where K, the broad beam attenuation coefficient, is less than α since the former includes scattered components. K only achieves a value independent of depth for large depths where the asymptotic scattering distribution has been established and the light due to the filtering effect of water has become almost monochromatic, when K is roughly equal to a, the absorption coefficient.

Figure 4 shows the order of magnitude of illuminance (lumens/m^2) as a function of depth. The wide variation between oceanic and coastal waters (see Table 1) indicates the impossibility of giving generally valid numerical data. For the present purposes the depth at which the illuminance is sufficient for viewing by eye and TV camera, or photographic recording is of interest. Human photopic vision, viewing by vidicon television tubes and photography with acceptable time exposures are limited to about 1 lux. This level is achieved for the most turbid waters indicated in Fig. 4 at 50 m depth while scotopic vision possible to about 2×10^{-4} lux, is limited to 100 m depth. Electro-optic low light viewing devices can extend this limit to 150 m at least.

It is clear that for much of man's exploration of the sea the natural light levels are too low and too variable. Artificial illumination is therefore normally required.

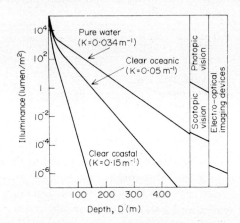

FIG. 4. Greatly simplified illustration of the variation of illumination with depth. The values of K for coastal waters is especially subject to variation. The depth ranges at which the unaided human eye and electrical imaging devices can detect target shapes is indicated.

7. Artificial Light Sources

The requirements of lights for close range differ from those for extended range viewing. In the latter case, selective absorption is responsible for the loss of light energy with wavelengths separated from the absorption minimum. Thus for efficient use of available energy the spectral output of the light source should be limited to the minimum absorption region.

A. Incandescent Filament Lamps

The most commonly used light source is the incandescent filament. The spectral output from such lamps is broad, with much of the energy lying in the red and infra-red region. Figure 5 shows the spectral distribution of such a lamp, together with the spectral distribution after the light has passed through 10 m of clear water.

B. Discharge Lamps

Discharge lamps, however, have spectral output characteristic of the elements present in the arc discharge. Thus the common mercury arc discharge has prominent spectral lines at 545–580 nm (Fig. 5). This spectral output can be modified with other metallic additives. The addition of thallium which yields a dominant line at 530 nm has been specially emphasized as an underwater light source (Harford, 1968). Discharge lamps can offer a more compact light source than the filament types of equivalent power, and this is advantageous in designing lamps with narrow beams as required for projection through appreciable distances. The higher luminous efficiency (lumens per watt

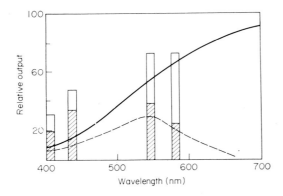

FIG. 5. Comparison of spectral output of typical filament lamp (continuous spectra) and a mercury arc discharge lamp (discrete lines). The spectra after transmission through 10 m of clear water are shown by the dashed curve for the filament lamp and the shaded region of the discrete spectra. The removal of red and infra-red wavelengths results in loss of about 90% of the filament lamp's energy output.

electrical power consumed) of the discharge lamps has also been emphasized (50–100 for discharge lamps compared with 20–30 lumens per watt for tungsten-halogen filament lamps). For short range viewing and colour photography where the colour quality of the target is not totally lost by selective absorption, the limited spectral range of some discharge lamps is disadvantageous. However, recently developed discharge lamps, including for example dysprosium, offer a wide spectral output.

The use of short, intense pulses and shuttered receivers can be especially advantageous in applying special viewing techniques (Sections 14–21) and to minimize back scatter due to the presence of natural light. In addition, the imaging of moving targets is possible. Most discharge lamps can be operated in a pulsed mode, the xenon discharge lamp being familiar in electronic flash apparatus for photography. Mertens (1970) gives a full account of both bulb and electronic flash apparatus in underwater application.

C. Lasers

Kornstein and Wetzstein (1968) have discussed early development of lasers for in-water use and their application in range gating. Lasers containing noble gases offer suitable wavelengths, but relatively low optical efficiency (about 0.01%). Argon gas lasers yield lines at 488 and 514 nm and continuous optical output of the order of watts is usual.

Solid state lasers offer higher optical efficiency but normally yield longer wavelength radiation (694.3 nm for ruby lasers, 1060 nm for neodynium doped glass or yttrium aluminium garnet). The use of unharmonically vibrating crystals such as potassium dihydrogen phosphate (KDP) allows optical frequency doubling, that is the wavelength is halved. A neodymium doped laser with a KDP frequency doubler yields 530 nm with about 10% of the energy of the natural line at 1060 nm. A mean power of 1 W at 530 nm is difficult to achieve, although power will be increased with further developments. The expected practical power limit for solid state lasers is due to optical distortion by heating of the laser rod. For this reason liquid lasers currently being developed, in which the medium is continuously circulated and cooled, promise higher mean power output and even tunable wavelengths.

Although lasers offer lower mean power than can be provided by conventional lamps, two special features have application in the viewing systems discussed in Sections 16–18. The small diameter beam with natural divergence less than 0.1° is valuable in image scanning methods, while the relative ease with which short, intense pulses are produced, for example 10 ns pulses of 10 megawatt peak power, is being applied in range gating.

While the straight, narrow beam from lasers has been used for metrology in underwater survey, the possibilities arising from the coherent nature of the light have been little explored.

8. Imaging Receivers

It is here assumed that of the three dominant imaging receivers namely the human eye, photographic cameras and electro-optic imaging (television) tubes, the last is least familiar to the reader. Hence an outline of the distinguishing features of these tubes, with some indication of the considerable variety of modes of use is given. Only those features of relevance in the present context are discussed, detailed operation of these devices being covered by Vine (1965), Jensen (1968) and Soule (1968). This is followed by comparison of performance of these tubes with photographic film and the human eye; for more detailed discussion see also Jensen (1968) and Soule (1968).

A. Image Intensifiers

In image intensifiers the optical image is focused onto a photo-emissive surface. The emitted electrons, having the spatial and intensity distribution of the optical image, are accelerated to fall on a phosphor screen where light is re-emitted. Given sufficient accelerating voltage, electron multiplication at the phosphor results in overall light amplification. Electrostatically focused intensifiers yield a light gain of order 50. Several such intensifiers may be joined in cascade to yield overall gains of 10^5. This order of gain can also be achieved with the use of channel plate intensifiers which offer an extremely compact device. Image intensifiers have been applied for direct viewing by eye (Schagen *et al.*, 1961) and with photographic cameras in low light or very high speed photography.

B. Television Tubes

In electro-optic imaging tubes the optical image is similarly converted to electronic form but this electronic image is stored on a target which is scanned by an electron beam to yield a time varying output voltage. This signal can then be transmitted by wire, radio or even acoustic carrier to a remote monitor where it is used to control the intensity of an electron beam which excites light emission from a phosphor screen. The scanning of the target in the tube and the scanning of the read-out electron beam are maintained in synchronism.

In the compact and rugged vidicon tube used in most closed circuit television applications, the photo-electric converting surface is a photo-conductor. The output signal is derived by the scanning electron beam effectively sensing the conductivity variation across this layer.

A photo-emissive surface is used in the image orthicon and some more sensitive versions of the vidicon. The emitted electrons are focused onto a separate storage layer where the resulting charge pattern is sensed by the scanning electron beam. Tube sensitivity can be increased by electron multiplication before storage or, when return beam scanning is used, by electron multiplication of the return beam. Such high sensitivity tubes include the EBItron, using Electron Bombardment Induced conductivity, the Secondary Electron Conduction (SEC) vidicon and the Silicon Intensifier Tube (SIT). The image isocon is a development of the already sensitive orthicon and yields improved signal to noise ratio and dynamic range (Mouser, 1969). The sensitivity of imaging tubes may be further increased by adding image intensifiers.

The flexibility of the electro-optic method is often overlooked, since inexpensive, standard systems have become so widely used. These systems, based on broadcast television standards, use linear horizontal scanning with 625 lines per picture height, complete scanning of one picture occupying 1/25th of a second (UK standard). The scanning electron beam both reads the stored image information and erases it. Normally the image falls continuously onto the tube so that the exposure time is near 1/25th of a second. The eye viewing the presented image, however, integrates scans over about 0.2 seconds.

Reproduction of moving targets with continuous exposure is limited by finite capacitance of the target which results in incomplete erasure of the image with a single electron scan. Typically 10 scans are required to reduce the signal to 5% of its initial value. As a result, a moving target is imaged with lower resolution than when static. Examples of static and dynamic resolutions are given in an extensive set of data by Towler and Swainston (1972).

The possibility of separating the expose, read and erase functions is frequently overlooked, since simultaneous expose-read-erase is standard. The exposure time can be selected within relatively wide limits, as with photographic cameras, while the read-out scanning of the image and, separately, erase can follow exposure. Increase of exposure time can be used to improve image quality at low light levels (Section 13) while reduced exposure using a shutter can be essential to record moving targets without blurring. Increase of exposure is limited by charge leakage. A static image can be usefully integrated for about one second with a standard vidicon, but high resistance storage layers have been developed to allow integration over several hours.

Schade (1970) describes the use of a high resolution vidicon camera with shuttered time exposure followed by signal read-out. Residual charges remaining after read-out are erased by the flash of a high intensity light. Alternatively a visual display of each exposure can be maintained for more than 30 s, using fast scan multiple read-out utilizing the large storage capacitance of the tube.

Slow scan read-out has been used when the bandwidth of the transmission link is limited, as for example when television pictures are transmitted from deep space or through water using an acoustic carrier.

C. *Spectral Response*

The maximum sensitivity and spectral variation of sensitivity of an imaging tube is determined by the photo-electric converting surface.

Figure 6a shows the spectral response of the commonly used antimony trisulphide photo-conductor. The spectral responses of one of the wide range of photo-emitting surfaces is shown in Fig. 6(a), with corresponding data for monochrome film.

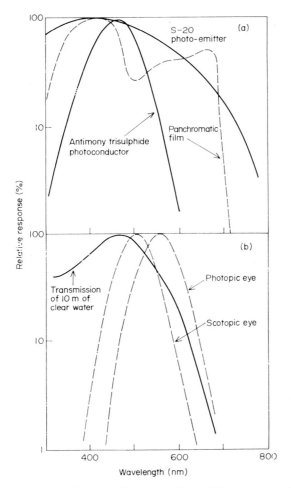

FIG. 6. (a) Comparison of relative spectral response of typical high-speed panchromatic film, a standard vidicon photo-conductor (antimony trisulphide) and a high efficiency photo-emitter (EIA type No. S-20). (b) Relative spectral response of photopic and scotopic eye compared with the relative transmission of 10 m of water.

The more extensive spectral sensitivity, particularly in the red and infra-red regions, of some artificial sensors (Fig. 6(a)), is not useful in water except at close range. Indeed, when viewing distant targets through water, high sensitivity to the more strongly absorbed wavelengths can be disadvantageous, since light received in this region must have originated from short range and thus is unwanted background to the required image. Examples of situations where it could be advantageous to restrict the spectral sensitivity of the detector using spectral filters are when viewing a relatively distant object in shallow water by natural light, or when using a filament lamp, which has high intensity in the red, relatively close to the receiver.

The relative spectral sensitivity of the human eye is shown in Fig. 6(b). Comparing these data with the transmission curve of water also shown, one sees that the human eye is relatively well adapted to seeing through water. Lythgoe (1972) discusses the different spectral responses of the eyes of fish in relation to their light environment.

D. Contrast Reproduction

The general response curve (Fig. 7) applies for electro-optical receivers and photographic film, the responses being electric current and density of developed film respectively. The threshold level shown is the inherent electron noise in the former case, and film base plus fog level

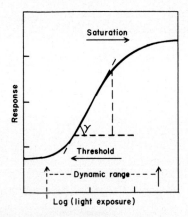

Fig. 7. The general response curve for a detector.

for developed film. The dynamic range of the detectors is given by the range of exposure between the threshold and saturation levels, while the slope of the curve in this region (commonly designated γ) determines the reproduction of received image contrast; a high gamma leading to high output contrast. Detector output is further modified in presenting the image on the television monitor or the photographic print or projected image.

The commonly used vidicon with antimony trisulphide photoconductor has γ between 0.6 and 0.8 depending on exposure. Photoemissive detectors have linear response ($\gamma = 1$). Television display tubes have markedly non-linear response (in this case light output for a given voltage input) with effective $\gamma \simeq 2$; this compensates for low γ reproduction of the standard vidicon, although non-linear amplifiers allow modification of γ. In underwater applications higher contrast reproduction should be advantageous to compensate for low contrast images. Wide variation of γ for film is possible in the development process, and a range of photographic printing papers with different γ is readily available.

For a well illuminated target subtending at least 1° at the eye a normal contrast perception of 2% is common. This is not readily achieved by film or electronic methods. However, human contrast perception falls with target size, edge blurring (Thomas and Kovar, 1965) and at low light levels. Lingrey (1968) records improved target detection range by television compared with divers' eyes. Such results could be accounted for by the high γ response possible with television compared with the degraded performance of the eye with typical underwater conditions.

E. Resolution

At high light levels resolution is limited by the optical quality of the detecting system. This limit may however not be achieved with insufficient image irradiance. These limitations are discussed separately in this and the following sections respectively.

Limiting resolution is commonly quoted, being the maximum number of alternating black and white lines per unit distance which can be distinguished. For photographic film, 100 line pairs per mm (lpm) on the negative is considered high, and less than 50, low resolution. The fundamental limitation of resolution (about 1000 lpm) is given by the grain size. A large sensitive area is chosen for maximum resolution. The wide range of film sizes available is well known.

For imaging tubes with linear horizontal scanning it is usual to quote limiting resolution as the maximum number of lines per picture height which can be resolved. Vertical resolution must be less than N, the number of electronic scan lines used to cover the image. Jensen (1968) gives $\dfrac{N}{2\sqrt{2}}$ as the number of horizontal black and white line pairs which can be resolved. Horizontal resolution, determined by the bandwidth of the video amplifiers and transmission system, is chosen to be equal to vertical resolution. The common 625 line scan thus should resolve 220 line pairs, much less than a low resolution film of 35 mm format whose 14 mm height at 50 lpm should resolve 700 line pairs. Although closed circuit television systems have usually been designed to broadcast standards, this is by no means essential. Thus in special systems, when extremely high resolution is required, up to 5000 lines per picture height have been used (Schade, 1970).

The human eye has the remarkable ability to accommodate to scene illuminances over a range 10^4–10^{-5} lux. Above 10^{-2} lux colour sensitive, photopic vision is used and an angular resolution of at least 1 minute of arc is normal. At optimum viewing distance this resolution is equivalent to about 15 line pairs per millimetre. Thus a high resolution photographic negative can be magnified about 6 times linearly before its limiting resolution approaches that of the eye. At greater magnification the observer tends to increase the viewing distance so that the finest detail resolved on the photograph subtends about 1 minute of arc at the eye. Similarly in viewing a television monitor the preferred viewing distance is chosen so that adjacent scan lines subtend about 1 minute of arc at the eye. A maximum visual resolution in TV lines per picture height is then equal to the number of scan lines.

F. Sensitivity

As the light intensity falling on an imaging detector is reduced, the required signal becomes increasingly difficult to distinguish against the inherent noise. The concept of signal to noise ratio is of central importance in discussion of sensitivity. Increasing image resolution implies detection of a signal from increasingly smaller picture elements with correspondingly smaller signal to noise ratio. Hence resolution is limited by the available signal to noise ratio of the detecting system.

The output signal from a vidicon is low and electronic noise arises mainly in the high gain amplifiers required. In the more sensitive tubes electron multiplication yields a higher output signal at a given illumination level with far less noise generation. In fact the most sensitive systems can be used at light levels where the limitation becomes the statistical noise due to random arrival of the very limited number of light quanta (see Section 12). The resolution of some imaging devices as a function of photo-cathode illumination is shown in Fig. 8. Data from Jensen (1968) for a high speed monochrome film is also shown, an exposure time of 0.2 s equal to the integrating time of the human eye is assumed. The region with detector illuminance below 10^{-2} lux is only accessible to the film with extended exposure time.

A scene luminance scale for Fig. 8 is calculated using the standard relationship (Kingslake, 1965).

$$\text{Photo-cathode illuminance} = \text{scene luminance} \frac{\pi T_0}{4f^2} \qquad (5)$$

f being the equivalent aperture number and T_0 the transmission of the lens; using values $T_0 = 0.8, f = 1$ for the present illustration.

At scene illumination greater than 10^{-2} lux, the high resolution capabilities of modern film and of the unaided eye can be realized. Below 10^{-2} lux the eye adapts after some time to colour blind, scotopic vision, sacrificing angular resolution which falls with decreasing light intensity and target contrast. Acuity is improved with the use of binoculars which, having large entrance pupils, collect more target light (Fig. 9). This is however necessarily achieved with the loss of angular field of view. Vision can however be restored to the photopic

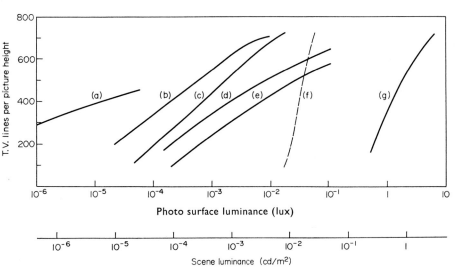

FIG. 8. The limiting resolution as a function of photo-cathode illuminance for:

(a) an RCA 4470 orthicon plus image intensifier (after Mertens, 1970)
(b) an RCA 4804 SIT tube
(c) an EEV Co. P880 Isocon
(d) an EMI 9777 Ebitron
(e) a Westinghouse WL 30691 SEC vidicon
(f) Kodak Royal X film developed to 1600 ASA (after Jensen, 1968)
(g) a typical photo-conductor vidicon.

Curves (b), (c), (d) and (e) taken from manufacturers' data.
The scene luminance scale is here calculated using equation (5) with lens transmission, $T_0 = 0.8$ and lens equivalent aperture number $f = 1.0$.

region, the field of view being maintained or varied at will with the use of image intensifiers (Fig. 9). (The second law of thermodynamics indicates that addition of energy is required in order to enhance the image brightness in an optical system.)

Using the result (Section 8 E) that the unaided eye viewing a television monitor can resolve a number of TV lines/picture height equal to the number of scan lines, the eye's acuity shown in Fig. 9 as 10 min at 10^{-4} cd/m² is equivalent to about 62 TV lines per picture height with a 625 line picture. With this scene luminance the isocon, curve (c) of Fig. 8, shows resolution of about 300 TV lines per picture height. At lower light levels the eye maintains resolution better than this isocon, but does not reach the performance of the intensifier orthicon shown in Fig. 8. Experience of low light television operating successfully where

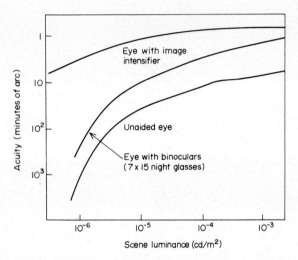

Fig. 9. Acuity of the human eye at low light levels (after Schagen et al. 1962). The image intensifier data is for a binocular viewing system proposed by these authors.

the dark-adapted eye sees nothing is impressive and becoming increasingly familiar.

9. Transmission of a Light Beam through Water

A light beam travelling through sea water suffers absorption and scattering. If a source emits energy L_0 into a solid angle ω, a detector at distance r, receives an irradiance of unscattered light, S_u given by

$$S_u = \frac{L_0}{\omega r^2} e^{-\alpha r} \tag{6}$$

Light is scattered from the beam predominantly at small angles and thus forms a halo of illumination around the main beam. At small path lengths (for $\alpha R < 0.1$) most of this light will be scattered only once, so that the angular distribution about the main beam is that given in Fig. 3. For greater distances the light will have been scattered several times so that the halo will become increasingly spread. The intensity observed in the main beam will be augmented by multiple scattered light, the total intensity increasing with the angle of acceptance of the detector (Fig. 10). An equivalent result is obtained if the detector acceptance

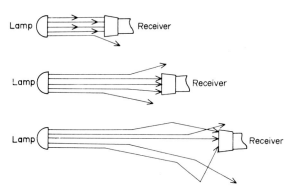

FIG. 10. Illustration of the addition of multiple scattered light to directly transmitted light as the distance from lamp (L) and receiver (R) is increased.

angle is narrow but the light is emitted over an increasing angle. The effective attenuation of the light intensity will then be considerably less that that given by equation 6. Indeed if all the scattered light could be collected by the detector, attenuation would result only from absorption (see Fig. 11).

Due to the considerable practical importance of this increased intensity in the design of devices aiming to transmit light through long distances under water, measurements have been made by Duntley (1963) and Okoomian (1966). Duntley shows that for a beam spread of 20° the ratio of multiple to single scattered intensity is 1, 10 and 100 after 1.4, 4 and 11 attenuation lengths respectively. Okoomian's measurements, using a laser source, extended to 38 attenuation lengths where, for a receiver angle of 26°, the ratio is 10^8. Duntley gives a semi-empirical formula for the total beam intensity.

A numerical approach to this problem has been made by simulating the possible trajectories of light rays in a computer (the Monte Carlo technique). Using results from such studies, Funk (1973) has calculated an effective attenuation coefficient α^* for light from a source of finite angular width for clear ocean water ($\alpha = 0.05$, $b = 0.03$ m^{-1}) and typical turbid water ($\alpha = 1.055$, $b = 0.760$ m^{-1}). He shows that, for the clearer water, the effective attenuation coefficient is independent of r so that the total beam intensity S_0 can be written

$$S_0 = \frac{L_0}{\omega r^2} e^{-\alpha^* r} \qquad (7)$$

with α^* a constant with value between α and a. Unfortunately corresponding information for the more turbid water is not given. Figure 11 shows how the effective attenuation coefficient, α^*, changes from α to nearly a as the angular beam width is increased.

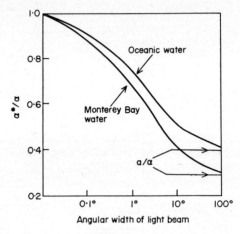

Fig. 11. The ratio of effective attenuation coefficient, α^* to narrow beam attenuation coefficient α as a function of the angular width of the light beam for clear and turbid waters (after Funk, 1971).

10. Transmission of an Image through Water

The light carrying the image of a target directly to the receiver also suffers absorption and scattering. While absorbed light merely reduces the intensity of the received image, that part of the scattered light which still falls upon the receiver, that is mainly the forward scattered component, results in an image displaced in position from the unscattered beam. As a result, an initially sharp image is received as a blurred image, that is, the resolution is degraded.

Gazey (1970) has made observations of this loss of resolution. Varying the turbidity by increasing addition of alumina power of mean diameter 2.5 μ, he recorded the minimum resolvable angle subtended by a bar chart by eye, photographic camera and standard TV vidicon camera. Observations over a 1 m path, with particle densities up to 6 milligramme/litre (7.5 10^{11} particles/m³), showed a loss of minimum resolvable angle from 0.01° in clear water to 0.1° in the most turbid when the "ideal geometry" (Section 11) of a self-luminous target was used. The further deterioration of resolution in the presence of ambient light illuminating the whole water path was also shown.

A. Resolution and the Modulation Transfer Function

The modulation transfer function gives a more comprehensive representation of resolution than the commonly used limiting resolution. If a target has dark and light bars of reflectivity ρ_1 and ρ_2 the modulation of this chart is $m_0 = \rho_2 - \rho_1$. An imperfect optical system would image this target with reduced modulation, m. The ratio $\dfrac{m}{m_0}$ is called the modulation transfer which, if measured for charts with differing numbers of bars per unit distance (spatial frequencies), describes the modulation transfer function (MTF). This function has been much used in recent years to describe the performance of optical components and systems. A perfect optical system has modulation transfer function of unity for all spatial frequencies, in a real optical system the MTF falls off at higher frequencies.

The MTF should rigorously be measured with targets of sinusoidally varying reflectivity. Since these are difficult to produce, the black and white bar target is normally employed and correction made in analysis.

Replogle (1966) presents measurements of the MTF of relatively clear water for ranges up to 18 m. For ranges between 5–15 m his data can be fitted with the Gaussian form

$$M_v = \exp - (\tfrac{1}{2}\sigma^2 \nu^2) \qquad (8)$$

where ν is the spatial frequency in bars per radian and $\sigma = 1.18\ 10^{-3}$ radians independent of range (Mertens, 1970). This data for relatively clear water cannot, however, be applied for other conditions.

Yura (1971) shows that the MTF resulting from particle scattering differs from that from refractive deterioration. Given the angular distribution function $\beta(\theta)$ and a method of handling the tedious calculation of multiple scattering, the MTF for a water path can be calculated from first principles. Wells (1969) has simplified the theoretical problem by considering only scattering at angles sufficiently small that $\sin \theta \approx \tan \theta \approx \theta$ and derives the necessary relationship in closed form. Yura (1973) has calculated the MTF for water in which refractive deterioration dominates. Thus adequate theoretical basis for calculation of MTF exists, but experimental data is lacking.

11. Back Scatter and Light Source/Receiver Geometry

Figure 12 shows a typical arrangement of an optical receiver viewing a target by artificial illumination. In addition to light reflected from the target, suspended particles scatter some light back toward the receiver, the intensity increasing with the quantity of suspended particles which are both illuminated and seen by the receiver. This common

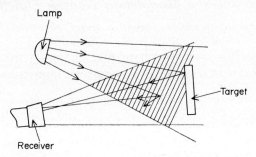

Fig. 12. In a typical arrangement of lamp and receiver relative to a target, light is scattered by particles in the shaded region and so reduces the target contrast.

volume can be reduced by appropriate placing of the light relative to the receiver and photographers working in smoky atmospheres are familiar with the need to displace the illumination from the camera. Nevertheless underwater cameras are frequently designed and marketed with a lamp fixed firmly within a few centimetres of the lens. Their use in turbid waters is thus limited to very short ranges.

It is worth noting that the ideal geometry for reducing back scatter is to place the lamp immediately at the target, so that little of the space between target and receiver is illuminated. A familar example occurs when driving in fog, a preceding car being more readily seen by its tail lights than with the use of the following car's headlamps; indeed this is a strong argument for very bright rearlights for use in fog.

The methods discussed later to remove back scatter can only approach and not surpass the performance of the ideal geometry.

A. Contrast Loss by Back Scatter

Again consider the target consisting of dark and light regions of equal area reflecting irradiances S_1 and S_2 into the detector. The inherent target contrast, C, is then

$$C = \frac{S_2 - S_1}{S} = \frac{\rho_2 - \rho_1}{\rho} \tag{9}$$

where $S = S_2 + S_1$ and $\rho = \rho_2 + \rho_1$.

Back scatter reduces the inherent contrast, C, of a target to an apparent contrast, C_A (Duntley, 1962). If an irradiance S_0, reflected by this target, arrives at the detector together with a back scattered irradiance, B, then

$$C_A = \frac{(S_2 + B/2) - (S_1 + B/2)}{(S_2 + B/2) + (S_1 + B/2)} = \frac{S_2 - S_1}{S + B} = \frac{C}{(1 + B/S)}$$

The ratio $R_c = C_A/C$ is called the contrast ratio and

$$R_c = (1 + B/S)^{-1} \qquad (10)$$

The apparent contrast is also reduced by imperfect optical resolution, the reduction factor being the MTF at the spatial frequency considered. The final apparent contrast is then

$$C_A = C \cdot R_c \cdot M_v \qquad (11)$$

Where M_v is the modulation transfer function at the spatial frequency v.

12. Signal to Noise Ratio, Detectability and Limiting Resolution

The ratio of signal to r.m.s. noise is of central importance in discussing the sensitivity of all radiation detectors. Although in the present section the discussion is for an electro-optic receiver, the concept has been applied to the human eye, photographic film and other detectors (Jones, 1959; Schade, 1964). Hodara and Marquedant (1968) discuss the signal to noise ratio concept in underwater optics in some detail.

The accuracy with which a given signal can be detected is limited by noise, which may be an inherent property of the detector, independent of any signal, or can be created in the process of signal detection. In the case of an electro-optic detector these components include amplifier noise, dark current and electron beam noise.

Following modern convention, the performance of real detectors is discussed by comparison with an ideal detector which although unattainable, is defined as one which has no self generated noise. Such an ideal detector is still limited in detection accuracy by the statistical fluctuations in the arrival of the discrete quanta of radiation; one watt of light of wavelength λ contains $\frac{\lambda}{hc}$ photons/s e.g. $2.67 \; 10^{18}$ photons per second or 1 lumen = $4.54 \; 10^{15}$ photons per second for $\lambda = 540$ nm.

The responsive quantum efficiency, η, is defined as the fraction of incident photons which result in a detected response (photo-electrons, developed film grains or neural response for photo-optic, film and eye respectively). If q photons are detected repeatedly the number detected on each occasion is subject to statistical fluctuation whose root mean square value is shown in standard text books to be $\sqrt{q}$.

If an irradiance, S, is converted to photo-electrons at a receiver of area A, the photo-current, i, is then

$$i = \left(\frac{e}{t}\right)q = e\frac{SA\eta\lambda}{hc} \qquad (12)$$

where e is the charge of an electron and t the time over which the detector integrates the signal. The root mean square noise current Δi for an ideal detector is given by

$$\Delta i^2 = \left(\frac{e}{t}\right)^2 q = \frac{ei}{t}.$$

The signal to noise ratio, k, defined as $\frac{i}{\Delta i}$ is given by

$$k^2 = \frac{t}{e}i = \frac{SA\eta\lambda t}{hc} \qquad (13)$$

or for a detector having inherent r.m.s. noise current i_D

$$\Delta i^2 = \frac{ei}{t} + i_D^2$$

when

$$k_0^2 = k^2\left[1 + \frac{t}{e}\frac{i_D^2}{i}\right]^{-1} \qquad (14)$$

where k_0 is the observed SNR for the detector and k is the SNR for the detector without its inherent noise, that is the equivalent ideal detector. We define a factor,

$$Q = \frac{k_0^2}{k^2}. \qquad (15)$$

From Eqn. (14)

$$Q = \left[1 + \frac{t}{e}\frac{i_D^2}{i}\right]^{-1} \qquad (14a)$$

Q is clearly not a single numerical factor for each detector, but varies with detector illuminance (Eqn. 14a). Since i_D is also a function of i, the dependence of Q on i it is usually complex (Towler and Swainston, 1972).

The signal to noise ratio concept is now extended to the viewing of an image to show that, for a given image illuminance, the need for adequate signal to noise ratio limits the resolution which can be achieved

Imagine, for simplicity, an image consists of dark and light elemental areas with resulting detector currents i_1 and i_2 respectively, and contrast C. The signal to noise ratio, k, is then given in terms of the corresponding detected currents, i_1 and i_2 as

$$k^2 = \frac{(i_2 - i_1)^2}{\Delta i_2{}^2 + \Delta i_1{}^2} \tag{16}$$

then with a total image area, A, comprised of n elements, one can readily show that

$$nk^2 = C^2\left(\frac{t}{e}\right)i = C^2\left(\frac{SA\eta\lambda t}{hc}\right) \tag{17}$$

where S is the total image irradiance at the detector.

For standard TV images formed by linear scans having N lines per picture height, and equal horizontal and vertical resolutions with aspect ratio $\frac{4}{3}$, $n = \frac{4N^2}{3}$.

Introducing the inherent noise current as before again yields Eqn. (14). Towler and Swainston (1972) discuss extensively the noise characteristics of a range of imaging tubes, and show a valuable set of comparisons of limiting resolutions over a wide range of illuminance. For the present purpose the purely formal representation defined by Eqn. (15) is used. Thus

$$\frac{4}{3}(Nk_0)^2 = nk_0{}^2 = C^2 Q \frac{SA\eta\lambda t}{hc} \tag{18}$$

A. Inspection, Recognition and Detection

Equation 18 indicates how, as the image is increasingly divided into finer resolution elements, the decreasing number of electrons available in each element results in increasing statistical noise. For a chosen k_0 value the limiting resolution can thus be calculated.

The values of N and k_0 required of an image depends on the visual task undertaken. Thus detailed inspection requires high resolution capability. As N is reduced, the point is reached where the target can only be recognized, that is distinguished from another set of possible targets. Further reduction allows only the detection of a target. The presence of noise further inhibits the ability to detect, recognize or inspect the target.

Several studies have aimed to give the required values of N and k_0 to achieve inspection, recognition and detection of a variety of objects. Waynant (1971) states that N should be 6 to 8 for recognition, and 2 to 4 for detection of an object, and that a value k_0 near 3 is required. Wagenaar and Van Meeteren (1969) find 50% correct identification of military targets with 6.4 ± 1.5 TV lines per object, and that this is relatively insensitive to k_0 for $k_0 > 2$.

B. Extensions for Through-water Viewing

As explained in Section 11, when S is accompanied by a background (e.g. back scatter) irradiance, B, the inherent target contrast C should be replaced by the apparent contrast C_A. Further, statistical fluctuations of the current arising from the background, Δi_B must be included in calculating the SNR which becomes

$$k^2 = \frac{(i_2 - i_1)^2}{\Delta i_2^2 + \Delta i_1^2 + \Delta i_B^2}$$

whence with obvious substitutions and using Eqn. (10)

$$\frac{4}{3}(Nk)^2 = nk^2 = C_A^2 R_C \frac{SA\eta\lambda t}{hc} \qquad (19)$$

Introducing the factor Q and substituting for C_A from Eqn. (11), Eqn. (19) becomes

$$\frac{4}{3}(Nk_0)^2 = nk_0^2 = C^2 R_C^3 M_N^2 Q \frac{SA\eta\lambda t}{hc} \qquad (20)$$

Here the dependence of M on resolution is recognized with the subscript N.

A further source of statistical fluctuation arises from the variable number of scattering particles in the viewed region. This source of noise is currently being evaluated by the author.

13. System Performance in Terms of its Components

An image will be focused on an area A of the detector's photosurface with optics (lens or mirror) whose effective aperture number is f. Then S is given in terms of S_0, the irradiance falling on the target, using Eqn. (5) as

$$S = \pi \frac{S_0 \rho(\varepsilon)}{4f^2} T_0 T_p \qquad (21)$$

Here T_0 and T_p are the transmissions of the optical system and path from target to detector, $\rho(\varepsilon)$ is the reflection coefficient per unit solid angle at angle ε to the normal to the reflecting surface. Thus

$$\tfrac{4}{3}(Nk_0)^2 = nk_0^2 = \tfrac{1}{4}F_L F_T F_W F_D \tag{22}$$

where
$$F_L = \frac{S_0 \lambda}{hc} \tag{22a}$$

$$F_T = \pi C^2 \rho(\varepsilon) \tag{22b}$$

$$F_W = R_c^3 M_N^2 T_p \tag{22c}$$

$$F_D = \frac{T_0}{f^2} \eta t A Q \tag{22d}$$

The four F factors are those determined respectively by light source, target, water path and detector.

The increase of Nk_0 with target irradiance and reflectivity are perhaps obvious, the stronger dependence on target contrast is less generally realized. The importance of high contrast markings on a target which may be the object of underwater search is thus emphasized. Varnado and Hessel (1973) show theoretically that detection range can be extended with the use of a photo-luminescent target.

The factor F_W, in effect, summarizes the through-water viewing problem, the three components representing contrast, resolution and intensity losses. We use the methods already given to calculate the factor F_W and hence the limiting resolution of some through-water viewing systems in Section 14. The detector dependent factor F_D is discussed in the following section.

A. Performance of Receivers

The factor F_D could be split into two, the first $\dfrac{T_0}{f^2}$, being a function of the optical system, while the remaining factors A, t, η, and Q depend on the particular detector.

A primary requirement of a sensitive imaging system is efficient collection of light, that is, high transmission optics of wide acceptance aperture (low f-number). The progress towards ever smaller f-number lenses has been much stimulated with the use of computers in lens design. Especially high aperture optics designed for use with low light viewing systems has been extensively discussed by Soule (1968); f values as low as 0.5 are currently designed. When wide angle of view is also required, as in search applications, the lens can become very large.

The maximum photo-sensitive area, A, is normally fixed by restrictions on size, weight and expense of the lens-receiver. Limitations on the camera size usually restricts the use of the larger formats. Television receivers are most commonly based on the 25 mm diameter imaging tube with photosensitive area 9.8 × 12.6 mm being compatible with the use of lenses designed for the 16 mm cine-photography format.

The maximum integrating time, t, is normally determined by the need to "freeze" any relative movement of target and receiver. Image motion compensation devices are used to retain a maximum t in airborne receivers. Where image motion is negligible, the advantage of maximizing t can be simply demonstrated by comparing individual frames of a cine film with the same scene presented by normal projection. Marked improvement in the latter case arises from integration of frames in the eye over a period of about 0.2 seconds.

η may be determined for film from the number of developed silver grains per incident photon and for the human eye from measurements of the probability of perceiving light flashes as a function of a number of photons in the flash. Soule states that values $\eta \simeq 0.01$ are found for both film and the human eye. The higher sensitivity of electro-optic devices arises largely from the considerably higher responsive quantum efficiency for photo-emissive surfaces, values of greater than 0.2 being normally achieved.

Jones (1959) has defined "detective quantum efficiency" Q_D as

$$Q_D = \eta Q. \tag{23}$$

Jones points out that due to ambiguity in defining η for most detectors, the separation into two factors η and Q is less meaningful than consideration of Q_D. He presents data on Q_D for the human eye, film, vidicons and image orthicons. Q_D plotted against exposure shows maxima for each detector. Maximum values for Q_D are 6%, 1%, 1% and 0.1% for an image orthicon, Royal-X film, human vision and a vidicon respectively. Introducing measured value of η, Jones concluded that $Q \simeq 0.3$ for the orthicon. More recent developments of low light imaging tubes have probably improved on this already impressive performance, thus imaging devices are within a factor less than 3 of the ideal noiseless detector performance. The dark-adapted human eye has apparently a comparable value of Q, but the effective η is at least a factor of 10 lower. Nevertheless the human eye is remarkable in adapting to the needs of low light level viewing, increasing its pupil size and effective integration time, and also integrating the received image over a larger area of the retina. Thus the eye makes every effort to increase the SNR by reducing f, increasing t, and decreasing the number of resolution elements, n.

14. Calculation of Back Scatter

Figure 13 serves to define the geometrical parameters in a lamp-target-receiver arrangement. This of course is not the only nor always the most advantageous geometry. In some circumstances it is practicable to advance the light to illuminate the target from close range. Indeed, other methods of minimizing back scatter can only approach this "ideal geometry" as a method of reducing back scatter. Patterson (1970) discusses the advantages of placing the light behind the camera when a very wide angle view is required.

Back scattered irradiance for a chosen geometry can be estimated

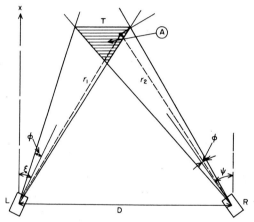

FIG. 13. Definition of geometrical parameters in an arrangement of lamp (L), target (T), and receiver (R). The shaded region is both illuminated and viewed by the receiver.

using a number of simplifying assumptions. Assumptions arise principally because in situations where back scattered irradiance is of importance, multiple scattering is the dominant component, this being notoriously tedious to calculate.

Here the method of Funk (1971) is used, that is, single scattering analysis is extended with the use of an effective attenuation coefficient. Funk discusses the appropriate attenuation coefficients for incident light, light reflected by the target and back scattered light. For simplicity of representation these coefficients are assumed equal here. (The reader is referred to Funk (1971 and 1972) for further discussion.)

With our simplifying assumptions and the geometry of Fig. 13 the back scattered irradiance, B, at the detector is

$$B = \frac{L_0}{\omega} \int^v \beta(\theta) \exp - [\alpha^*(r_1 + r_2)] r_1^{-2} r_2^{-2} \cos \psi^{-1} dV \qquad (24)$$

The integral extends over that volume marked "A" in Fig. 13 which is both illuminated and viewed and $\omega = 2\pi(1 - \cos\phi)$.

If the target covers the whole of the illuminated area irradiance, S, at the detector from reflection by the target is

$$S = \frac{L_0}{r^2} \rho(\varepsilon) \exp(-2\alpha^* r) \cos\xi \qquad (25)$$

In Eqn. (25) the use of an effective attenuation coefficient α^* implies that the forward scattered light as well as the unscattered is included in the target irradiance (see discussion in Section 9).

For a Lambertian reflector with total reflection coefficient ρ

$$\rho(\varepsilon) = \rho \frac{\cos\varepsilon}{\pi} \qquad (26)$$

and for a distant target, $\varepsilon \approx 0$ $\rho(\varepsilon) \approx \rho/\pi$.

Using Eqns. (24) and (25) the contrast ratio can be estimated. clearly an exact calculation requires more comprehensive treatment of the multiple scattering; little theoretical or experimental effort has been devoted to this problem.

In order to give numerical examples of S, B and R_C, we must choose appropriate values for optical parameters. The angular distribution function $\beta(\theta)$ is relatively flat for θ between 120° and 180° and for θ near 180° may be taken as a constant, β (180). Values of $\beta(180)$ between 3.10^{-4} and 2.10^{-3} for oceanic and lake waters have been quoted by Jerlov (1968). For coastal waters data is not available, here value $\beta(180) = 5.10^{-3}$ m^{-1} sterad^{-1} is used to represent turbid water. $\rho = 0.16$ is chosen to represent a target of average reflectivity. Since α^* can vary so widely, a range of values is examined (in practice, the back scatter coefficient increases with α^* rather than remaining constant as is here assumed).

In principle, the back scattered intensity, B, is a function of ϕ and of D separately. However, calculation shows that within the range of interest and with error of less than 3%, B is a function of the ratio D/ϕ. This result is valuable in summarizing the results. Fig. 14 shows S/L_0 and B/L_0 as a function of range for several values of D/ϕ.

So far we have considered the average contrast loss over the target, which would be important in target detection. When more detailed viewing of features inside the target area is required, the contrast variation across the target is of interest. For example, in Fig. 13 the left hand side of the target, being viewed through a longer illuminated path, suffers more contrast loss than the right hand side. A more even distribution of back scatter can be achieved with the use of a pair of lights illuminating the target, the receiver being in the centre. An evenly distributed background is advantageous since in electro-optic receivers it is relatively simple to subtract constant background level. In photographic processing it is also possible to minimize the disturbing effects of constant background haze. The back scattered intensity from inside the illuminated region of Fig. 13 is calculated with appropriate modification of the volume of integration in Eqn. (24). Funk (1971), Duntley

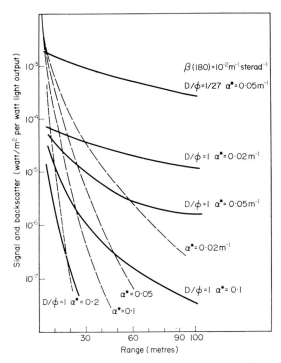

FIG. 14. The target irradiance (broken curves) for effective attenuation coefficients $\alpha^* = 0.02$, 0.05, 0.1, and 0.2 m^{-1}. Back scatter irradiance for geometries defined by the $D/\phi = 1$ (full curves) together with one example of the effect of reducing D (upper curve). Other parameters used are defined in the text.

(1966) and Angelbeck (1966) give more comprehensive formulae for this case.

A further source of unwanted background haze could arise from light, initially reflected by the target in a direction which would have avoided the receiver, being scattered by suspended particles into the receiver. A rough estimate for plane targets suggests that this source of background is comparable with that from back scattering by particles. Although Replogle (1966) gives data which confirms this expectation, no further experimental work has been reported, Funk (1972) gives a method of estimation.

The volume of integration in Eqn. (24) and thus the back scattered intensity, B, are reduced either by increasing the separation of source and receiver, D, or reducing the angles, ϕ. Since the effect is to isolate the region of detectable light to that near the target, this procedure is called geometrical isolation.

15. Geometrical Isolation

This method, requiring only attention to the angular spread of the light beam, angle of view and relative placing of light and detector, is perhaps the simplest method of maintaining adequate image contrast in conditions of appreciable scattering. We use the data of Fig. 14 to illustrate the potential improvement of contrast and the need for sensitive imaging receivers to take advantage of the resulting range improvement which becomes possible.

The upper curve shows how the back scattered irradiance exceeds that from the signal at very short ranges when small separation and

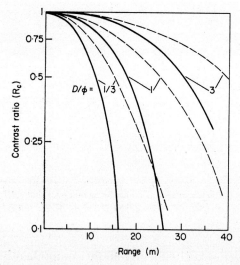

FIG. 15. The fall-off of target contrast with range for $\alpha^* = 0.2$ (full curves) and $\alpha^* = 0.05$ m^{-1} (broken curves) with values 3, 1 and $\frac{1}{3}$ for the geometrical parameter D/ϕ.

wide angle illumination is used. Reduction of the ratio D/ϕ by a factor of 27 results in decrease of B by two orders of magnitude. The range at which $B = S$ for constant geometry is also shown to decrease with α^*. Figure 15 shows the fall of contrast ratio with range.

The rapid decrease of contrast ratio from 0.25 to 0.1 over a range of only 1 m with $D/\phi = 1/3$, $\alpha^* = 0.2$ m^{-1} is consistent with the observed rapid disappearance of targets as noted in the Introduction. The range at which the contrast ratio reaches 0.1 is increased from 15–26 m by increasing the ratio D/ϕ threefold ($\alpha^* = 0.2$ m^{-1}). This valuable extension of potential viewing range is, however, accompanied by a marked decrease in image irradiance due mainly to attenuation in the water path.

The image irradiance at the photosensitive surface of the detector, S, can be written using Eqns. (21), (25) and (26) as

$$S = \frac{L_0}{\omega} \frac{T_0 \rho}{4f^2} \frac{e^{-2\alpha^* r}}{r^2} \cdot \cos \xi \tag{27}$$

Using $\phi = 3°$, $\omega = 0.87 \ 10^{-2}$ sterad. Then with $T_0 = 0.8$, $f = 1.0$, $\rho = 0.16$, $\cos \xi = 1$

$$S = 3.68 L_0 \frac{e^{-2\alpha^* r}}{r^2} \tag{28}$$

Thus with $\alpha^* = 0.2$ m^{-1}, $S = 4.06 \ 10^{-5} L_0$ W/m^2 at 15 m and $1.66 \ 10^{-7} L_0$ W/m^2 at 26 m. A powerful (1 kW) discharge lamp with thallium additive would yield 100 watts optical power or 64 000 lumens at 530 nm (Section 7 B), the resulting image illuminances then being 2.7 and 1.1×10^{-2} lux at 15 and 26 m. Using the data of Fig. 8, and image irradiance 2.7 lux would allow use of a standard vidicon with resolution of approximately 400 TV lines/picture height. At the longer range to retain this resolution, a more sensitive tube such as the Ebitron or SEC vidicon is required.

The effect of decreasing image irradiance on limiting resolution with a given detector can be calculated, using Eqn. (20) with Eqn. (23)

$$\frac{4}{3}(Nk_0)^2 = (nk_0)^2 = C^2 R_c^3 M_N^2 \cdot AQ_D \frac{t\lambda}{hc} \cdot S \tag{29}$$

using $A = 1.2 \ 10^{-4}$ m^2 (9.8×12.6 mm format) $t = 0.2$ s (eye integration time), $k_0 = 2$ (see Section 12 A).

$\frac{\lambda}{hc} = 2.67 \ 10^{18}$ photons/s per W, $C = 1$ (a maximum contrast target)

with Eqn. (28) for S.

$$N = 0.665 \ 10^6 \ M_N R_c^{3/2} Q_D^{1/2} \frac{e^{-\alpha^* r}}{r} \tag{30}$$

In the absence of scatter in the water path, $R_C = 1$, and $M_N = 1$, limiting resolution thus varies with range as $\dfrac{e^{-\alpha^* r}}{r}$. With scattering, the reduction of R_C and M_N with range further emphasizes the rapid loss

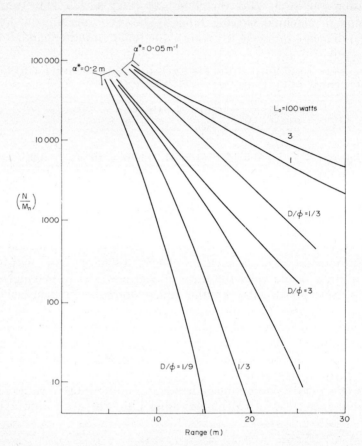

FIG. 16. Limiting resolution (for a TV picture with the linear scanning) divided by the modulation transfer function as a function of range, calculated from Eqn. (30). As the geometrical parameter D/ϕ is varied, the contrast ratio and hence the limiting resolution are changed. Here the contrast ratios from Fig. 15 are used. In practice, at short range the limiting resolution is determined by the limitations of the receiver and would not exceed 400 to 600 TV lines/picture height for a normal quality television receiver.

of resolution in laden waters. A thorough treatment requires consideration of the variation of Q_D with S (Immarco and Oigarden, 1966; Towler and Swainston, 1972). For the present illustration we use a constant value $Q_D = 0.01$, to represent the broad maximum value found by Jones (1959) for high speed film and the eye, and within a factor 5 of the maxima for the best low noise imaging tubes. Since data for calculation of M_N is not available, the ratio $\dfrac{N}{M_N}$ calculated from Eqn. (30) is plotted in Fig. 16.

Preliminary trials of viewing system using the geometrical isolation method have been conducted (Cocking, 1974) using the submersible "Pisces". Using $D = 3$ m and $\phi = 3°$, improved contrast compared with that found by natural illumination was demonstrated. A low light television receiver (Ebitron) was used and a target at 10 m range could be examined in relatively turbid waters when the target was barely detectable using a standard vidicon camera or direct viewing. The value of a zoom lens, allowing a variable angle of view, was demonstrated.

16. Target Scanning Methods

In geometrical isolation, the contrast ratio is improved by increasing D, and decreasing ϕ. The maximum separation distance, D, is limited by practical considerations, while for conventional lamps collimation angle of $\phi = 1°$ is near the limit which can be achieved with reasonable

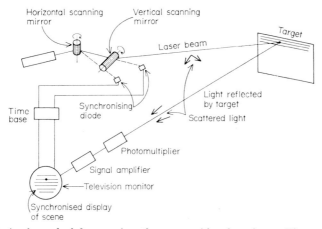

Fig. 17. A simple method for scanning of a target with a laser beam. The scanning of a television monitor is synchronised with the scanning of the target and the reflected intensity, detected by the photomultiplier, modulates the intensity of the monitor's canning spot. The whole target is viewed continuously in this simple scheme.

sized reflectors or lenses, and a more generally practicable figure is $\phi = 3°$. Divergence angles less than $0.1°$ are usual with lasers. With such small divergence, a small element of the target is illuminated, for example a 2 cm diameter spot at 10 m range. The whole target area is then covered by scanning with the light beam. A variety of ways to achieve this have been described (Wall and Coleman, 1968; Angelbeck, 1966). Figure 17 shows one proposal using rotating mirrors for horizontal and vertical scanning. A simple photo-detector is used as the receiver, the time varying signal being applied to a standard TV monitor. The electron scanning of the monitors is synchronised with the scanning light.

Back scatter is minimized if the optical receiver scans the target in synchronism with the illuminated region. This could, in principle, be achieved with a further set of rotating mirrors linked to those controlling the light spot. Alternatively the received light from the target can be

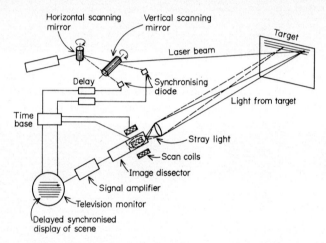

Fig. 18. The laser beam target scanner as in Fig. 17, modified so that the viewed region of the target is synchronously scanned, in this case using an image dissector tube (Papp, 1965).

scanned by applying signals from the light scanning mirrors to the scan coils of an image dissector (Papp, 1965), see Fig. 18.

In principle back scatter can be reduced indefinitely by reducing the angles of light emission and receiver view. However, in practice, as we have seen, scattering results in an angular spread of the light beam and, in effect, a spread of the viewed area of the target (Duntley, 1966). Thus the water medium limits the minimum value of ϕ which can be usefully employed.

The optical power available from standard lasers is of the order 1 watt compared with 100 W from conventional lamps. The smaller angular spread of the laser, for example 0.1° compared with 3° yields comparable target irradiance over the small illuminated target region. Thus an acceptable SNR can be achieved despite the low power in the illuminated region, the use of a synchronised receiver being especially advantageous (Angelbeck, 1966). The information rate from the whole scanned scene is, however, still determined by the mean optical power available.

Martin (1966), Waynant (1971) and Angelbeck (1966) have discussed the laser scanning method. Angelbeck has analyzed the system performance and given limited results from tests in which a 3 mW argon laser was used. He employed a standing acoustic wave in a liquid-containing cell to deflect the light beam. Two such cells placed at right angles were used to scan the target. No full scale tests of laser scanning have apparently been reported.

17. Time Dependent Isolation: Range Gating

The range gating method restricts the volume both illuminated and viewed to a region near the target as in the geometrical isolation method, but uses the finite propagation velocity of light to achieve discrimination. The method, being analogous to radar, has been called lidar (LIght Detection And Ranging).

If a light source is operated for a period Δt of the order of nanoseconds, the light emitted will travel to a target at range r and return to the receiver near the light source in a time $\frac{2r}{c}$, where c is the velocity of light in water. If the receiver is open only for a short period, the range, r, from which light can be received can be selected by choosing the delay $\left(\frac{2r}{c}\right)$ after the light pulse. If light pulse and receiver gate are each equal to Δt the total travel path $2r$ is determined to within a maximum error $\Delta r = 2\Delta tc$.

The back scattered illuminance with range gating can be calculated using Eqn. (24) by further restricting the volume of integration to that defined by

$$(2r + \Delta r) > (r_1 + r_2) > (2r - \Delta r).$$

It is also necessary and simple to include the duty cycle function discussed by Wall and Coleman (1968).

Figure 19 shows the contrast ratio obtained using range gating with $\Delta t = 20$ and 40 ns which can be readily achieved using pulsed lasers and gated electro-optic receivers, or with Kerr cells (electrostatic light switches). The volume of integration and hence the contrast ratio, largely determined by Δt, is almost independent of D and ϕ, thus the receiver can be placed close to the light without loss. With increasing range, Fig. 19 shows that the contrast ratio at first falls and then maintains an almost constant value, since the volume of illuminated light remains effectively constant once the light pulse is fully visible to the receiver. The data in Fig. 19 may be directly compared with that for geometrical isolation, since water and target conditions have been assumed the same. It is clear that for short ranges, that is less than $2c\Delta t$, the range gating method achieves less than the basically simpler geometrical method. However, at greater ranges considerable and often impractical separations, D, would be needed to achieve the same contrast ratio.

Range gating can, in principle, be used to reduce back scattered intensity effectively to zero, since techniques to achieve pulses as short

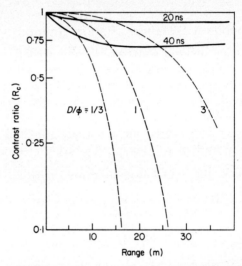

FIG. 19. Contrast ratio as a function of range achieved with range gating (full curves) compared with those from geometrical isolation (dashed curves). Light pulse and receiver gate in the former method are assumed equal to 20 and 40 ns and the geometrical parameter D/ϕ are indicated for the latter case. Attenuation coefficient $\alpha^* = 0.2$ m^{-1}.

as 10^{-12} s have been demonstrated. As with laser scanning the practical limitation is the available optical power. In range gating the available power is collected in short intense pulses with the result that the SNR during these pulses can be high. The relatively low mean optical power, which determines the target information rate, has not prevented extensive development of the method in the USA.

Heckman (1966) demonstrated through-water range gated photography using a laser yielding blue-green (550 nm) light from a neodymium doped glass rod coupled to a KDP crystal for frequency doubling. A rotating prism Q-switch produced a pulse of 20 ns duration with 0.5 millijoules optical energy. The receiver was polaroid type–410 film with an equivalent ASA rating of 10 000. An image converter tube was used as a gate over the detector. Range gated photographs using 20 ns laser pulses and receiver gate at 10.3 m (four attenuation lengths) was demonstrated. This work was the fore-runner to development of a more powerful laser source enclosed for underwater use, and a highly sensitive gated imaging receiver. Detailed calculations on the proposed system are given by Immarco and Oigarden (1966) and experimental results from the completed system by Keil, Immarco and Kerpchar (1968). An yttrium aluminium garnet (YAG) laser rod produced a 500 kW peak, 10 ns pulse which was frequency doubled to 530 nm with a lithium niobate crystal; pulse rates up to 15 pulse/s were possible. The receiver, an image orthicon with an image intensifier added, could be gated by applying a 14 ns voltage pulse to the image intensifier. The improvement possible with range gating compared with continuous lighting was demonstrated at ranges up to a maximum of 39 m (corresponding to 4.5 attenuation lengths). Results from range gating were compared with those using the ideal geometry, that is of illumination immediately at the target. At the greatest range shown the range gated results fell far short of those using the ideal geometry. The authors state that available power and not back scatter limited the range capability of this system.

Rattman and Smith (1972) have described briefly the application of pulsed lasers for underwater viewing using range gating and also for rapid depth sounding in shallow water from a helicopter. In the former mode, pulses with a repetition rate up to 50 pulse/s with peak power 40 mW at 530 nm were used, recognizable images at ranges of 20 m, being 4–5 times the range for unaided vision, are stated by the authors. The application for depth measurement is analogous to the usual sonic methods. The intense reflection from the water surface and the weaker bottom reflection are timed to determine water depth.

18. Continuous Time Dependent Isolation

A method of using the finite velocity of light to isolate scattering in the region of the target while continuously viewing has been discussed by Wall and Coleman (1968) and Waynant (1971). The basic system for scanning a target has already been described in Section 16. Since the travel time of the light is $\frac{2r}{c}$, the receiver would need to view in a direction delayed by a small angle $\psi = \frac{2r}{c}\Omega$ where Ω is the angular velocity of the scanning beam. Thus the viewed range, r can be chosen by the delay angle, ψ. In effect, the scanning is being used as a timing clock to achieve a time delay as in the pulsed method. Clearly for the method to be effective, ψ must be greater than ϕ, a condition requiring narrow angle beams and angle of view, and very high scanning speeds, e.g. for $r = 10$ m, $\dfrac{2r}{c} = 90$ ns, thus $\psi > (\phi = 0.1°)$ needs $\Omega > 10^6$ °/s. Perhaps understandably, the method has not yet been demonstrated in water.

19. Application of Polarizing Filters

Light, when scattered usually acquires some degree of polarization (Kerker, 1969). For example, Rayleigh scattering by density fluctuations in the upper atmosphere gives rise to strong polarization of skylight, as can be observed by viewing through a linear polarizing filter. The polarization of natural light under water arising from scattering in the water was first observed by Waterman (1954). Increased target contrast with the use of such polarizing filters in the sea has been demonstrated by Lythgoe and Hemmings (1967).

Artificial light, since it is readily completely polarized, can be used to emphasize more strongly the difference in scattering properties of wanted targets against unwanted scattering by suspended particles. For example, back scattering by small particles causes no change in the plane of polarization, whereas light is largely de-polarized on reflection from a matt surface due to the many reflections which such light suffers at a rough (when measured in terms of light wavelengths) surface. Thus when viewed through a linear polarizing filter orientated at right angles to the plane of the source, light scattering by the small particles would be removed, while one half of a totally depolarized beam would be transmitted. Circular polarized light may also be used to differentiate scattering by small particles and by matt surfaces (see Mertens, 1970).

Gilbert and Pernicka (1967) used circular polarizing filters to demonstrate a considerable improvement in contrast in relatively turbid water (attenuation length 1.08 m). Their photographic evidence of targets at up to 8.7 m range shows an impressive removal of the back scattered light which, without filters, obliterated the target image. Further, measurements with a telephotometer showed an average contrast improvement for white targets of 15.1 and for black targets of 13.6. If such improvements could be consistently obtained, this simple technique could be applied in all circumstances, including for individual divers. Unfortunately, Gilbert (1970) reported that such observations by divers had, in practice, shown negligible improvement. Since the degree of polarization of back scattered light is strongly dependent on particle size and refractive index (Kerker, 1969), Gilbert undertook a study of the effect of a particle size on contrast improvement by polarization discrimination. Contrasts were determined as a function of diameter and concentration of polystyrene spheres suspended in water. Circular polarization of source and receiver improved contrasts of matt targets for all concentrations of scattering spheres with diameters less than 0.8 microns. With spheres of diameter 1 μ contrast improved only with concentrations giving α between 0.25 and 1 m^{-1}; for spheres of diameter 6 to 100 μm, circular polarization always degraded contrast.

Reduction of back scatter using polarization techniques thus appears to give improvement or deterioration strongly dependent on the particular particulate suspension. Nevertheless the technique due to its extreme simplicity is almost always worth testing. When image quality is limited by available light intensity rather than by back scatter, the intensity loss of at least a factor of four can, however, prove disadvantageous.

20. Image Processing

The output signal from an optical receiver can be modified by subsequent treatment. A familiar example is the modification of picture contrast possible with photographic printing with selection of printing papers. Image processing comprises a wide variety of such modifications which can be used to diminish or enhance particular features of an image. Of special interest in the present context are reduction of back scatter and enhancement of detail reproduction.

Signal processing is well known in communications engineering, and mathematical treatment in terms of Fourier synthesis is well developed. Image information is relatively more complex, consisting of two-dimensional Fourier components in place of the more usual one-dimensional signal of communication engineering.

Television scanning techniques allow two-dimensional images to be represented as a one-dimensional signal. Processing of this signal is possible, but image information from different parts of the picture are separated in time. Thus complete image processing is only possible if the full image information is stored with immediate access. Digital computers have been used to process images but powerful computers are required for this work. Billingsley (1970) describes and presents examples of image modifications made by such digital techniques.

Simultaneous processing of the whole image, either in its optical form or converted to an electronic image, is possible. Mazurowski (1969) discusses operations on an optical image, Mertens (1970) illustrates simple photographic techniques to improve image sharpness and Munsey (1968) describes the use of an electronic image storage tube, and demonstrates the removal of back scatter due to atmospheric haze from aerial photographs.

Where two or more separate images of the same subject are available, correlation between these separate images can give further improvements. Intra-frame correlation can be used with optical, electronic or digital images to improve detectability or image quality. Further, known target shapes can be selectively sought and relative movement of particular regions of the image can be detected.

In terms of two-dimensional Fourier components, back scatter represents an unwanted addition at low spatial frequencies. Resolution loss is described by loss of high frequencies, in fact by modifying the frequency spectrum describing the image by the modulation transfer function, which falls at high frequencies. Image processing of through-water images should therefore diminish low frequency and enhance high frequency components. The limitation on low frequency diminuation is simply that low frequency components of the wanted image are reduced, while enhancement of high frequency components is limited by simultaneous enhancement of image noise.

This discussion suggests that an image, degraded by through-water transmission, could be improved by passing through a frequency dependent filter having minimum transmission at low spatial frequencies, and increasing at high frequencies. A simple technique called "diffuse masking" to create such a filter photographically is described by Mertens (1970). An alternative technique using coherent illumination (a laser or monochromatically illuminated pinhole) allowed direct viewing of the processed image. This method in which operations are made on the Fourier components of the image is based on the fact that when an image placed in the first focal plane of a lens is illuminated by coherent light, the Fourier transformed image is formed in the second focal plane. Required modification in the Fourier space can then be performed with the mask at this plane. Further transformation, using a second lens, allows the modified image to be displayed. Since a given spatial frequency is represented in Fourier space by a point, chosen frequencies may be diminished or removed with relatively simple masks (Mazurowski, 1969).

Some image processing occurs in any artificial viewing system, often without the operator's realizing it, for example, in adjustment of the contrast control on a television monitor. The black level and contrast response, γ, can often be selected in standard television cameras. Reduction of an even background haze and enhancement of image contrast would benefit if these were immediately under the control of a knowledgeable operator. Experienced photographers, in fact, make use of their ability to modify these parameters by exposing film to use the strongly non-linear toe of the response curve to reduce even haze and by developing film for high gamma. Edge gradient may be enhanced by choice of developer and development techniques. More advanced image processing techniques touched on here have apparently not yet been applied to in-water viewing. An important exception is the human eye-brain combination which performs the most sophisticated image processing, including time correlation. A unique advantage of the eye-brain is its ability to adapt to the needs of the particular viewing environment. The principal advantages over direct viewing of artificial techniques namely, remote viewing and image storage are thus gained at some loss.

21. Concluding Remarks

An adequate discussion of visual perception for normal in-air viewing would be lengthy and complex; with the complication of the imperfect optical properties of natural water it is still more illusive. The present chapter has aimed only to indicate the major physical phenomena which bear on this problem, to summarize the relevant literature and finally to indicate in outline the areas where potential for development of solutions to the practical problems lie. The emphasis in the latter enquiry has been on electro-optic devices, a field of considerable development in the present decade. Their potential in low-light viewing and enhancement of required image features with simultaneous viewing by a remote observer promise major improvements in through-water viewing.

With the expected expansion in under-sea civilian activities, improved underwater viewing will be required in survey and inspection of the sea-bed to choose routes for cables and pipes and search over wide areas for lost or missing objects. High resolution viewing will be necessary in underwater inspection of ships' hulls, oil rigs, docks and harbour walls, and many other structures often in very turbid waters. In the improvement of fishing techniques inspection of fish trawls in actual operation and examination of spawning grounds would benefit from improved viewing capabilities. Direct observation and also time lapse recording are already being used to check on changes of flora and fauna as a monitor of the effect of pollution (McIntyre, 1971).

In addition to these applications which involve transmission of an image through the water, a further class of applications requires a light beam or light pulses for the purposes of measurement, geometrical survey or information transmission.

Whether the potential improvements in all these areas is realized will be largely determined by how clearly the potential can be spelt out and its benefits quantified. The analyses and data, summarized in this review, are not yet adequate for such firm conclusions. Instead, a wealth of interesting technical, experimental and theoretical studies in the field of hydro-optics appears. A rapid survey of the material in the present article revealed 15 separate subjects deserving further attention, and offering fruitful fields of study; just one example is given here. The quantification of the optical properties of waters requires development of measuring instruments for field use, in particular transmissometers, scatter meters and instruments to quantify resolution loss. Given these instruments a survey of the optical properties of coastal waters could provide information on the factors which affect visibility, such as current and sea state. The correlation of these observations with measured particulate content and thermal inhomogeneity of the water would yield data for analysis and comparison with basic scattering theory. This data if correlated with visual observation of standard targets would yield information of practical application.

With increasing interest in universities in problems in applied science and the stimulus resulting from the use of diving as a research tool (Wood and Lythgoe, 1971), one may hope for some thorough experimental and theoretical studies leading to more complete basic data and methods to employ these data in quantifying the potential of optical and visual methods in underwater imaging.

References

Angelbeck, A. W. (1966). Application of a laser scanning and imaging system to underwater viewing. *In*: "Underwater Photo-optics: Seminar Proceedings", pp. B-VII-1-12. Society of Photo-optical Instrumentation Engineers, Redondo Beach, Calif.

Ashley, L. E. and Cobb, C. M. (1958). Single particle scattering functions for latex spheres in water. *J. Opt. Soc. Amer.* **48**, 261–268.

Beardsley, G. F. Jr., Pak, H. and Carder, K. (1970). Light scattering and suspended particles in the eastern equatorial Pacific Ocean. *J. Geophys. Res.* **75**, 2837–2845.

Billingsley, F. C. (1970). Applications of digital image processing. *Appl. Opt.* **9**, 289–292.

Cocking, S. J. (1974). Scattering by suspended particles and through water viewing. *Underwater J.* (to be published).

Deirmendjian, D. (1969). Electromagnetic Scattering on Spherical Polydispersions. Elsevier, New York.

Duntley, S. Q. (1962). Underwater visibility. *In*: "The Sea" (Ed. M. Hill), vol. 1, pp. 452–455. Wiley, New York.

Duntley, S. Q. (1963). Light in the sea. *J. Opt. Soc. Amer.* **53**, 214–233.

Duntley, S. Q. (1966). Principles of underwater lighting. *In*: "Underwater photo-optics:

Seminar Proceedings", pp. A-I-1-7. Society of Photo-optical Instrumentation Engineers, Redondo Beach, California.

Funk, C. J. (1971). Computer simulation of the performance of advanced underwater optical viewing systems. *In*: "IEEE 1971 Conference on Engineering in the Ocean Environment", pp. 74–84. Institute of Electrical and Electronic Engineers, New York.

Funk, C. J. (1972). Energy constrains on underwater optical and acoustical imaging systems. Doctoral thesis, University of California.

Funk, C. J. (1973). Multiple scattering calculations of light propagation in ocean water. *Appl. Opt.* **12**, 301–313.

Gazey, B. K. (1970). Visibility and resolution in turbid waters. *Underwater Sci. Technol. J.* **2**, 105–115.

Gilbert, G. D. (1970). The effects of particle size on contrast improvement by polarization discrimination for underwater targets. *Appl. Opt.* **9**, 421–428.

Gilbert, G. D. and Pernicka, J. C. (1967). Improvement of underwater visibility by reduction of backscatter with a circular polarization technique. *Appl. Opt.* **6**, 741–746.

Harford, J. W. (1968). Underwater lighting: a status report. *Marine Sci. Instrument.* **4**.

Heckman, P. J. Jr. (1966). Underwater range-gated photography. *In*: "Underwater Photo-optics: Seminar Proceedings", pp. B-IX-1-9. Society of Photo-optical Instrumentation Engineers, Redondo Beach, California.

Hodara, H. and Maquedant, R. J. (1968). The signal to noise ratio concept in underwater optics. *Appl. Opt.* **7**, 527–534.

Hodkinson, J. R. (1966). Particle sizing by means of the forward scattering lobe. *Appl. Opt.* **5**, 839–844.

Immarco, A. and Oigarden, T. (1966). Underwater detection and classification system. *In*: "Underwater Photo-Optics: Seminar Proceedings", pp. B-X-1-12. Society of Photo-optical Instrumentation Engineers, Redondo Beach, California.

Jensen, N. (1968). "Optical and Photographic Reconnaissance Systems". Wiley, New York.

Jerlov, N. G. (1968). "Optical Oceanography". Elsevier, Amsterdam.

Jerlov, N. G. (1970). Oceanic light scattering properties related to dynamic conditions. *Soc. Photo-opt. Instrument. Eng. J.* **8**, 89–93.

Jones, R. C. (1959). Quantum efficiency of detectors for visible and infrared radiation. *In*: "Advances in Electronics and Electron Physics" (Ed. L. Marton), vol. 11, pp. 87–183. Academic Press, New York and London.

Kain, J. M. (1971). Measurement of underwater light. *In*: "Methods for the Study of Marine Benthos" (Ed. N. A. Holme and A. D. McIntyre), pp. 53–58. Blackwell Scientific Publications, Oxford.

Kalle, K. (1966). The problem of gelbstoff in the sea. *Ann. Rev. Oceanogr. Marine Biol.* **4**, 91–104.

Kampa, E. M. (1970a). Underwater daylight measurements in the Sea of Cortez. *Deep-Sea Res.* **17**, 271–280.

Kampa, E. M. (1970b). Underwater daylight and moonlight measurements in the eastern North Atlantic. *J. Marine Biol. Assoc. U.K.* **50**, 397–420.

Keil, T., Immarco, A. and Kerpchar, M. (1968). Recent underwater range-gated measurements. *In*: "Underwater Photo-optical Instrumentation Applications: S.P.I.E. Seminar Proceedings", vol. 12, pp. 21–32. Society of Photo-optical Instrumentation Engineers, Redondo Beach, California.

Kerker, M. (1969). "The Scattering of Light and Other Electromagnetic Radiation". Academic Press, New York and London.

Keith-Walker, D. G. (1970). The attenuation of collimated light by sea water.

Havant, Hants., Plessey Electronics. (Paper presented at AGARD-NATO meeting "Electromagnetics of the sea", Paris, 22–26 June, 1970.)

Kingslake, R. (1965). Illumination in optical images. *In*: "Applied Optics and Optical Engineering" (Ed. R. Kingslake), vol. II: "The Detection of Light and Infrared Radiation", pp. 195–228. Academic Press, New York and London.

Kornstein, E. and Wetzstein, H. (1968). Blue-green high powered light extends underwater visibility. *Electronics* **41,** 140–150.

Lee, A. J. and Folkard, A. R. (1969). Factors affecting turbidity in the southern North Sea. *J. Conseil Internat. pour l'Exploration de la Mer*, **32,** 291–302.

Lingrey, J. L. (1968). A study of underwater colour saturation recording. *In*: "Underwater photo-optical instrumentation applications: S.P.I.E. Seminar Proceedings", vol. 12, pp. 51–61. Society of Photo-optical Instrumentation Engineers, Redondo Beach, California.

Lythgoe, J. N. (1971). Vision. *In*: "Underwater Science" (Ed. J. D. Woods and J. N. Lythgoe), pp. 103–139. Oxford University Press, London.

Lythgoe, J. N. (1972). The adaptation of visual pigments to the photic environment. *In*: "Handbook of Sensory Physiology", vol. VII/1: "Photochemistry of Vision" (Ed. H. J. A. Dartnell), pp. 566–603: Springer-Verlag, Berlin.

Lythgoe, J. N. and Hemmings, C. C. (1967). Polarized light and underwater vision. *Nature (Lond.)* **213,** 893–894.

McIntyre, A. D. (1971). Photography and television. *In*: "Methods for the Study of Marine Benthos" (Ed. N. A. Holme and A. D. McIntyre), pp. 59–70. Blackwell Scientific Publications, Oxford.

Martin, R. E. (1966). Extended range underwater viewing systems. *In*: "Underwater Photo-optics: Seminar Proceedings", pp. B-VIII-1-4. Society of Photo-optical Instrumentation Engineers, Redondo Beach, California.

Mazurowski, M. J. (1969). Mathematics with light. *New Scientist* **44,** 636–639.

Mertens, L. E. (1970). "In-water photography: Theory and Practice". Wiley Interscience, New York.

Morrison, R. E. (1970). Experimental studies on the optical properties of sea water. *J. Geophys. Res.* **75,** 612–628.

Mouser, D. P. (1969). The image isocon—a low light television tube. *IEEE Transactions on Broadcasting*, *BC*-15.

Munsey, C. J. (1968). An electro-optical underwater search and visibility enhancement technique. *In*: "Underwater Photo-optical Instrumentation Applications: S.P.I.E. Seminar Proceedings", vol. 12, pp. 175–182. Society of Photo-optical Instrumentation Engineers, Redondo Beach, California.

Okoomian, H. J. (1966). Underwater transmission characteristics for laser radiation. *Appl. Opt.* **5,** 1441–1446.

Papp, G. (1965). On a novel application of the image dissector. *J. Soc. Motion Picture Television Eng.* **74,** 782–783.

Patterson, R. B. (1966). A wide angle camera for photographic search of the ocean bottom. *In*: "Underwater Photo-optics: Seminar Proceedings", pp. C-XII-1-8. Society of Photo-optical Instrumentation Engineers, Redondo Beach, California.

Patterson, R. B. (1970). Underwater lighting techniques. *In*: "Marine Technology 1970", vol. 2, pp. 1303–1313. Marine Technology Society 6th Annual Conference, June 21–July 1, Washington, D.C.

Rattman, W. and Smith, T. (1972). Lasers for depth sounding and underwater viewing. *Hydrospace*, **5**(1), 57–59.

Rebikoff, D. I. (1966). Mosaic and strip scanning photogrammetry of large areas underwater regardless of transparency limitations. *In*: "Underwater Photo-optics:

Seminar Proceedings", pp. C-XIII-1-11. Society of Photo-optical Instrumentation Engineers, Redondo Beach, California.

Replogle, F. W. (1966). Underwater illumination and imaging measurements. *In*: "Underwater Photo-optics: Seminar Proceedings", pp. A-V-1-10. Society of Photo-optical Instrumentation Engineers, Redondo Beach, California.

Ross, H. E. (1971). Spatial perception underwater. *In*: "Underwater Science" (Eds. J. D. Woods and J. N. Lythgoe), pp. 69–101. Oxford University Press, London.

Schade, O. H. (1964). An evaluation of photographic image quality and resolving power. *J. Soc. Motion Picture Television Eng.* **73,** 81–119.

Schade, O. H. (1970). High-resolution return-beam vidicon cameras: a comparison with high-resolution photography. *J. Soc. Motion Picture Television Eng.* **79,** 674–705.

Schagen, P., Taylor, D. G. and Woodhead, A. W. (1962). An image intensifier system for direct observation at very low light levels. *Advances in Electronics and Electronic Physics*, **16,** 75–84.

Soule, H. V. (1968). "Electro-optical Photography at Low Illumination Levels". Wiley, New York.

Stanley, E. M. (1971). The refractive index of seawater as a function of temperature, pressure and two wavelengths. *Deep-Sea Res.* **18,** 833–840.

Thomas, J. P. and Kovar, C. W. (1965). The effect of contour sharpness on perceived brightness. *Vision Res.* **5,** 559–564.

Towler, G. O. and Swainston, P. (1972). Assessing the performance of low light-level camera tubes. *Advances in Electronics and Electron Physics* **33B,** 961–978.

Tyler, J. E. and Smith, R. C. (1970). Measurements of spectral irradiance underwater. Gordon and Breach, New York.

Varnado, S. G. and Hessel, K. R. (1973). Analysis of underwater optical detection systems using photoluminescent targets. AEC report SLA-73-364.

Vine, Benjamin H. (1965). Electro-optical devices. *In*: "Applied Optics and Optical Engineering" (Ed. R. Kingslake), vol. II: "The Detection of Light and Infrared Radiation, pp. 229–278. Academic Press, New York and London.

Visser, M. P. (1970). The turbidity of the southern North Sea. *Deutsche Hydrogr. Zeitschr.* **23,** 87–117.

Wagenaar, W. A. and van Meeteren, A. (1969). Noisy line-scan pictures. *Photogram. Eng.* **35,** 1127–1134.

Wall, M. R. and Coleman, K. R. (1968). Underwater viewing. *In*: "Underwater Photo-optical Instrumentation Applications: S.P.I.E. Seminar Proceedings", vol. 12, pp. 21–32. Society of Photo-optical Instrumentation Engineers, Redondo Beach, California.

Waterman, T. H. (1954). Polarization patterns in submarine illumination. *Science, N.Y.* **120,** 927–32.

Waynant, R. W. (1971). Application of a scanned-laser active imaging system to atmospheric and underwater viewing environments. Naval Research Laboratory, Washington (NRL Report 7287.)

Wells, W. H. (1969). Loss of resolution in water as a result of multiple small-angle scattering. *J. Opt. Soc. Amer.* **59,** 686–691.

Williams, J. (1970). "Optical properties of the sea". U.S. Naval Institute, Annapolis, Maryland.

Woods, J. D. and Lythgoe, J. N. (1971). "Underwater science: an Introduction to Experiments by Divers". Oxford University Press, London.

Yura, H. T. (1971). Small-angle scattering of light by ocean water. *Appl. Opt.* **10,** 114–118.

Yura, H. T. (1973). Imaging in clear ocean water. *Appl. Opt.* **12,** 1062-1066.

Appendix

h	Planck's constant
e	electronic charge

Optical Parameters of Water

c	velocity of light in water
μ	refractive index
b	total scattering coefficient
$\beta(\theta)$	angular distribution function for scattering at angle θ
a	absorption coefficient
α	attenuation coefficient
α^*	effective attenuation coefficient
K	broad beam attenuation coefficient
M_ν, M_N	modulation transfer function (MTF) at spatial frequency, ν or T.V. resolution N
T_P	transmission of water path

Parameters of the Light Source

L_0	optical energy of light source
λ	wavelength of light
ω	solid angle of emitted light

Parameters of the Target and of Backscatter

C	inherent target contrast
C_A	apparent target contrast
R_C	contrast ratio $\left(= \dfrac{C_A}{C}\right)$
ρ_1, ρ_2	total reflectivity of dark and light areas of the target
S_1, S_2	reflected irradiance from dark and light areas of the target
$\rho(\varepsilon)$	reflection coefficient per solid angle at angle ε to target normal
$S(= S_1 + S_2)$	total irradiance from target reaching the detector
S_0	radiance falling on target
B	backscatter irradiance reaching the detector

Parameters of the Detector

T_0	transmission factor of lens
f	equivalent aperture number of lens
γ	slope of response curve for detector
η	responsive quantum efficiency
t	integration time of detector
A_e	area of photo sensitive surface
i, i_B	detected current from target and backscattered light
$\Delta i, \Delta i_B$	fluctuation of i and i_B, respectively
i_D	inherent noise current of detector
k, k_0	signal to noise ratio (SNR) for ideal and real detectors respectively
Q	$\left(= \dfrac{k_0^2}{k^2}\right)$
Q_D	$(= \eta Q)$ detective quantum efficiency

Depth Estimation by Divers

HELEN E. ROSS
Department of Psychology, University of Stirling, Scotland

and

SAMUEL S. FRANKLIN
Department of Psychology, Fresno State College, Fresno, California, U.S.A.

1. Introduction	191
2. Experiment 1	192
A. Subjects	192
B. Method	192
C. Results and discussion	193
3. Experiment 2	194
A. Sites and subjects	195
B. Method	195
C. Results and discussion	196
4. Conclusions	197
Acknowledgements	197
References	197

1. Introduction

It is important for a diver to know his depth. If he is at a greater depth than he thinks, he may be in danger of narcosis, of exhausting his air supply, or of requiring extra decompression time on the ascent. If he is at a more shallow depth than he thinks, he may attempt to decompress at too shallow a depth, or surface unintentionally due to increased buoyancy at the end of a dive. For reasons of safety and scientific accuracy divers normally rely on a depth gauge; but it is interesting to know how accurate their estimates are without such an aid.

There are various types of information the diver may use in estimating his depth. He may use visual cues, such as the appearance of the surface or seabed; or he may watch the apparent displacement of an anchor rope or underwater cliff while gauging the vertical distance of an ascent or descent; or he may notice changes in light intensity. In addition to visual cues, he may rely on proprioceptive estimates of the vertical distance swum; or on cues more specifically related to depth, such as changes in suit thickness and buoyancy, changes in the flow of air from the demand valve (which varies with different types of valves—Williams, 1964), changes in the taste of the air (sometimes claimed to be "metallic" at depths below 120 ft, 36.6 m), the "drunken" feeling of narcosis at depths below about 120 ft, changes in pressure on the eardrum, and changes in temperature at steep thermoclines (though their location may change over a few hours—Woods and Fosberry, 1967). Divers making visual estimates of the surface or seabed in clear water can become fairly accurate with practice, though they have a tendency to underestimate the distances both upwards and downwards (Ross et al., 1970b). In water of moderate or low visibility the visual cues are reduced, and the diver must rely more on the other types of cues. The present experiments were designed to investigate depth estimation in water of moderate visibility (about 15 ft–25 ft, 4.6–7.6 m).

2. Experiment 1

The first experiment was concerned with the accuracy of depth estimates with and without the sight of a rope. It was expected that the presence of the rope would improve performance, by giving a visual cue to the rate of descent.

A. Subjects

The subjects were 10 divers (9 male, 1 female), with diving experience ranging from 1–15 years. They were university staff and students, and members of a Los Angeles advanced Scuba course.

B. Method

The experiment took place from a boat anchored off Santa Catalina Island, California, in about 100 ft (30.5 m) of water. The visibility was about 25 ft (7.6 m), and the surface temperature about 15°C. The divers carried out the experiment in pairs, one being the subject and one the experimenter. The experimenter carried a depth gauge (oil-filled Scuba Pro, accurate to 5%), and a pencil and record card. The subject attempted to descend to 15 ft, 30 ft or 45 ft (4.6, 9.1 or 13.7 m) without looking up at the surface, and then signalled to the experimenter who recorded the depth of the subject's head. The subject then attempted the other two depths, both divers surfacing between each attempt. This procedure was followed twice, once in view of a white rope suspended from the boat, and once out of sight of the boat or any other landmark. Half the subjects performed first with the rope in sight, and half first without the rope. The order in which the three depths were attempted was varied between subjects. When the first subject had completed his tests, the experimenter and subject reversed roles and repeated the procedure. The subjects were given no knowledge of results until after the dive.

C. Results and Discussion

The mean depth estimates for nine of the divers are shown in Table 1, together with the standard deviations. (One subject was dropped from the analysis, as he showed differences from the mean up to six times the standard deviation of the group. He was an experienced diver, but claims to have been taking drugs at the time.)

TABLE 1. Mean depth estimates (ft) for 9 subjects with and without a rope. Standard deviations are shown in brackets.

	Attempted depth		
	15	30	45
Rope	17.2 (3.02)	31.9 (2.95)	43.0 (5.12)
No rope	20.4 (4.11)	35.2 (7.51)	39.0 (5.14)

The mean results suggest that divers are tolerably accurate at estimating depth over the range tested. The means differ by about 2 ft–6 ft (0.6–1.8 m) from the attempted depth—an error no worse than that of many depth gauges. However, the variability of the estimates (indicated by the standard deviations), shows a more serious situation. Some of these divers could have been in trouble if they had relied on their subjective depth estimates for decompression stops or other purposes. The results show that divers tend to swim too deep (or underestimate their true depth) at 15 ft and 30 ft (4.6 and 9.1 m), and swim too shallow (or overestimate their depth) at 45 ft (13.7 m). These trends were significant on a 2-tailed binomial test for the "No Rope" condition at 15 ft ($p < 0.04$) and 45 ft ($p < 0.02$). The slight tendency to overestimate depth when deep is clearly not dangerous; but the tendency to underestimate it when shallow could be dangerous for decompression stops—particularly when the error is so great that 28 ft (8.5 m) can be mistaken for 15 ft (4.6 m), as happened to one subject. Similar errors of underestimation tend to occur when looking up and judging the height of the surface (Ross et al., 1970b), or when judging the distance swum in a horizontal plane (Ross et al., 1970a).

Divers who dispense with depth gauges should beware of this type of error near the surface.

The presence of a rope appears to provide useful visual information. It can be seen from Table 1 that the mean errors (whether of over or under estimation) are smaller with the rope present. The difference between the conditions was significant at $p = 0.02$ (1-tailed binomial test), comparing the total arithmetic (unsigned) errors over all three depths. The differences in the standard deviations show a similar trend, being smaller with the rope present. This difference was significant only at the 30 ft (9.1 m) condition ($F = 6.476$, $p > 0.01$). A rope with no distinguishing features cannot provide an absolute cue to depth, but it can enable a diver to gauge his rate of descent or ascent, and take into account passive movement due to positive or negative buoyancy. A diver is often unaware of passive rising and sinking (Ross and Lennie, 1968); and even if he is aware of it, he may have difficulty in maintaining himself at a certain depth against incorrect buoyancy. For example, Andersen (1968) found that divers who were 9 lb (4.1 Kg) overweight tended to sink about 5 ft (1.5 m) below their intended level (when using a depth gauge). A rope, or other visual reference, may help the diver by providing immediate visual feedback about active or passive changes of depth. In empty or murky water, quite large passive changes can go unnoticed.

The tendency to overestimate depth (or failure to swim deep enough) when attempting 45 ft (13.7 m) is interesting. It is a reversal of the more predictable underestimation found at shallow depths, and needs some explanation. One possibility is that the 45 ft estimates were influenced by the presence of a strong thermocline at about 40 ft (12.2 m). Another possibility is that there is a "starting effect", in which the diver fails to notice the first few feet of his descent. Alternatively, there may be a change in the quantity or quality of depth cues at different depths; or divers may be afraid of venturing unnecessarily deep without a depth gauge, and may start to err on the cautious side with increasing depth. The second experiment was designed to investigate some of these possibilities.

3. Experiment 2

In this experiment several diving sites were used, to eliminate the effects of any local thermoclines. The "starting effect" hypothesis was tested by asking subjects to descend for 15 ft and 30 ft (4.6 and 9.1 m) starting from the surface, and starting from 20 ft (6.1 m) below the surface. If there were no difference between the distances swum from the two starting points, this would tend to support the "starting effect" hypothesis. If, on the other hand, subjects swam for a shorter distance when starting from 20 ft than when starting from the surface, this would tend to support the theory that deeper depth judgements are influenced by anxiety or by a change of depth cues. The latter effect was predicted, since some subjects had complained of anxiety at depth, and since there was no evidence from a horizontal swimming experiment (Ross et al., 1970a) to suggest that there was any "starting effect" peculiar to the first 30 ft (9.1 m).

A. Sites and Subjects

Ten subjects were tested off Santa Catalina Island, California, in about 100 ft (30.5 m) of water, when the visibility was about 25 ft (7.6 m). There was a thermocline at about 40 ft (12.2 m). The subjects were 9 males and 1 female, from the same diving population as in Experiment 1 (though they did not take part in that experiment).

Six subjects were tested in Cow Bay, Jamaica, in over 100 ft (30.5 m) of water, when the visibility was about 20 ft (6.1 m). There was no thermocline, the temperature being 26°C at all depths. The subjects were 5 males and 1 female, members of the Jamaica Sub-Aqua Club (Kingston Branch), with experience ranging from 15 dives to 5 years.

Four subjects were tested at sites in or near Oban Bay, Scotland, in about 90 ft (27.4 m) of water, when the visibility was about 15 ft (4.6 m). There was no obvious thermocline, the temperature being about 13°C. The subjects were four males, members of the United London Hospitals Diving Group and the Dunstaffnage Marine Research Laboratory, with diving experience ranging from 2-10 years.

B. Method

The method was essentially similar to that of Experiment 1, except that no rope was present, and all tests took place out of sight of the boat. In Cow Bay, Jamaica, no boat was used, as the experiment could be run from the steeply-shelving shore. The divers worked in pairs, one acting as subject and one as experimenter. The subject attempted to swim to a depth of 15 ft or 30 ft (4.6 or 9.1 m), either starting from the surface, or starting from 20 ft (6.1 m) (in which case he attempted true depths of 35 ft and 50 ft, 10.7 to 15.2 m). Both divers returned to the surface or to 20 ft between trials, the experimenter indicating the 20 ft level when necessary. The order of the four trials was varied between subjects. A Scuba Pro depth gauge was used at Santa Catalina, and an S.O.S. T70 or a Spirotechnique P15 at Cow Bay and Oban.

C. Results and Discussion

The problem of whether a thermocline affected the results was examined by comparing the data from Santa Catalina (thermocline at 40 ft, 12.2 m) with that from Cow Bay and Oban (no thermocline). The mean depth estimates of these two groups of subjects are shown in Table 2. There were no significant differences between the groups, so the thermocline cannot have been an important variable. A thermocline could not be a useful depth clue unless it was stable, and its location known to the divers. At Santa Catalina only the subjects who had previously acted as experimenters knew its precise depth, but their estimates were no different from those of the first subjects.

TABLE 2. Mean depth estimates (ft) for 10 subjects in water with a thermocline, and 10 in water with no thermocline. The depths in the table represent depths below the starting point—true depth is obtained by adding 20 ft to the values below the 20 ft start.

	Attempted depth from starting point			
	Surface start		20 ft start	
	15	30	15	30
Thermocline	17.0	34.8	15.6	29.2
No thermocline	17.7	30.9	15.0	28.1
Combined mean	17.4	32.9	15.3	28.7
Standard deviation	3.38	7.49	5.74	14.05

Since there was no difference between the groups, their results were combined. The combined means and standard deviations are shown in Table 2. They are similar to the results from Experiment 1. When starting from the surface there was again a tendency to go too deep when estimating 15 ft and 30 ft (4.6 and 9.1 m), which was significant only at 15 ft ($p = 0.015$, 1-tailed binomial test). When starting from 20 ft (6.1 m), subjects tended to go less deep than they had from the surface. This tendency was apparent for both the 15 ft and 30 ft estimates, but was significant only at 30 ft ($p = 0.002$, 1-tailed binomial test). The mean estimates from 20 ft show less error than those from the surface, since the tendency to swim too deep has been reduced. However, the variability of the estimates has *increased* (as indicated by the standard deviations shown in Table 2). The difference in variability was significant at the 30 ft level ($F = 3.5$, $p < 0.05$) but not the 15 ft level.

It can be concluded from this experiment that there is a tendency to swim less deep when estimating greater depths (below about 35 ft,

10.7 m); and that this is not due to local thermoclines, or to a "starting effect". The overestimation of true depth may be related to the reduction of cues at greater depths. Some depth cues, such as the rate of change of pressure, buoyancy and light intensity, are more marked near the surface than at depth. Anxiety is, however, probably a more important factor, especially since the method of *reproduction* rather than *verbal estimation* of depths was used. If the subject were lead to a certain depth and then asked to estimate the depth, he would not achieve any reduction of apparent danger by overestimating his true depth; whereas when he is leading the descent he can reduce his apparent danger by stopping too shallow. This method of depth estimation is therefore biased by an element of risk, which is not present in horizontal distance reproduction (Ross *et al.*, 1970a), or in verbal depth estimates (Ross *et al.*, 1970b). Since anxiety is probably present, this rather than the reduction of cues with depth, could account for the increased variability at depth. Anxiety is known to cause a deterioration of performance at some tasks under water (Baddeley, 1967), and it may contribute to the more variable performance of novice divers at perceptual estimates (Ross *et al.*, 1970a).

4. Conclusions

Divers tend to swim slightly too deep when attempting to estimate depths shallower than about 35 ft (10.7 m), and they do not swim deep enough when attempting dives to greater depths. The former error also occurs in horizontal distance estimates. The latter error is peculiar to greater depths, and may be due to fear. Both types of error are reduced when additional visual cues are present.

Acknowledgements

We should like to thank the Office of Naval Research, Dr. G. Weltman, Dr. G. Egstrom, Mr. F. Gasser and members of Occidental College and U.C.L.A. for their assistance in California; members of the Kingston Branch of the Jamaica Sub-Aqua Club and the Edinburgh University Jamaican Diving Expedition for their assistance in Jamaica; and members of the United London Hospitals Diving Group and the S.M.B.A. Dunstaffnage Marine Research Laboratory for their assistance at Oban. The work in Oban and Jamaica was supported by the Medical Research Council.

References

Andersen, B. G. (1968). Diver performance measurement: Underwater navigation; depth maintenance; weight carrying capabilities. Tech. Rep. U-417-68-030 General Dynamics, Electric Boat Division, Groton, Conn. 06340, U.S.A.

Baddeley, A. D. (1967). Diver performance and the interaction of stresses. *Underwater Ass. Rep.* 1966–67, 35–38.

Ross, H. E., Dickinson, D. J. and Jupp, B. P. (1970a). Geographical orientation under water. *Human Factors* **12,** 13–23.

Ross, H. E., King, S. R. and Snowden, H. (1970b). Size and distance judgements in the vertical plane under water. *Psychol. Forsch.* **33,** 155–164.

Ross, H. E. and Lennie, P. (1968). Visual stability during bodily movement under water. *Underwater Ass. Rep.* 1968, 55–57.

Williams, S. (1964). Progress towards the ideal demand valve. *Triton* **9(4)**, 24–31.

Woods, J. D. and Fosberry, G. G. (1967). The structure of the thermocline. *Underwater Ass. Rep.* 1966–67, 5–18.

An Investigation into Colour Vision Underwater

CHARLES W. FAY

*Department of Natural Philosophy, University of Edinburgh**

1. Introduction 199
2. Apparatus 199
3. Experimental Procedures 200
4. Colour Classification 201
5. Results 202
6. Conclusion 206
Acknowledgements 207
References 207

1. Introduction

Most divers are aware of the change in colours (in particular the rapid loss of reds) as they descend underwater. This is due to the non-uniform spectral absorption of light by water. The absorption differs from place to place depending on the type of water (Jerlov, 1968) and the particles suspended therein, making it difficult to predict exactly what colour an object will appear. It is also noticed by many divers that, in deep dark waters, the beam of light from a torch has a red tint and likewise objects viewed by it. This colour adaptation has been investigated on land in simulated condition (Kinney, 1967).

An investigation underwater into these two aspects of colour perception was carried out by members of the Edinburgh University Sub-Aqua Club Jamaican Diving Expedition (1970) in co-operation with volunteers from the Jamaican Sub-Aqua Club.

2. Apparatus

Torch. A Spirotechnique Super-Aquaflash was used.
Colour charts. Two small squares of $\frac{1}{8}$-inch (0.3 cm) perspex (a and b) each contained four coloured squares (a red, yellow, green and a blue), taken from a Winsor and Newton Winton Oil Colour Chart (ref. no. 58) and cemented on with liquid perspex. Six larger perspex plates (c)

* Presently at Institute for Marine Environmental Research, Plymouth, England.

each contained 21 colours selected from the Winton Chart and randomly laid out; each chart had the same colours.

Underwater Spectrometer. This instrument was specially designed and built to measure the spectral distribution of light underwater, incident from above. Housed in a small Eddystone dyecast box ($4\frac{1}{2}$ in × $3\frac{1}{2}$ in × 2 in) (11.43 × 8.89 × 5.08 cm) with a $\frac{1}{4}$-in (0.64 cm) perspex watertight lid, it consisted of an ORP 12 CdS photoconductive cell with an infra-red filter (Chance CB2), in front of which could be placed any one of six Kodak Wratten monochromatic filters (70, 72B, 73, 74, 75, 98). The signal from the cell was amplified and displayed as shown in Fig. 1. The spectrometer had four sensitivity ranges and a provision for testing the internal battery. There were two watertight controls through the perspex, sealed with "O" rings; one was for the

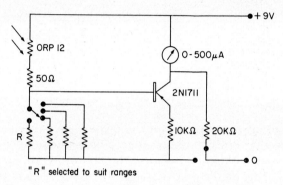

Fig. 1. Circuit diagram of the signal amplifier for the CdS cell used in the measurement of natural daylight.

"range" and "test" switch and the other for the disc on which were mounted the monochromatic filters. The intensity sensitivity was calibrated in Light Numbers against a Weston Master V lightmeter and the spectral sensitivity was calibrated using a photoflood lamp (3200°K) of known spectral distribution. A Formica sheet on which to record the readings was attached to the casing. The instrument was carried on a strap round the diver's neck like a camera, and has been successfully tested to a depth of 60 metres.

3. Experimental Procedures

Two persons were required to conduct the experiment with no more than four subjects at a time. One person stayed on shore with the set of large colour charts (c) and a notebook for recording the results; the other, with colour charts (a) and (b), torch and depth gauge, acted as "dive-leader" for the four subjects—volunteers from the Jamaican Sub-Aqua Club of 3rd Class B.S.A.C.* rating and above.

After being briefed, the five SCUBA divers descended to a depth of 30 m where they were shown colour chart (a) illuminated by the torch. Remembering these colours, they returned to the surface and swam ashore (approximately 30 m) where each was presented with two large colour charts (c) from which they were asked to select the colours nearest to those seen at the bottom. These were noted by the shore controller. The divers went down again to 30 m and this time were shown colour chart (b) viewed in natural light; they returned to shore and performed the same routine for selecting the colours apparent at that depth. This was repeated with colour chart (b) at 18 m, 9 m, 0.5 m (i.e. head just below the surface), and finally on dry land (as a control), with a 5 min interval between viewing and colour selection—all with colour chart (b).

This procedure allowed maximum information to be obtained from the subjects without running them into decompression times—a point to be carefully considered when asking divers to go to such depths on repeat dives.

An alternative routine was used to check the perception of the red colours. The subjects were taken to each of the required depths, starting at the deepest and working upwards, and permitted to write down on a Formica slate a description of the apparent colours. This allowed a greater number of colours to be remembered and subsequently identified, with a shorter diving time.

On completion of testing the subjects, the shore-controller (with "buddy"-diver) dived to measure the spectral distribution of light incident from above at 30, 18, 9, and 0.5 metres depth and on shore. This sometimes proved difficult due to passing clouds causing variations in the light intensity. The most consistent results were obtained on a cloudless day at about noon.

4. Colour Classification

The colours selected from the Winton charts were analysed on a Unicam Spectrophotometer SP800; these graphs were normalized to the spectral response of the human eye and the colours classified according to dominant wavelength. To further determine the actual physical colour in uneven spectral lighting (e.g. torchlight and, at depth, natural light) the distributions were normalized to these lighting conditions and reclassified according to dominant wavelength. Table 1 shows the classification of colours in air.

* British Sub-Aqua Club.

5. Results

The results of the "torch-test" are displayed in Figs. 2 and 3. The upper graphs give the spectral distribution of light as measured by the spectrophotometer. The histograms beneath give the average colour selections made by the subjects. The four colours (33, 43, 26, 04) seen in the water are marked in boxes on the abscissa; and below are their equivalent colours when illuminated by torchlight. The first group of 4

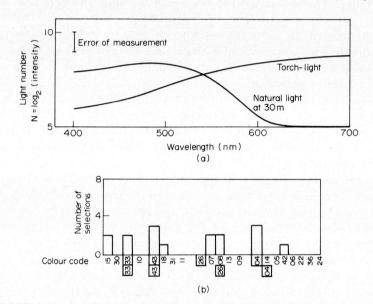

FIG. 2.(a) Relative spectral distribution of the ambient daylight at 30 m depth and of torchlight used to obtain the results in Fig. 2(b). (b) The appearance of colours seen underwater but illuminated by torchlight as they were remembered on return to the surface. Four colours were used in the test (Nos. 33, 43, 26 and 04). The hues that these colours appeared when illuminated by torchlight on land are given in the boxes on the lower line. The histograms show the number of times each of the comparison colours was "matched" to the four test colours. Four subjects were used in this experiment.

subjects (Fig. 2) viewed these colours with the natural light intensity of the same order as the torchlight. Under these conditions the subjects were able to estimate colours accurately as if viewed in air. The second group of 15 subjects (Fig. 3) viewed the colour chart in a natural light whose intensity was lower than that of the torch. Statistical analysis (the binomial test) on their estimation of colours shows that only in their description of the orange was there any significant deviation in perception from the physical colour. The wavelength of the Cadmium Orange (04) viewed in torchlight was 596 nm, equivalent to

Chrome Orange (14), but it was seen by subjects as Cadmium Red (05) whose wavelength is 604 nm. This could be due to an enhanced sensitivity of the eye to red light in the blue/green environment, causing the red components of the orange colour to stand out above the orange or yellow components.

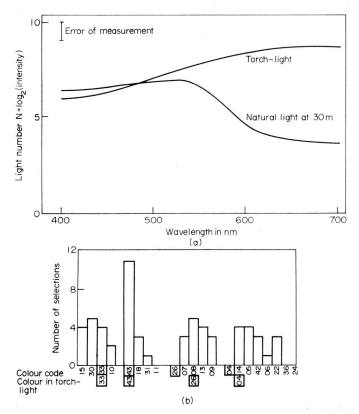

FIG. 3(a, b). A similar experiment to the one whose results are shown in Fig. 2(a, b), except that 15, not 4, subjects were used.

The results of the "natural-light" test are displayed in Fig. 4b. Figure 4a gives a typical outline of the spectral distribution of light at various depths in the sea around the coast of Jamaica. The histograms in Fig. 4b represent the average colour selections made by the subjects. These show that under such lighting conditions, the perception of blues and greens remained similar to that on land; this is to be expected since the spectral distribution for these wavelengths (450 nm–550 nm) is constant. At 100 ft (30 m) the yellow colour showed a significant shift towards green, but this is accounted for by the change in the spectral distribution in this region (550 nm–580 nm); so that no perceptual distortion is apparent. It is the changing appearance of red colours which is most striking as subjects descended deeper. This was due to the progressively increasing absorption of red light relative to blue. This absorption is not as rapid as found in British waters so that even at 30 m there is still sufficient red light to stimulate perception. Specifically all that can be said is that down to 9 m Cadmium Red (05) becomes

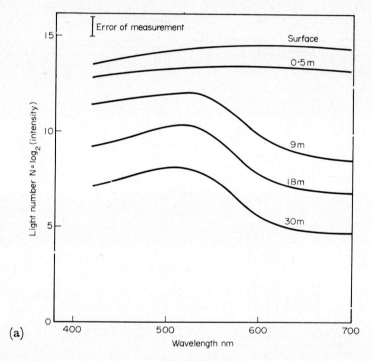

FIG. 4(a, b). Relative spectral distribution of daylight at the surface and at four depths under water during the experiments whose results are shown in Fig. 4(b). (b) The appearance of the four test colours when viewed at four depths under water as they were remembered on return to the surface. In the "surface" experiment the four colours were viewed just beneath the water surface. The four test colours were 30, 18, 08 and 05, printed here in bold type. The histograms above the line represent the number were "matched" to the four test colours.

deeper in colour and appeared as Cadmium Red Deep (06); thereafter at 18 m it took on a light brown colour—Light Red (27)/Indian Red (23)—and at 30 m a darker, greyer brown—Indian Red (23)/Burnt Sienna (02).

Photographs of the colour chart were taken at each of the depths and the Cadmium Red (05) appears much blacker on film than as seen by

COLOUR VISION UNDERWATER 205

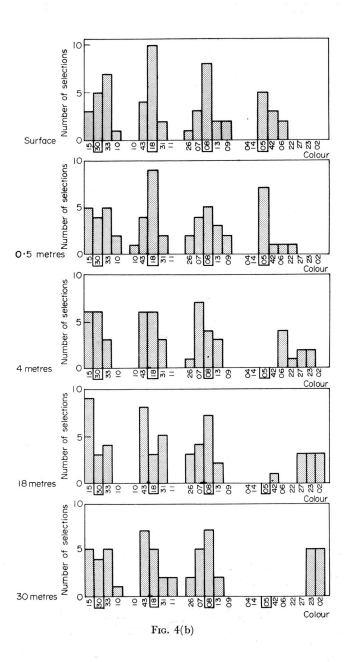

Fig. 4(b)

the subjects *in situ*; this indicates an increased sensitivity of the eye to red light in the blue/green environment as found in the torch test.

TABLE 1. Colour Classification.

Colour Name	Code	Dominant Wavelength (nm)
Cobalt Blue	15	528
Oxford Blue	30	530
Prussian Blue	33	532
Cerulean Blue	10	535
Viridian	43	536
Emerald Green	18	538
Oxide of Chromium	31	541
Chrome Green	11	550
Lemon Yellow	26	556
Cadmium Lemon	07	559
Cadmium Yellow Pale	08	563
Chrome Yellow	13	567
Cadmium Yellow	09	580
Cadmium Orange	04	590
Chrome Orange	14	594
Cadmium Red	05	604
Vermilion	42	607
Cadmium Red Deep	06	610
Geranium Lake	22	620
Rose Madder	36	630
Ivory Black	24	—
Light Red (Brown)	27	—
Indian Red (Brown)	23	—
Burnt Sienna	02	—

6. Conclusion

Colour vision underwater is dependent upon the particular spectral distribution of light found and upon the human adaptation to it. This adaptation takes the form of an increased sensitivity in perception of those wavelengths of light which are reduced in intensity, thus causing a distortion in the perception of the true physical colour as shown in the "torch-test" and the "natural light test". A rough measurement of the time for adaptation to occur was measured and found to be of the order of 30 seconds. Adaptation may continue after this period of time, but it was not detectable in this preliminary measurement.

Accurate descriptions of colours underwater would be facilitated by the diver carrying a waterproof colour chart against which to compare colours *in situ*. A possible alternative is to develop a light source (e.g. torch) whose intensity is automatically controlled by a photo-sensor to redress the imbalance of the spectral distribution—allowing for the effect of colour adaptation.

Acknowledgements

The author wishes to thank the members of the Jamaican Sub-Aqua Club, especially Miss A. Westwater, for their co-operation; also Dr. Helen Ross for her useful advice and Mr. D. Borthwick for his invaluable assistance.

This work was carried out with the aid of grants from the Medical Research Council, Edinburgh University Weir Fund and Messrs Kodak Ltd. Underwater watches were kindly supplied by the Omega Watch Company.

References

Jerlov, N. G. (1968). Colour of the sea. "*Optical Oceanography*", Elsevier, Amsterdam.

Kinney, J. Ann S. (1967). Adaptation to a homochromatic visual world, Sub. Med. Res. Lab. Report No. 499.

Narcosis and Visual Attention

HELEN E. ROSS
and
M. H. REJMAN

Department of Psychology, University of Stirling, Scotland

1. Method	210
2. Results and Discussion	212
Acknowledgements	216
References	216

The narcotic effect of breathing compressed air at depth has been noted for some time, although scientific study of the condition has been mainly confined to the last thirty years (see Bennett, 1966). Most research on the effects of narcosis on performance has been concerned with manual dexterity and intellectual tasks (e.g. Shilling and Willgrube, 1937; Baddeley, 1966; Baddeley *et al.*, 1968), and has shown that intellectual tasks generally suffer more impairment than manual tasks at depth. Little research has been carried out on performance at visual tasks, despite the fact that divers sometimes report a "narrowing of the visual field" at depth.

Weltman and Egstrom (1966), Weltman *et al.* (1971) and Baddeley (1967) have indicated the importance of the anxiety which may be induced by the potential danger of diving or a pressure chamber. Anxiety alone may worsen performance, or may aggravate the effects of narcosis. Weltman *et al.* (1971) report that responses to peripheral visual signals were reduced for novice subjects performing the test in a simulated pressure chamber, when they believed themselves to be at depth but were actually at atmospheric pressure. They found a similar decrement (Weltman and Egstrom, 1966) for novice divers at a depth of about 20–25 ft (6.1–7.6 m) in the sea.

Other authors have shown that different types of stress can improve or worsen visual performance, and the use of complex tasks has made possible a more precise analysis of these effects. Hockey (1970a) used a two-task situation to study the effect of loud noise on performance. His task comprised a (primary) tracking task in the centre of the visual field, with a (secondary) array of lights which the subject had to monitor. He found that under noise—an arousing stress—performance on the primary task improved, while scores on the secondary task showed complex changes without such overall benefit. Hockey concluded that stress produced a shift in the distribution of attention over the various components of the task. The same task was used by Hamilton and Copeman (1970) to examine the effects of alcohol and noise on attention. Alcohol was found to have a similar effect to noise in that it focused attention on the important aspects of the task while degrading the rest, but the overall level of performance was lower than in normal conditions. The authors concluded that alcohol has an arousal-like effect on the profile of attention, but that there must also be some decrease in the rate of information transmission through the human operator.

Deep diving normally involves both narcosis and anxiety. Narcosis probably acts as a depressant and anxiety as an arouser, but their combined effect may not be simple. Broadbent (1963) points out that when stresses are combined they frequently interact in a complex rather than an additive manner—especially if the stresses are operating on different mechanisms.

The following experiment was carried out to see whether narcosis produced a simple "narrowing of the visual field", or some more complex effect.

1. Method

Subjects

The eleven subjects used in the experiment came from the Royal Naval diving team at H.M.S. Neptune, Faslane, Helensburgh, and their ages ranged from 18–40 years. There were six Clearance Divers (CDs) capable of depths down to 180 ft (54.9 m), and five Ships Divers (SDs) capable of depths down to 120 ft (36.6 m).

Apparatus

The pressure chamber used in the experiment was located outside the Diving Section HQ. Its dimensions were 10 ft (3 m) in length and 6 ft (1.8 m) in diameter, and it was capable of holding eight men. The two portholes ($5\frac{1}{2}$ in, 14 cm diameter) at either end of the chamber were used for viewing the experimental display, which was placed outside the porthole.

The display consisted of five lights attached to a 2 ft (6 m) long horizontal bar on a black background. Four red lights and one central white light were spaced at 5 in (12.7 cm) intervals. The display was mounted on a stand and adjusted for height until it was level with the centre of the porthole, and was positioned 12 in (30.5 cm) from the porthole. The subject sat inside the chamber, and placed his nose against the glass of the porthole while viewing the display. The distance from his eyes to the centre of the display was approximately 15 in. (38.1 cm). The entire display subtended a visual angle of 100°, and completely filled the field of view through the porthole.

The lights were 14 V 3/4 W ("peanut" bulbs), and the applied voltage was about 12 volts. The experimenter controlled the onset of the

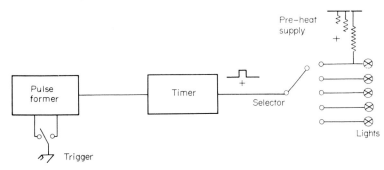

FIG. 1. Block diagram of display apparatus.

lights from outside the chamber by a selector switch and other equipment (Fig. 1). The duration of each flash was between 15–20 ms.

Procedure

The experimenter flashed the lights in a predetermined random order at irregular intervals, with an average of one every 4 s. The central white light was flashed forty times, and the peripheral red lights twenty times each, making a total of 120 signals in each test. The test took 8 min to complete.

The subjects were instructed to tap on the inside of the chamber whenever they detected a signal. They were asked to pay attention primarily to the central light, but to detect as many other lights as possible. The experimenter recorded the presence or absence of a response, and noted any false positives. Each subject was given a short practice run before the first test.

The subjects were normally tested in pairs, one at each end of the pressure chamber. They were tested simultaneously by two experimenters, using identical sets of apparatus. Simultaneous testing was employed to make maximum use of the available manpower and pressure chamber time. One SD lacked a partner, and was tested on his own. Each pair of divers underwent two pressure chamber sessions, normally on two consecutive days. Each session comprised two or three test runs at different depths. The length of a session varied between 30 and 50 min, depending on the number and depth of the test runs. The sequence of tests for each subject is listed below in Table 1.

TABLE 1. Sequence of depths (ft) at which tests were conducted (1 ft = 0.3048 m).

Subjects	Session 1	Session 2
CDs 1 and 2, 5 and 6	0, 180	120, 10, 0
CDs 3 and 4	0, 120, 10	180, 0
SDs 1 and 2	0, 120	60, 10, 0
SDs 3 and 4, 5	0, 60, 10	120, 0

2. Results and Discussion

The number of correct detections for each of the five lights was counted for each subject and each test run. The number of false positives was too small to be worth analyzing. There were no consistent differences in performance on the first and last tests (both of which were always at the surface), so the scores for these two tests were combined to give a more reliable estimate of surface performance. Scores for the left and right lights were also combined for the outer peripheral lights and the mid peripheral lights, giving three locations (central, mid, peripheral) with 40 trials for each. The mean percentage detection scores for the three locations at all depths are shown for the SDs in Fig. 2 and the CDs in Fig. 3. Analyses of variance for the two groups of divers are shown in Tables 2 and 3. The variance ratios are compared with the remainder variance in all cases.

The effect of the light locations was significant for both groups of divers, showing the expected performance decrement towards the periphery. The variance between subjects was also significant for both groups. This was to be expected, since individual differences are commonplace in both psychological and narcosis experiments (see Bennett, 1966). Both groups showed a significant interaction between depth and locations, implying that depth does not have the same effect at all locations. Depth as a main factor was not significant for either group, when compared with the above interaction. The SDs also showed significant interactions between all the main factors. These interactions make it difficult to give a simple interpretation of the results. It is clear, however, that depth does not produce a straightforward "narrowing of the visual field", or an overall decrement in performance. The results seem to suggest that at 180 ft (54.9 m) there is an overall decrement, but more data would be necessary to confirm this. If this is the case, it would imply that the effect of narcosis is not important until 180 ft: other factors, such as anxiety, may cause other performance changes at shallower depths.

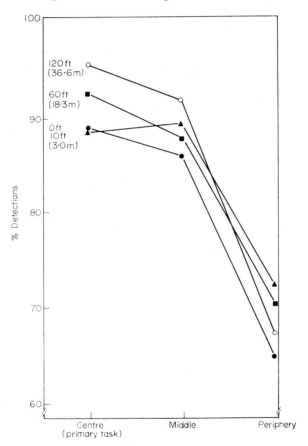

Fig. 2. Percentage detection scores for Ships Divers.

It can be seen from Figs. 2 and 3 that depth appears to affect the primary task scores (central light) of each group of divers in opposite directions. SDs show superior performance on the primary task at depth, while CDs show the opposite. Although it is unlikely that the trained divers used in this study were "anxious" in the way that Weltman et al.'s (1971) novice subjects were, it may be fair to say that being in the chamber is "arousing". This would be more likely to be true of the slightly less experienced SDs, and might therefore explain their better primary task performances (Hockey, 1970a). The secondary task performance of SDs showed complex effects and may be partly confounded by differences in strategy. For example, a pilot study

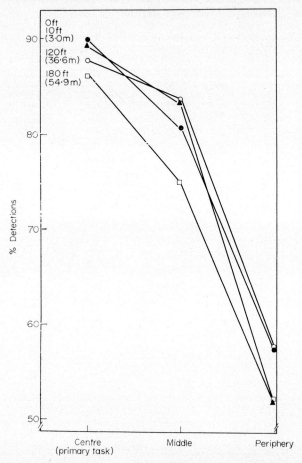

FIG. 3. Percentage detection scores for Clearance Divers.

showed that subjects who fixated the centre of the display detected more peripheral stimuli than those who scanned. It is possible that the SDs concentrated on the primary (central) task, due to the arousing situation, and scored more on the peripheral signals almost by accident. Sanders (1963) also found that highly motivated subjects performed much better at large angle displays when they did not move their heads.

The CDs show a rather different pattern. Their performance at the primary task shows a gradual deterioration with increasing depth, while changes in performance at the secondary task were inconsistent.

"Narrowing" is a seductive term, and many researchers have used it to imply an actual contraction of the visual field. Hockey (1970b)

TABLE 2. Analysis of variance of scores for the Ships Divers.

Source	S.S.	d.f.	M.S.	V.R.	p
Depths	118.08	3	39.36	6.48	<.01
Locations	1257.81	2	628.91	103.61	<.001
Subjects	2764.98	4	691.25	113.88	<.001
D × L	132.16	6	22.03	3.63	<.025
D × S	225.49	12	18.79	3.10	<.01
L × S	263.66	8	32.95	5.43	<.001
Remainder	145.87	24	6.07		
Total	4908.05	59			

TABLE 3. Analysis of variance of scores for the Clearance Divers.

Source	S.S.	d.f.	M.S.	V.R.	p
Depths	40.29	3	13.43	0.08	n.s.
Locations	2620.95	2	1310.48	7.76	<.01
Subjects	4611.81	5	922.36	5.47	<.001
D × L	3400.72	6	566.79	3.36	<.01
Remainder	9277.48	55	168.68		
Total	19951.25	71			

argues convincingly that this is unlikely to be the case. Instead some funnelling of *attention* seems to occur under certain stresses. In the diving situation, where more than one stress is involved (and probably more than one mechanism), it would appear useful to employ sophisticated tasks which can be subdivided and analyzed precisely. Such tasks would give more information than the gross measures of performance which have normally been used in pressure chambers or the sea. There are, of course, severe difficulties in using electrical or other complex equipment in such environments.

Due to limitations on the type of equipment that could be used and the number of subjects available, we cannot draw many firm conclusions from this experiment. Comparisons between the two groups of divers may well be unreliable. We can, however, state that a simple "narrowing of the visual field" does not appear to occur under mild narcosis: subjects may redeploy their attention in different ways, and the redeployment may vary with the task and the depth, and with the diving experience.

Acknowledgements

We are very grateful to Commander P. G. Hammersley for permission to use the facilities at H.M.S. Neptune, and to the diving officer Sub-Lieut. J. Cook for advice and help in organizing the trials, and to the members of the diving team who willingly took part in the experiment. We should also like to thank Mr. John Bevan for suggesting improvements to the manuscript. This work was supported by a grant from the Medical Research Council.

References

Baddeley, A. D. (1966). Influence of depth on the manual dexterity of free divers: a comparison of open sea and pressure chamber testing. *J. Appl. Psychol.* **50,** 81–85.

Baddeley, A. D. (1967). Diver performance and the interaction of stresses. *Underwater Assoc. Rep.* 1966–67, 35–38.

Baddeley, A. D., De Figueredo, J. W., Hawkswell-Curtis, J. W. and Williams, A. N. (1968). Nitrogen narcosis and performance underwater. *Ergonomics* **11,** 157–164.

Bennett, P. B. (1966). "The Aetiology of Compressed Air and Inert Gas Narcosis". Pergamon, Oxford.

Broadbent, D. E. (1963). Differences and interactions between stresses. *Q.J. Exp. Psychol.* **15,** 205–211.

Hamilton, P. and Copeman, A. (1970). The effect of alcohol and noise on components of a tracking and monitoring task. *Brit. J. Psychol.* **61,** 149–156.

Hockey, G. R. J. (1970a). Effect of loud noise on attentional selectivity. *Q.J. Exp. Psychol.* **22,** 28–36.

Hockey, G. R. J. (1970b). Signal probability and spatial location as possible bases for increased selectivity in noise. *Q.J. Exp. Psychol.* **22,** 37–46.

Sanders, A. F. (1963). The selective process in the functional visual field. *Report for Institute of Perception*. RVO-TNO. Soesterberg, Netherlands.

Shilling, C. W. and Willgrube, W. W. (1937). Quantitative study of mental and neuromuscular reactions as influenced by increased air pressure. *U.S. Nav. Med. Bull.* **35,** 373–380.

Weltman, G. and Egstrom, G. H. (1966). Perceptual narrowing in novice divers. *Human Factors* **8,** 499–506.

Weltman, G., Smith, J. E. and Egstrom, G. H. (1971). Perceptual narrowing during simulated pressure chamber exposure. *Human Factors* **13,** 99–108.

Diver Performance—Nitrogen Narcosis and Anxiety

J. P. OSBORNE
and
F. M. DAVIS*

United London Hospitals Diving Group and Royal Naval Physiological Laboratory, Alverstoke, Hants.

1. Introduction	217
2. Methods	218
A. Experimental design	218
B. Experimental tasks	219
C. Physiological measurements	220
3. Results	221
A. Performance tests	221
B. Plasma cortisol	221
C. Electrocardiographic (ECG) and heart rate monitoring	223
4. Discussion	223
Acknowledgements	224
References	225

1. Introduction

In the simulated conditions of a dry pressure chamber, diver performance is impaired by the effects of nitrogen narcosis (Keisling and Maag, 1962). Two experiments conducted in open water (Baddeley, 1965, 1966) showed that there was a much greater reduction in diver efficiency at depth than would be anticipated from the pressure chamber studies, although a third experiment (Baddeley et al., 1968) did not show this difference. In his examination of these results, Baddeley (1967) suggested that anxiety may be the cause of the extra impairment of performance.

Those experiments were carried out in the Mediterranean. The present experiment, carried out under the more strenuous conditions of British coastal waters by the United London Hospitals Diving Group, was designed to see whether diver performance at depth was impaired to an extent similar to that in the Mediterranean, and if anxiety at depth could be demonstrated by physiological measurements.

The physiological response to anxiety is mediated through the

* Present address: 3500 90th Ave. S.E., Apt. 257, Mercer Island, Washington, 98040, U.S.A.

sympathetic nervous system and the endocrine organs, particularly the pituitary and adrenal glands (Selye, 1957). It results in changes in heart rate, blood pressure, respiration, metabolism, etc., which can be easily measured on dry land but present greater problems during diving. However, the activity of the endocrine glands themselves can be assessed with modern techniques by measuring the levels in body fluids of the hormones they secrete. In this experiment, plasma levels of cortisol, secreted by the adrenal cortex, and its urinary excretion products were measured, and electrocardiography (ECG) was performed during the dives.

2. Methods
A. Experimental Design

Each subject performed 4 psychometric tasks once at 3 m depth and once at 30 m in open water, all on the same day. Some delay between dives was caused by the subject having to remove all his diving equipment except dry suit for the collection of physiological data. The depth at which the subject would be tested first was selected randomly and subjects informed the evening before. The subjects practised each task 3 times on dry land and once in the sea at 3 m depth in order to minimize learning effects during the experiment.

Procedure. The tests were carried out from the island of Kerrera, off Oban on the west coast of Scotland, over a two week period in the summer of 1969. The 3 and 30 m sites were both within 80 m of the shore. Polythene sheets four metres square were spread over the bottom at each site to prevent divers disturbing the sand (at 3 m) or mud (at 30 m) and reducing visibility. Temperature, depth and visibility were recorded before each "run". Average water temperature throughout the experiment was 11°C with the shallow site usually 1°C warmer than the deep one; tidal range was 3–4 m with a mean depth of 30 m at the deep site; visibility varied between 3 and 10 m (mean 5 m) at the deep site.

During the "runs" the subject sat on a rigid metal frame with a flat surface in front of him on which the tasks were performed, and with a 30 lb weight belt over his lap for stability. Although artificial lighting was not thought necessary at the shallow site, the test bench at 30 m was always illuminated (sealed beam lamp connected by cable to 2 × 12 V accumulators on shore).

In order to reach the sites, divers descended directly down "into the blue", a procedure generally considered more alarming than following the seabed down. Subjects were then tested by an experienced buddy diver who was in contact with the surface via a DUCS set. Tasks were timed by the buddy diver with a stop-watch in a pressure-proof perspex case, and the surface was informed simultaneously. The tester signalled the beginning and end of tasks to the subject by visual signals or by removal of the task. The total time required to complete all four tasks was a maximum of 25 min from leaving the surface, thus necessitating a 5 min decompression stop at 3 m after the 30 m dives.

Subjects. The subjects were 10 young male divers from the United London Hospitals Diving Group; all except one were medical students. Three had their first open water diving experience during the experiment, two had several years diving experience, and the rest were in their first or second year of diving.

B. Experimental Tasks

All subjects performed the following four tasks:

Sentence comprehension

This was measured with a reasoning test comprising a series of sentences claiming to describe the order of the two letters A and B which followed the sentence as a pair—either AB or BA. The subject had to decide whether the sentence described the letter pair correctly, i.e.:

	True	False
1. A follows B—AB		✓
2. B is preceded by A—AB	✓	
3. A does not follow B—BA		✓

The subject completed as many items as possible in three minutes; at each testing he was given a different test form on which the same items occurred in a different random order.

Manual dexterity

A modification of the Hand Tool Dexterity Test (Bennett, 1965) was used in which nuts and bolts were transferred from one end of a brass plate to the other in a specified manner using a set of spanners. One of the three rows of nuts and bolts was omitted so that the task was completed within the time allocated during the dive. Time taken for complete transfer was recorded.

Simple arithmetic

The subject was given two periods of 30 seconds in which to do as many simple addition sums as possible. Each sum consisted of 5 randomly selected digits and the score was taken as the number of sums correctly completed in the time allowed.

Memory

Each subject was required to memorize a list of 10 words selected randomly from the Teacher's Word Book of 40,000 Words (Thorndike and Large, 1944). This was done on four occasions using a different list each time, and on each occasion 30 s were allowed in which to memorize the list and 45 s in which to recall and write them down (this was ample time to write all 10 words down if the subject could remember them). However, on the second and third occasions the periods of memorizing and recall were separated by the 30 seconds arithmetic described above. Thus, the first and last occasions tested the combined effect of long- and short-term memory whilst the second and third tested only long-term memory.

C. Physiological Measurements

Attempts were made to monitor a number of physiological parameters, but only plasma cortisol estimations and ECG monitoring will be considered here.

Plasma cortisol

The maximum levels of cortisol occur 30 min after the causative stress, so venous blood samples were taken in heparinized syringes immediately upon the subject's return to the shore after each dive. A control sample was also taken prior to diving and a further control was taken at a later date, after 30 min rest at the same time of day as the dive had been conducted. The samples were stored frozen and subsequently measured using the fluorimetric analysis of Mattingley (1962).

ECG monitoring

A very simple ECG technique was used in which two stick-on electrodes were placed at either end of the subject's sternum. These were connected to a short cable passing out of the dry suit (which ensured that the electrodes remained dry) through a seal at the right hand shoulder. This cable terminated in one half of a McMurdo Water Mate underwater connector. On reaching the site the subject was connected via this plug to a cable running to the shore where the signals were collected by a Honeywell Cardioview battery operated ECG machine with a hot-wire pen recorder.

Even with over 100 m of cable, some of it ordinary lighting flex, the signal at the surface was perfectly adequate and required no further amplification. Meticulous attention to electrode application with thorough cleansing of the skin was important. An earth electrode was placed outside the suit on the divers cheek; a poor earth connection resulted in useless recordings.

3. Results

A. Performance Tests

Each task was examined for deterioration of performance at depth by a direct comparison between the scores of each individual diver at the two depths; these were then averaged. Three out of the four tests showed significant deterioration of performance at depth—sentence comprehension, manual dexterity and arithmetic; these changes are

TABLE 1. Average deterioration in performance of 3 tasks between 3 m and 30 m depth.

Test	Deterioration
1. Sentence Comprehension	16%
2. Arithmetic	16% errors 6% to 14%
3. Manual Dexterity	22%

set out in Table 1. The memory test showed no significant deterioration at depth either for short-term or long-term memory. Short-term memory was obtained from the difference between immediate recall (short + long) and delayed recall (long only).

B. Plasma Cortisol

These estimations show a number of interesting changes, and the values obtained are shown in Table 2. Pre-dive levels in all subjects were raised compared with control levels, whether the divers were about to dive to 3 m or to 30 m first. Post-30 m dive levels were raised well above both control levels and the levels obtained after diving to

3 m. Only subjects who had yet to dive to 30 m were levels after the 3 m dives significantly above pre-dive values (Table 3).

TABLE 2. (a) Plasma cortisol levels (microgrammes per 100 ml plasma) in divers diving to 30 m and then 3 m. (b) Plasma cortisol levels (mg per 100 m; plasma) in divers diving to 3 m and then 30 m.

Method error = ±2 mg/100 ml plasma.

(a) Subject	Pre-dive	Post 30 m	Post 3 m	Control (av. of 2)
A.H.	14.5	22.5	21.0	12.5
R.B.	9.0	14.5	7.0	13.75
J.W.	22.0	20.0	10.0	19.5
I.G.	17.0	27.0	16.0	9.5
T.M.	21.0	20.5	15.0	10.0

(b) Subject	Pre-dive	Post 3 m	Post 30 m	Control (av. of 2)
J.C.	19.5	19.5	20.0	14.0
P.C.	7.5	17.5	21.5	9.5
J.A.	22.5	14.0	14.0	10.25
R.S.	12.5	20.25	20.75	11.0
R.C.	15.5	15.0	23.5	10.0

TABLE 3. Comparison of plasma cortisol levels between different diving situations.

Comparison	% Difference	Statistical analysis $n = 10$
1. Pre-dive v. control	+48%	$t = 3.10$, $p\ 0.01$
2. Pre 3 m v. pre 30 m	insignificant	($p\ 0.05$)
3. Post 30 m v. control	+87%	$t = 5.39$, $p\ 0.01$
4. Post 30 m v. post 3 m	+31%	$t = 3.59$, $p\ 0.01$
5. Post diving v. control	insignificant	($p\ 0.05$)
6. Post 3 m/pre 30 m v. control	+71%	$t = 6.98$, $p\ 0.01$

C. Electrocardiographic (ECG) and Heart Rate Monitoring

Where ECG monitoring was not possible, heart rate was measured at the right wrist by palpation of the radial artery. No particular pattern of change was noted in these divers, and heart rates were not raised significantly during dives to either depth even in novice divers.

Only intermittent ECG monitoring was achieved. During these periods no abnormalities of cardiac rhythm were demonstrated at any time. Although some excellent ECG tracings were obtained, the data are insufficient for detailed analysis. What has been shown is that with simple equipment adequate tracings from working divers at depth can be obtained.

4. Discussion

The deterioration of 22% in manual dexterity is similar to that found in Mediterranean boat diving (20% at 30 m) and far greater than that found under simulated conditions in a pressure chamber (Baddeley, 1965). The deterioration of 16% in sentence comprehension is comparable to the 15% found by Baddeley et al. (1968) in the same experiment in which deterioration of manual dexterity was only 3.5%. Arithmetic errors follow the same trend as seen by Baddeley (1966) in boat dives to 60 m—6% at 3 m to 21% at 60 m. Thus performance in British waters seems to be impaired to an extent similar to boat diving in Mediterranean conditions.

One possible interpretation of the discrepancy between manual dexterity and sentence comprehension results is that the manual dexterity task was clearly made more difficult by being performed in the water (Baddeley, 1965) whereas this is not the case for sentence comprehension (Baddeley et al., 1968). This implies that the manual dexterity task was more sensitive underwater because of the increased difficulty which therefore allows it to exhibit a larger effect of anxiety/narcosis.

There is no correlation between the increase in plasma cortisol and the decrease in performance; this would be expected since it is known that a rise in cortisol level is a non-specific response to a psychological stress and is not correlated with the degree of anxiety of the individual (Bridges et al., 1968). However, the increase in cortisol associated with the 30 m dives is most likely to be due to an anxiety response to the deep diving. Unfortunately, the influence of cold stress on the cortisol levels cannot be ruled out in this experiment as it is known that exposure to a cold environment also results in raised plasma cortisol levels

(Keatinge, 1969). Only oral temperatures were taken with any regularity, and this is undoubtedly inadequate for a reasonable assessment of the cold stress experienced in this series of dives. Accurate body temperature measurements, though difficult to obtain in divers, are a necessity for further studies.

The results presented here fulfil the aims of this initial field study in showing a deterioration in some aspects of diver performance over and above that demonstrated in dry pressure chamber work or in non-stressful open water diving. In addition, changes in at least one physiological parameter—plasma cortisol—known to be influenced by anxiety, were demonstrated. Of interest is the difference in the results of the manual dexterity test and the reasoning test performed in stressful and non-stressful diving conditions. This suggests that different factors are influencing different aspects of diver performance at depth.

Interpretation of these results can only be tentative as a number of variables have been inadequately assessed, most particularly body temperature and the tightness of diving procedure control.

Acknowledgements

We are indebted to Dr. A. Baddeley, Dr. P. Cole, Dr. H. V. Hempleman, and Dr. C. Wilton-Davis for their invaluable advice and assistance. We would also like to thank our sponsors: The British Sub-Aqua Club, British Petroleum, The British Oxygen Company, The Gilchrist Educational Trust, and the Deans of King's College, St. Bartholemew's and St. Thomas's Hospital Medical Schools.

References

Baddeley, A. D. (1965). The influence of depth on manual dexterity of free divers: a comparison between open sea and pressure chamber testing. *J. Appl. Psychol.* **50**, 81–85.

Baddeley, A. D. (1965). The relative efficiency at depth of divers breathing air and oxyhelium, *Underwater Assoc. Symp. Malta*, 1965, **13**.

Baddeley, A. D. (1967). The interaction of stresses and diver performance. *Underwater Assoc. Rep.* 1966–1967, **35**.

Baddeley, A. D., De Figueredo, J. W., Hawkswell Curtis, J. W. and Williams, A. N. (1968). Nitrogen narcosis and underwater performance. *Ergonomics* **11**(2), 157–164.

Bennett, G. K. (1965). "Manual of Directions, Hand Tool Dexterity Test". The Psychology Corporation, New York.

Bridges, P. K., Jones, M. T. and Leak, D. (1968). A comparative study of four psychological concomitants of anxiety. *Archiv. Gen. Psychiat.* **19**, 141–145.

Keatinge, W. R. (1969). "Survival in Cold Water". Blackwell Scientific Publications, Oxford.

Keisling, R. J. and Maag, C. H. (1962). Performance impairment as a function of nitrogen narcosis. *J. Appl. Physiol.* **46**, 91–95.

Mattingley, D. (1962). *J. Clin. Path.* **15**, 374.

Selye, H. (1957). "The Stress of Life". Longmans Green, London.

Body Temperature Monitoring during Diver Performance Experiments

F. M. DAVIS,*
J. BEVAN,†
J. P. OSBORNE
and
J. WILLIAMS

United London Hospitals Diving Group and Royal Naval Physiological Laboratory, Alverstoke, Hants.

1. Introduction	225
2. Diving Procedure	226
3. Temperature Monitoring	227
A. Thermistor temperature probes	227
B. Temperature-sensitive radio pill transmitters	229
4. Results	230
5. Discussion	231
Acknowledgements	235
References	235

1. Introduction

In 1967, Baddeley in an analysis of three open water experiments in the Mediterranean suggested that anxiety may be a crucial factor in the exaggerated drop in efficiency of divers in the open sea. In 1969, the United London Hospitals Diving Group carried out a diver performance experiment in British waters, during which plasma cortisol levels as a physiological measure of anxiety were estimated. The results of this experiment are presented in the previous chapter.

In that experiment, plasma cortisol levels were significantly increased in association with 30 m dives, but this result was complicated by three factors. The most important of these was that severe cold stress is known to influence the levels of cortisol in the body in a similar manner to anxiety (Hardy, 1961), and body temperature was inadequately monitored during the dives. In addition, consecutive dives on the same day were likely to produce carry-over effects from one dive to the next, and, as the various blood samples were taken at different times of the day, they were likely to be affected by the natural diurnal variation

* Present address: 3500 90th Ave. S.E., Apt. 257, Mercer Island, Washington, 98040, U.S.A.
† Presently with Comex Driving Ltd., Fairburn Industrial Estate, Dyce, Aberdeenshire, Scotland.

in plasma cortisol level. These factors would tend to reduce the usefulness of the estimations. Therefore the experimental design was altered to improve the validity of hormone estimations, and methods of monitoring body temperature of divers in open water were tested. A further series of experiments was carried out in 1970. In this chapter the diving procedure and the temperature monitoring methods are described in detail, and some aspects of the temperature information are discussed.

2. Diving Procedure

The dives were again conducted from the island of Kerrera, off Oban, on the west coast of Scotland. A test bench, consisting of a small metal table, with a metal chair welded to it back-to-front, was bolted on the centre of a 4-m square angle-iron frame covered with a thick sheet of polythene. This was lowered to the bottom on two sites at depths of 3 m and 30 m, both within 80 m of the shore. The polythene prevented the divers disturbing the bottom, and thus reducing their visibility. Two cables were laid from the shore to each site, and an independent lighting unit (SCUL 200, Inner Space Design Systems) was fixed on the deep site. Water temperature was 12°C average (cf. 11°C in 1969); the shallow site was 1°C warmer than the deep one. Visibility on both was similar, being 3 to 8 m.

The subject commenced to kit-up about 1.5 h before the dive. Temperature probes were positioned, a radio pill swallowed (see below), and then a wet suit was put on. A urine sample was collected before he was helped into a dry suit by two assistants. A blood sample was then taken from the left arm, his aqualung fitted by his dressers, and he was taken to the diving site by boat. Exercise was limited as much as possible both before and during the dives.

Diving was in pairs, one diver being connected to the shore by a diver's communication set (DUCS), and he acted as dive leader and conducted the performance experiments on the subject. Both divers wore twin 60 cu ft sets with new, single-hose two stage regulators so that air supply was more than adequate for the 30 m dives. The divers were made to dive straight down following the shot line. On reaching the site, the subject sat astride the chair with an extra 15 lb weight belt across his lap for stability. One of the cables was connected to a lead coming from the right shoulder of his suit, and the other cable (terminating in an encapsulated ferrite rod aerial) was tucked under a waistband (Fig. 1). The subject then performed four tasks as directed by his companion: a sentence comprehension test; a simple arithmetic test; a memory test; and a manual dexterity test. These are described in the previous chapter.

When the tests were completed the subject was "unplugged", and the divers returned to the surface, carrying out decompression, as instructed by the dive marshall over the DUCS, on a short vertical line suspended below the shot line buoy. Thirty metre dives were limited to 25 min bottom time, even if all the tests had not been completed in that time.

FIG. 1. The external aerial to pick up signals from a temperature-sensitive radio pill transmitter in the gastro-intestinal tract can be seen positioned under the diver's waist strap.

On surfacing, the subject was immediately returned to shore where further blood and urine samples were obtained, and the divers were debriefed.

3. Temperature Monitoring

A. Thermistor Temperature Probes

A thermistor is a temiconducting element made of a compressed bead of a heavy metal oxide which has a large negative temperature coefficient of resistance. The bead is positioned in the tip of a probe, and is usually fused around the wires. Thermistors have a rapid response time and exhibit a large change in resistance with a small change in temperature, which is relatively simple to amplify to give a direct readout of temperature. This relationship, however, is logarithmic, and thermistors tend to vary in their temperature/resistance characteristics, each requiring individual calibration with the amplifier/display system to be employed. Matched thermistors which do not require individual calibration are now available, and this allows considerable flexibility of use, since probes can be readily interchanged. In this experiment, YSI series 400 skin and rectal thermistor probes were used with a YSI Model 41 Telethermometer (Shandon Southern, Ltd). This system proved extremely rugged and reliable with a rapid signal response and no loss of accuracy over the 100 m cables used on the two sites.

The probes were positioned at four sites on the body: (1) on the medial aspect of the right calf; (2) on the right loin; (3) on the back, between the shoulder blades, and (4) in the rectum. The latter was inserted 10–12 cm, and was remarkably well tolerated by the subjects. The leads from the probes were connected to jack plugs, and then to a multi-way junction inside the dry suit. From this a cable passed through the dry suit at a seal on the right shoulder to terminate in one half of a Watermate underwater 8-pin connector. This was plugged into a 7-core screened cable running from the underwater site to the shore, where the temperature signals were recorded. Temperatures were taken before, at 5 minute intervals during, and once or twice after the dive.

This system of body temperature monitoring had the following advantages:

(1) Simplicity of operation.
(2) Accuracy to $0.1°C$, with rapid response times.
(3) Ruggedness of equipment.
(4) Knowledge of the exact position of the temperature sensing device.
(5) Multi-channel capabilities.
(6) No calibration whatsoever is required in the field.

The disadvantages include: (1) Once the diver was in his dry suit no recordings could be taken until he was "plugged in" on the bottom, and likewise, after the dive, until the dry suit was removed. (2) Shorting of the connections within the dry suit occurred initially as these were not waterproof, and neither was the original suit, which had to be replaced. Both these difficulties can easily be overcome by minor modifications to the system used. (3) The necessity of having the diver connected to the surface by a cable. (4) Positioning and fixing the probes takes up to 20 min.

B. Temperature-sensitive Radio Pill Transmitters

The use of radio pills sensitive to changes of temperature, pH, or pressure is now well established in clinical and research practice. Two types of temperature-sensitive pill have been used in the past. In the type used in this experiment, temperature changes are detected by a temperature-sensitive inductance core. As temperature changes, this

FIG. 2. A temperature-sensitive radio pill transmitter. The miniature mercury cell in the centre is inserted the opposite way about to that illustrated, and the cap screwed firmly on. The whole is sealed in a rubber finger stall for swallowing.

alters the tuning inductance of an oscillator operating between approximately 200 and 650 kHz. The aerial coil of the transmitter is separate from the tuning inductance and is connected to a low impedance section of the circuit. Power is supplied by a miniature 1.4 V Mallory cell (RM312H). The transmission frequency alters with temperature, and a sensitivity of 2.5% frequency change per degree Celsius can be obtained. The device is shown in Fig. 2.

Pills were calibrated with fresh batteries prior to each dive. The calibration consisted of immersing the pills in water at three standard temperatures whilst their output frequency was recorded. The results of temperature and frequency were plotted graphically as a calibration curve which enabled the pill temperature to be assessed to within 0.1° Celsius over the approximate 40 h life of the battery.

The procedure followed in preparing the pills for use was as follows: The battery previously used in the calibration of the pill was enclosed, and the pill's cap screwed on, thus ensuring an effective seal against a fitted nylon O-ring (Fig. 2). The pill, measuring 26 mm × 9 mm, was enclosed within a rubber finger stall and tied off as an additional safety and hygiene measure. The pill was swallowed by the subject approximately 1.5 h prior to his dive in order to afford sufficient opportunity for it to pass into the small intestine before the initiation of the dive.

This method of deep-body temperature monitoring (Bevan, 1971) had the following advantages: (1) Simplicity of operation. (2) No requirement for a direct hard-wire connection to the subject. (3) Accuracy to within 0.1 °C.

The disadvantages include: (1) Inconvenience to subject of recovery of the pills after passage through the gut. (2) If two or more pills are monitored simultaneously there is a problem of identification of the pills. (3) The effective range of the transmission is approximately 1.5 m. Therefore the ferrite rod aerial had to be positioned on or very close to the diver. (4) The need for a time-consuming calibration procedure in the field before each use of a pill if there are insufficient pills to cover the total number of experimental dives. (5) Slow response time in comparison to thermistor probes. (6) Lack of knowledge as to the exact position of the temperature-sensing device. Unexpected fluctuations in temperature can occur depending on the pill's position relative to abdominal organs, e.g. proximity to the liver which increases temperature or the anterior abdominal wall which lowers, sometimes markedly, the observed temperature (Lunn, personal communication). For this reason, some workers have suggested that the pill should be inserted into the rectum where its position is reliably known. This positional problem is of particular importance in a diving context as will be seen below.

4. Results

Examples of abdominal and rectal temperatures on the same subjects are shown in Figs. 3–8. Time 0 is the time at which the diver left the surface at the start of the dive. The limits of the dive are shown by the vertical dotted lines. In Figs. 6–8 peripheral skin temperatures are also shown.

From experience we found that 15–20 min were required from the time that the radio pill was swallowed until relatively stable readings were obtained. After this time no unusual temperature fluctuations were observed before the dives. This was due to the slow "warming up" of the pill to body temperature, the perspex case being a poor heat conductor. Accurate thermistor readings could be made within one minute of positioning the probes.

5. Discussion

During the dives, an initial slight rise in abdominal temperature was commonly seen. This could be due to proximity to the liver, but this is thought unlikely for two reasons: firstly, the time interval between ingestion of the pill and the start of the dive varied from diver to diver; and secondly, similar initial increases in rectal temperature were seen in a number of dives (Figs. 6–8). This initial rise was probably due to peripheral vasoconstriction occurring when the diver immerses, resulting in pooling of warm blood centrally. Such transient rises in

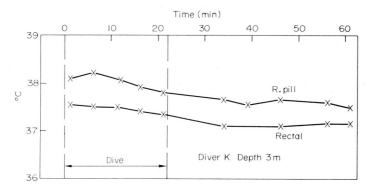

FIGS. 3–8. Typical temperature profiles obtained from divers using the two temperature monitoring techniques. The time limits of each dive are indicated by vertical dotted lines. The depth of each dive, and the site at which each temperature curve was measured are shown. All temperatures are in °Celsius.

rectal temperature have been demonstrated previously in divers working in colder waters off Norway (Skreslet and Aarefjord, 1968). Absolute temperature values measured rectally and in the small intestine differ by up to 1°C at times during these dives. This may be partly a genuine difference and partly due to small calibration errors for the radio pill—not every pill was checked against the rectal thermistor for exact matching of temperature readings. There was, however, very close correlation between the two techniques where changes in temperature during the dives were concerned.

After the initial period, as heat loss from the peripheries occurred (as shown by the steady drop in skin temperatures—Figs. 6–8—which, however, remain well above ambient water temperature), abdominal and rectal temperatures also started to fall slowly. Except in one 30 m dive, and two long 3 m dives, in all of which only wet suits were worn by the subjects, this fall was not marked, and the divers were not subjectively cold. Of particular interest was the continued slow drop in

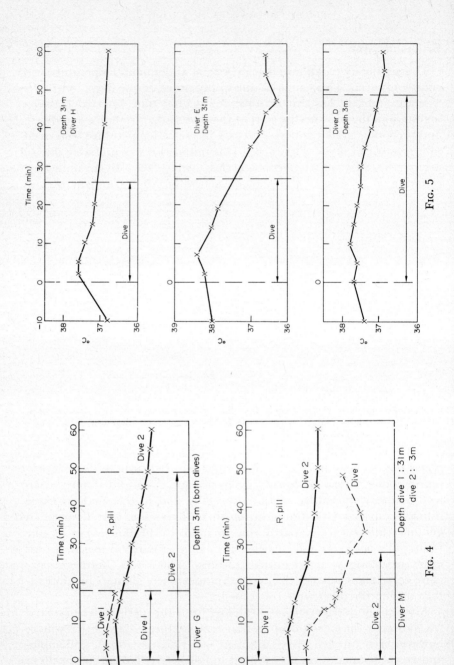

Fig. 5

Fig. 4

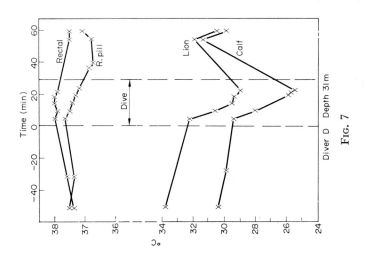

Fig. 7. Diver D Depth 31 m

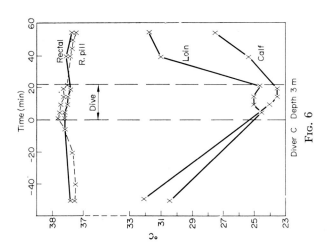

Fig. 6. Diver C Depth 3 m

abdominal and rectal temperatures after the dives, which was only reversed once the diver was fully clothed again. This "after-drop" is well known following immersion in cold water. It is due to the fact that skin temperature tends to remain low for a time, so that loss of heat from the deep tissues to the skin continues (Keatinge, 1969). This can be

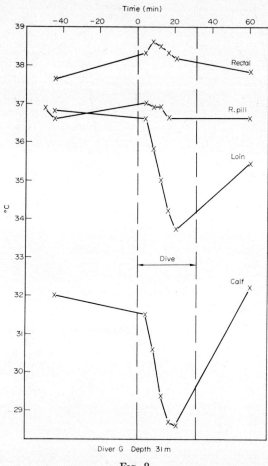

Fig. 8

accentuated by partial rewarming of a cold diver as this results in some increase in skin blood flow which further increases the rate of heat loss. Central body temperature in severe cases may thus be reduced to dangerously low levels. For this reason, rapid rewarming of a really cold diver by immersion in a hot bath (37–44°C) is probably the best course, as this at least minimizes the extent of the after-drop. This technique is often used for rewarming long-distance swimmers.

The evidence from these temperature observations suggests that, wearing a combination of a wet suit and a dry suit, the cold stress to which the divers were exposed during these dives was not marked, and was unlikely to be a major factor in the hormonal changes seen in either the 1969 or 1970 experiments. A few rectal temperatures were measured before and after some of the 1969 dives, and the changes seen were also minimal. As it is now known (Golstein-Golaire *et al.*, 1970) that severe cold stress is required to produce similar increases in plasma cortisol to those seen in the 1969 dives, one can conclude that the increased plasma cortisol levels seen in relation to the 30 m dives (see p. 217ff.) were indeed due to anxiety as was suggested at that time. The hormone results from the 1970 experiment (ULHDG Report 1970) in which the cortisol changes associated with the 30 m dives were much greater, and in which the additional evidence from adrenalin and noradrenalin excretion rates was available, add further support to this conclusion.

Acknowledgements

We wish to acknowledge the invaluable help and advice given to us by Dr. H. V. Hempleman, Mr. F. Lunn, and Dr. C. Wilton-Davis. We are most grateful for the loan of equipment from the Royal Naval Physiological Laboratory, the Royal Naval Personnel Research Committee, and the M.R.D. Research Laboratories, Hampstead. This work was made possible by grants from the Wellcome Trust, British Petroleum Ltd., The British Oxygen Company, and the Deans of King's College, Guy's, and St. Thomas's Hospital Medical Schools.

References

Baddeley, A. D. (1967). The interaction of stresses and diver performance. *Underwater Assoc. Rep.* 1966–67 **35**.

Bevan, J. (1971). The use of a temperature sensitive radio pill for monitoring deep-body temperature of divers in the sea. RNPL Report No. 12/71. Dept. of Naval Physical Research, M.O.D.

Davis, F. M. (1970). United Hospitals Diving Group Project. London.

Golstein-Golaire, J. (1970). Acute effects of cold on blood levels of growth hormone, cortisol, and TSH in man. *J. Appl. Physiol.* **29**(5).

Hardy, J. C. (1961). The physiology of temperature regulation. *Physiol. Rev.* **41**, 521.

Keatinge, W. R. (1969). Survival in cold water. Blackwell Scientific Publications, Oxford.

Skreslet, S. and Aarefjord, F. (1968). Acclimatization to cold in man induced by frequent scuba diving in cold water. *J. Appl. Physiol.* **24**, 177–181.

The Measurement of Respiration at High Ambient Pressures

J. B. MORRISON

*Royal Naval Physiological Laboratory, Alverstoke, Hants.**

---

1. Introduction 237
2. Review 237
3. Methods 244
4. The Measurement of Respiration at High Ambient Pressures . . . 248
References 250

1. Introduction

One of the main factors governing man's performance at high ambient pressures is respiratory function. Due to the increasing density of the gas breathed by the diver ventilatory capacity may become severely limited at depth. Added external resistance of breathing apparatus worn by the diver further reduces ventilatory capacity. There is as yet insufficient quantitative information available regarding the separate or cumulative effects of these limitations on a diver's ability to perform work. It is evident from the present studies however that alveolar carbon dioxide tension, P_{A, CO_2}, can become unnaturally high when performing work at depth. This suggests that the diver's ventilation is less than normal due to the increased respiratory effort required.

2. Review

A useful index of pulmonary function is maximum voluntary ventilation, MVV. MVV indicates a subject's maximum ventilatory capacity in litres/minute. It is normally measured over a period of 15 s of maximum breathing effort (Cotes, 1968; Lanphier, 1969). The variation of MVV with depth for air breathing is shown in Fig. 1. These curves by Lanphier are based on the results of several investigators (Seusing and Drube, 1960; Wood, 1963; Maio and Farhi, 1967;

---

* Presently at the Department of Kinesiology, Simon Fraser University, British Columbia, Canada.

Miles, 1969). The variation of MVV with depth was estimated by Miles (1969) to be

$$MVV \propto 1/\rho^{0.5}$$

where ρ = gas density

It can be seen from Fig. 1 that this relationship although not exact gives a reasonable approximation to the experimental results.

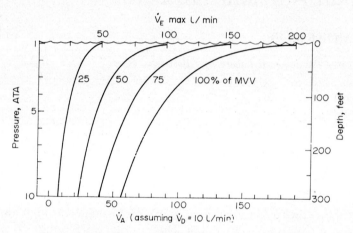

FIG. 1. Relationship of MVV to depth when breathing air.

Although there is no theory which provides an accurate description of gas exchange in the lungs, the pressure head required at the lungs to produce a given rate of gas flow may be approximated (Buhlmann, 1963; Rohrer (1915) by the relationship

$$P_{alv} = K_1 \dot{V} + K_2 \dot{V}^2 \qquad (1)$$

Where term $K_1 \dot{V}$ represents pressure drop in the regions of laminar flow and term $K_2 \dot{V}^2$ represents pressure drop in the regions of turbulent flow. More accurately, in regions of turbulent flow, for smooth walled tubes $P \propto \mu^{0.25}$, $\rho^{0.75}$, $V^{1.75}$ and at tube junctions or change in cross section, $P \propto \rho \dot{V}^2$ (Hughes and Brighton, 1967; Mead et al., 1967). In more recent theories a third term $K_3 \dot{V}^2$ has been added to represent pressure drop due to convective acceleration in the upper airways (Mead et al., 1967). Constant K_1 is dependent upon viscosity, μ, constants K_2 and K_3 are dependent on density. As depth increases therefore K_1 is unaltered in value, but K_2 and K_3 increase in proportion to the gas density. As term $K_1 \dot{V}$ is normally regarded as being small in value it can be approximated that for a given pressure head, P_{alv}, the gas flow rate $\dot{V}$ must therefore decrease with depth by a factor $1/\rho^{0.5}$.

Although derived from a somewhat different theory this relationship agrees with the prediction of Miles (1969). This formula tends to underestimate the value of MVV. A possible explanation is that the maximum pressure head in the lungs, P_{max}, which is partly dependent on pleural pressure, P_{pl}, produced by the respiratory muscles, is assumed independent of depth. As $\dot{V}_{max}$ decreases with depth, at the correspondingly reduced contraction rates, the respiratory muscles will be capable of greater forces.

The MVV of a diver is of little importance provided that he can attain the required level of ventilation to perform his work. It is normally

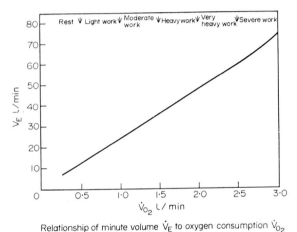

Relationship of minute volume $\dot{V}_E$ to oxygen consumption $\dot{V}_{O_2}$

FIG. 2. Relationship of Minute volume $\dot{V}_E$ to oxygen consumption $\dot{V}_{O_2}$.

assumed that for safety purposes a diver must be capable of maximum effort. In the design of breathing apparatus this condition is normally defined as $\dot{V}_{O_2} = 2.5$–3.0 litres/minute (R. N. Diving Manual, 1964; Riegel and Harter, 1969; Williams, 1969). Figure 2 shows the relationship of respiratory minute volume to oxygen consumption and the corresponding approximate work rate. The curve shown is based on measurements of respiration during exercise at the surface (Kao, 1963). From Fig. 2 it can be seen that an oxygen consumption of 3.0 l/min implies an expired minute volume $\dot{V}_E$ of 70 l/min for the diver. It is estimated from experimental measurements at the surface that an individual can maintain a work rate requiring a ventilation, $\dot{V}_E$, of 50–75% of MVV (Zocche et al., 1960; Shephard, 1967). It is predicted that use of breathing apparatus in underwater work would however very much reduce this value, and although no figures are available a value of about 25% of MVV is suggested (Lanphier, 1969), this figure being dependent on the characteristics of the particular apparatus. Combining the information of Figs. 1 and 2 and assuming the conditions available to the diver wearing breathing apparatus are $\dot{V}_{E\,max} = 25\%$ MVV, the diver breathing air is limited to moderate work at depths in excess of 30 ft (9.1 m) and light work at 100 ft (30.5 m) and deeper.

This conclusion based on predictive grounds does not agree with the practical evidence given by divers who claim to do relatively heavy work at depths in excess of 100 ft (30.5 m). There are very few accurate experimental measurements of underwater work available to support

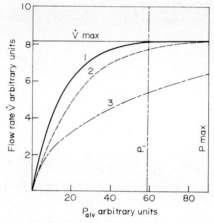

Fig. 3. Relationship of Expiratory flow to alveolar pressure.

either of these statements but it would appear from available estimates that the above prediction underestimates the work capability of the diver.

In order to explain the wide margin of discrepancy between the predicted value and the claims of divers it is necessary to look at other aspects of respiratory physiology.

Recent research work has shown that the flow rate of gas exchange in the lungs is not only limited by the parameters of Eqn. 1. There would appear to be an upper limit to expiratory flow rate $\dot{V}$ above which further increase in pleural pressure, P_{pl}, by increased muscular effort produces no corresponding increase in flow (Mead and Whittenberger, 1953; Mead et al., 1967). This is explained by the condition that above a certain value the pleural pressure induced by muscular effort will equal the internal pressure at some point in the downstream section of the thoratic airway. This point is known as the equal pressure point, EPP.

Thus downstream of the EPP the pleural pressure will tend to compress the thoratic airway and limit flow. Further increase in pressure head in the lungs therefore will result in increased resistance of the restricted airway and much of the increased respiratory work is in fact wasted. This concept is shown in Fig. 3. At a given lung volume the relationship will have a form similar to the curve 1. Any increase in P_{alv} above the value P_1 at which $\dot{V}_{max}$ is reached represents wasted effort. This fact may be incorporated into the theory by representation of Eqn. 1 as a discontinuous equation.

If we now consider the effect of adding external resistance (i.e. breathing apparatus), assuming turbulent flow, the equation becomes

$$P_{alv} = K_1 \dot{V} + (K_2 + K_3) \dot{V}^2 + K_4 \dot{V}^2 \qquad (2)$$

As term $K_4 \dot{V}^2$ represents resistance downstream of the thoratic region any increase in P_{pl} to overcome this resistance will result in a corresponding increase in the internal pressure at the EPP and therefore will not alter the maximum flow rate. The shift in the pressure-flow relationship due to this effect is indicated by curve 2 in Fig. 3. Hence pressures in excess of P_1 can be utilized to overcome the resistance of the breathing apparatus. The peak expiratory flow rate of the diver using apparatus would only be reduced if the resistance of the equipment at peak flow value $\dot{V}_{max}$ were to exceed $P_{max} - P_1$. This case is represented by curve 3 in Fig. 3. It should be noted that Fig. 3 is a schematic diagram to illustrate the theory and is not based on experimental results. This argument applies only to expiratory flows. As there is no such mechanism inhibiting inspiratory flow any added external resistance must represent an added burden on the respiratory system and hence reduce inspiratory flow rates.

In Fig. 2 the curve shown is based on measurements of respiration during exercise at surface. There is evidence that a diver may in fact perform work at a lower $\dot{V}_E/\dot{V}_{O_2}$ ratio than normal by tolerating a high P_{A,CO_2} (Lanphier, 1963; Hamilton, 1967). For example at an oxygen consumption of 2.5 l/min a diver may have a reduced ventilation of 50 l/min. A predicted relationship of $\dot{V}_E/\dot{V}_{O_2}$ for the fully acclimatized diver at depth is presented by Williams (1969).

Assuming therefore that the most liberal conditions available to a diver wearing breathing apparatus are $\dot{V}_{E\ max} = 50\%$ MVV and $\dot{V}_{E\ max} = 50$ l/min at $\dot{V}_{O_2} = 2.5$ l/min, the limit of maximum performance, as defined by these values, occurs at a depth of about 100 ft (30.5 m) (see Fig. 1). These calculations based on the information available suggest that at depths greater than 100 ft if a diver is to be assured of maximum effort or is required to perform very heavy work air breathing may no longer meet the stated safety requirements and a lighter gas mixture should be adopted.

It must be stressed that the conclusions presented here are purely predictive and there would appear to be few comprehensive experimental measurements taken under these circumstances. A study of $\dot{V}_{E\,max}$, $\dot{V}_{O_2}$, and the related P_{A,CO_2} values at depth with and without

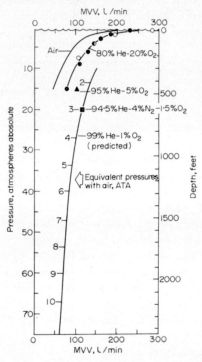

FIG. 4. Actual and predicted values of MVV when breathing helium-oxygen mixtures.

breathing apparatus would do much to improve the understanding of a diver's physiological condition whilst working under water.

As shown in Fig. 4 the use of helium-oxygen gas mixtures very much increases the safe working depth of a diver. Assuming the same conditions define maximum effort as for air diving (i.e. $\dot{V}_{E\,max} = 50\%$ MVV, $\dot{V}_{E\,max} = 50$ l/min, $\dot{V}_{O_2} = 2.5$ l/min). Figure 4, from Lanphier (1969), indicates that limitation of diver performance due to respiratory problems would be expected only at depths in excess of about 800 ft. At 1500 ft (457.2 m) predictions suggest that it should still be possible to obtain adequate ventilation during periods of heavy work ($\dot{V}_{O_2} \leqslant 2.0$ l/min). Although there is no information bearing directly on $\dot{V}_{E\,max}$

during deep saturation dives, experimental measurements have been made of divers' work performance during simulated chamber dives. These measurements refer to work performed under dry conditions with no breathing apparatus, but nevertheless offer a useful guide to what might be expected of a diver performing a work task during an excursion from a submerged chamber or habitat. Respiratory function

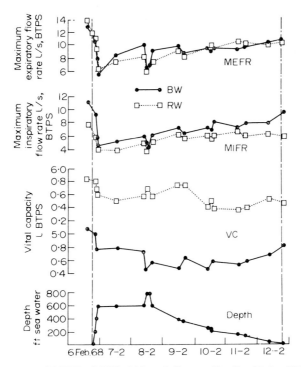

FIG. 5. Time course of MEFR, MIFR, VC and dive profile (depth) for 600 ft (182·9 m) saturation with excursions to 800 ft (243·8 m) (from Dougherty and Schaefer, 1969).

has been measured at exercise levels of 900 kg m/min (very heavy work) at 600 ft (Bradley *et al.*, 1968), 700 kg m/min (heavy work) at 1000 ft (Salzano *et al.*, 1970) and 300 kg m/min (moderate work) at 1500 ft (Morrison, 1971). The information from these experiments demonstrated no unexpected or gross changes in ventilation $\dot{V}$ or alveolar carbon dioxide tension P_{A,CO_2} and therefore it may be assumed that at these work loads and depths the physiological limits of performance were not attained.

In deep saturation diving following compression there is evidence of a time dependent relationship between respiratory mechanics and depth suggesting that the diver gradually becomes better acclimatized to his working depth as the dive progresses. This relationship as measured by Dougherty and Schaefer (1969) is shown in Fig. 5. The added complications of acclimatization, breathing apparatus and cold which must be considered mean that much more research is required before operational very deep saturation diving in the sea can be approached with confidence.

3. Methods

Experimental techniques designed to measure respiratory function at high ambient pressures have been developed and tested in the laboratory and under pressure. The basic requirements of the system were that detailed measurements of respiratory gas flows and gas analysis could be made during simulated saturation helium-oxygen diving and air diving

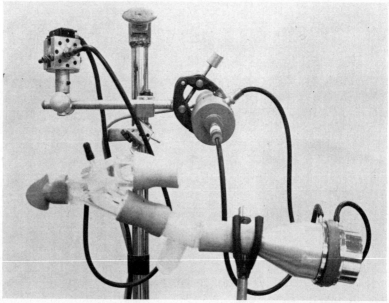

Fig. 6. Experimental set up of mouthpiece pneumotachograph and pressure transducers.

experiments in R.N.P.L. pressure chambers. Respiratory measurements were to be obtained of subjects at rest and at work under dry conditions.

The most common methods of measuring respiration are by means of body plethysmograph, a strain gauge round the chest (Miles, 1969), spirometer, electrical impedance plethysmograph, pneumotachograph and gas meter (Bradley *et al.*, 1968; Salzano *et al.*, 1970). Of these methods the strain gauge and electrical impedance plethysmograph do not offer sufficient accuracy and the theoretical principal of the latter has been challenged (Hill et al., 1967; Kinnen, 1969). From the remaining methods the pneumotachograph was selected as the most suitable instrument on the grounds of accuracy, linearity of output, response time and compactness of equipment, the last factor being of importance in the limited space of the pressure chamber.

The pneumotachograph consists of a series of small bore parallel tubes through which the respiratory gas passes under laminar flow conditions. Pressure drop across the region of laminar flow is measured by a differential transducer having a range of ± 2.4 cm H_2O. The

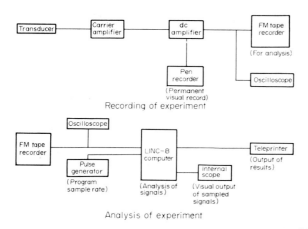

FIG. 7. Block diagram of instrumentation for recording and analysis of data.

signal output of the transducer is linearly related to gas flow rate. In order to avoid problems of temperature variation and condensation, the pneumotachograph is connected to the inspiratory side of a low resistance mouthpiece as shown in Fig. 6. Expired gas is exhausted by a separate valve to a collection bag from which a mixed expired gas sample can be taken. A second pressure transducer is fitted to a tapping at the mouthpiece to measure external resistance to respiration imposed by the equipment (see Fig. 6). The external resistance to flow at the mouthpiece is less than 1 cm H_2O under laboratory working conditions.

A block diagram of instrumentation used to record and analyze data is shown in Fig. 7. Signals from the transducers are amplified and recorded by an FM instrumentation tape recorder. Signal amplitudes are set to a suitable level for recording by means of oscilloscope display, and for reference purposes a permanent visual record is obtained by pen-recorder. In the laboratory, a calibration signal is recorded by connecting the pneumotachograph in series with a rotameter and passing a steady gas flow rate by means of a rotary blower. In the chamber, calibrations are performed after each test. The gas flow is passed through the chamber wall and measured by a rotameter at atmospheric conditions.

The data recorded on magnetic tape is replayed as input to a LINC 8 computer. As the computer is limited to a 4K memory bank it was necessary to develop programming in machine code form. The computer has been programmed to analyze the respiratory data as follows.

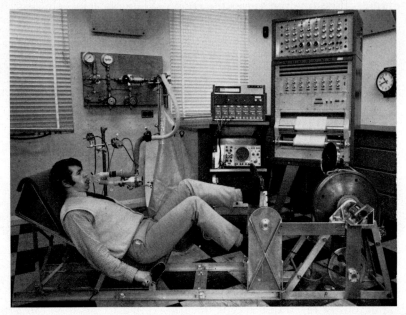

FIG. 8. Bicycle ergometer and instrumentation for measurement of respiration at rest and exercise.

The analog signal of inspired gas flow $\dot{V}$ is sampled and converted to digital form. Sample rate is set at 100/s by using a pulse generator to trigger the programme. The volume of each inspired breath is calculated by integration. At the end of each inspiration, time is sampled from an internal time clock in the computer. These values are stored in the computer and, at the end of the analysis, for each minute of the recorded experimental test, a print-out is obtained of the tidal volume of each breath, the average tidal volume per minute $\dot{V}_T$, the inspired volume per minute, $\dot{V}_I$, and the respiration rate.

In order that the diver's respiration could be measured during exercise at a steady work load a bicycle ergometer was designed and constructed in a form suitable for use in the limited space available in the saturation diving chamber. This machine, shown in Fig. 8, was constructed of lightweight materials and can be collapsed and stored under the chamber flooring when not in use in order to maximize diver living space. The machine is loaded by applying known frictional forces to the flywheel. By pedalling at a predetermined fixed speed the diver maintains a constant work rate.

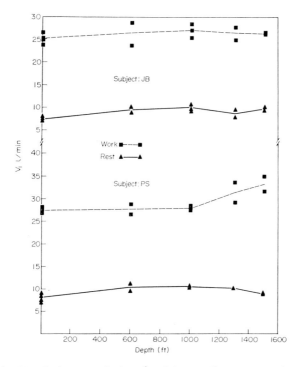

FIG. 9. Relationship of minute ventilation, $\dot{V}_I$, of the two divers, measured at rest and at a work load of 300 kg m/min, depth (from *J. Appl. Physiol.* **30,** 1971).

The experimental procedure adopted for the present series of deep dives is as follows. Respiration is measured at rest over a period of 20 min. Of this time the last 10 min are analyzed, the first 10 min being regarded as the time required by the subjects to reach a steady state. In order to keep conditions as constant as possible in the laboratory and in the chamber, measurements at rest are made with the subject in a relaxed position on the ergometer as shown in Fig. 8. Respiration is measured at a constant rate of exercise over a period of 10 min, the last 5 min of the exercise period being analyzed. As the dive series aimed at an eventual depth of 1500 ft (457.2 m) a moderate work load of 300 kg m/min was chosen as most suitable.

4. The Measurement of Respiration at High Ambient Pressures

A simulated dive was performed in a dry chamber complex. Two subjects, Sharphouse and Bevan, were compressed to successive depths of 600, 1000, 1300 and 1500 ft (182.9, 304.8, 396.2 and 457.2 m). Time

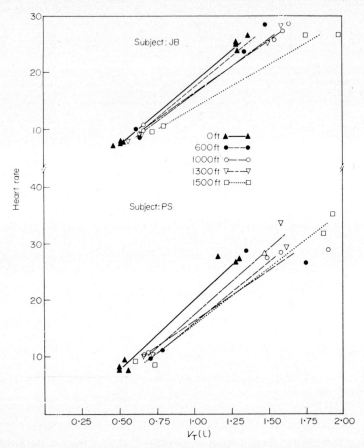

Fig. 10. Relationship of ventilation, $\dot{V}_1$, to tidal volume, V_T, at surface and under pressure (from *J. Appl. Physiol.* **30,** 1971).

spent at each depth was 24 h, 24 h, 22 h and 10 h respectively. The respiratory measurements outlined above were made at each depth.

The results obtained during the dive can be summarized as follows. Minute ventilation showed a slight increase at depth during both rest and moderate exercise (see Fig. 9). At 1500 ft (457.2 m) Sharphouse showed a marked increase in minute ventilation. Carbon dioxide elimination remained relatively constant at both rest and exercise except in Sharphouse where at 1500 ft a significant rise in $\dot{V}_I$ was accompanied by a corresponding increase in $\dot{V}_{CO_2}$. Resting values of alveolar carbon dioxide tension, P_{A,CO_2} were on average slightly lower at depth than at surface. During moderate exercise P_{A,CO_2} showed a small rise at depth in Bevan but remained unaltered in Sharphouse. The relationship of respiratory rate and tidal volume to minute ventilation was altered considerably as shown in Fig. 10. Respiratory

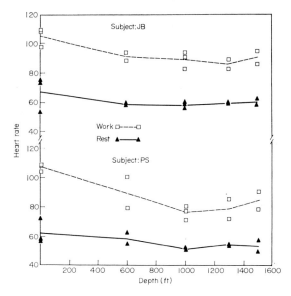

FIG. 11. Relationship of heart rate of the two divers, measured at rest and at a work load of 300 kg m/min, to depth (from *J. Appl. Physiol.* **30,** 1971).

rates were reduced and tidal volumes increased with depth. There was some evidence of bradycardia at rest and more significantly when exercising. As shown in Fig. 11 the extent of bradycardia did not alter significantly with increasing depth from 600–1500 ft (182.9 to 457.2 m).

In general it can be concluded from the experiment that man can exist with reasonable comfort in an oxygen-helium environment down to a depth of 1500 ft (457.2 m). When resting or performing moderate work under the conditions described there are minor changes in respiratory function but no significant respiratory problem.

Having established that work under pressure is possible to depths well in excess of 1000 ft (304.8 m), the main experimental goal is to obtain more detailed information of man's physiological conditions at different depths and exercise levels, and to determine to what extent performance is affected by breathing apparatus and the underwater environment. Although the main research programme involves the measurement of respiration during saturation dives breathing helium-oxygen mixtures, it is also intended to perform respiratory measurements at shallower depths under air diving conditions using the same experimental methods.

References

Bradley, M. E., Vorosmarti, J., Linaweaver, P. G. and Mazzone, W. F. (1968). U.S. Navy Deep Submergence Systems Project. Research Report 1–68.

Buhlmann, A. A. (1963). Respiratory Effects of Increased Pressure. *Proceedings of the 2nd Symposium on Underwater Physiology* (Eds. C. J. Lambertsen and L. J. Greenbaum, Jr.), pp. 98–108. Natl. Acad. Sci. Natl. Res. Council (Publ. 1181), Washington.

Cotes, J. E. (1968). Lung Function: Assessment and Application in Medicine, 2nd edn. Blackwell Scientific Publications, Oxford and Edinburgh.

Dougherty, J. H., Jr. and Schaefer, K. E. (1969). The Effect on Pulmonary Functions of Rapid Compression in Saturation-excursion Dives to 1000 feet. *Sub. Med. Res. Lab. Report No. 573*, publ. U.S. Navy Sub. Med. Center.

Hamilton, R. W., Jr. (1967). Physiological Responses at Rest and in Exercise During Saturation at 20 Atmospheres of He-O_2. *Proceedings of the 3rd Symposium on Underwater Physiology* (Ed. C. J. Lambertsen), pp. 361–374. William and Wilkins, Baltimore, Md.

Hill, R. V., Jansen, J. C. and Fling, J. L. (1967). Electrical impedance plethysmography: a critical analysis. *J. Appl. Physiol.* **22**, 161–169.

Hughes, W. F. and Brighton, J. A. (1967). Theory and Problems of Fluid Dynamics. Schaum Publ. Co., New York.

Kao, F. F. (1963). An Experimental Study of the Pathways Involved in Exercise Hypernoea Employing Cross Circulation Techniques. *The Regulation of Human Respiration* (Eds. D. J. C. Cunningham and B. B. Lloyd). Blackwell, Oxford.

Kinnen, E. (1969). A Defense of Electrical Impedance Plethysmography. *Med. Res. Eng.* **8**, no. 4, 6–7.

Lanphier, E. H. (1963). Influence of Increased Ambient Pressure Upon Alveolar Ventilation. *Proceedings of the 2nd Symposium Underwater Physiology* (Eds. C. J. Lambertsen and L. J. Greenbaum, Jr.), pp. 124–131. Natl. Acad. Sci. Natl. Res. Council (Publ. 1181), Washington.

Lanphier, E. H. (1969). Pulmonary Function. *The Physiology and Medicine of Diving and Compressed Air Work* (Eds. P. B. Bennett and D. H. Elliott), pp. 58–112. Bailliere, Tindall and Cassell, London.

Maio, D. A. and Farhi, L. E. (1967). Effect of gas density on mechanics of breathing. *J. Appl. Physiol.* **23**, 687–693.

Mead, J. and Whittenberger, J. L. (1953). Physical properties of human lungs measured during spontaneous respiration. *J. Appl. Physiol.* **5**, 779–796.

Mead, J., Turner, J. M., Macklem, P. T. and Little, J. B. (1967). Significance of the relationship between lung recoil and maximum expiratory flow. *J. Appl. Physiol.* **22**, 95–108.

Miles, S. (1969). Underwater Medicine, 3rd edn. Staples Press, London.

Morrison, J. B. (1971). R.N.P.L. Report, 1–71, pp. 34–60.

Riegel, P. S. and Harter, J. V. (1969). Design of Breathing Apparatus for Diving to Great Depths. *Amer. Soc. Mech. Eng. Conference*, New York.

Rohrer, F. (1915). Der Strömungswiderstand in den menschlichen Atemwegen und der Einfluss der unregelmässigen Verzweigung des Bronchialsystems auf den Atmungsverlauf verschiedenen Lungenbezirken. Pflügers. *Arch. Ges. Physiol.* **162**, 225–299.

Royal Navy Diving Manual (1964). B.R. 155.

Salzano, J., Rausch, D. C. and Saltzman, H. A. (1970). Cardiorespiratory responses to exercise at a simulated seawater depth of 1000 feet. *J. Appl. Physiol.* **28**, 34–41.

Seusing, J. and Drube, H. C. (1960). Die Bedeutung der Hyperkapnie für das Auftreten des Tiefenrausches. *Klin. Wschr.* **38**, 1088–1090.

Shephard, R. J. (1967). The maximum sustained voluntary ventilation in exercise. *Clin. Sci.* **32**, 167–176.

Williams, S. (1969). Underwater Breathing Apparatus. *The Physiology and Medicine of Diving and Compressed Air Work* (Eds. P. B. Bennett and D. H. Elliott), pp. 17–35. Bailliere, Tindall and Cassell, London.

Wood, W. B. (1963). Ventilatory Dynamics under Hyperbaric States. *Proceedings of the 2nd Symposium of Underwater Physiology* (Eds. C. J. Lambertsen and L. J. Greenbaum, Jr.), pp. 108–123. Natl. Acad. Sci. Natl. Res. Council (Publ. 1181), Washington.

Zocche, G. P., Fritts, H. W., Jr. and Cournand, A. (1960). Fraction of maximum breathing capacity available for prolonged hyperventilation. *J. Appl. Physiol.* **15**, 1073–1074.

The Design of a Lightweight Underwater Habitat

B. RAY

*Kingston Polytechnic, London.**

1. Introduction	253
2. The Design	253
3. Mooring	255
4. The Modular Life Support System	256
5. The Fuel Cell	257
6. Operation	257
Acknowledgements	259
References	259

1. Introduction

The expense and complexity of a habitat or submerged laboratory is largely due to two factors: the problem of placing the habitat on the seabed and the support requirements once it is operational. It was failure to find sufficient funding for these items that prevented the British "Kraken" underwater laboratory from going ahead. The present project, which grew from the pool of knowledge gained in the design of "Kraken", was an attempt to construct an inflatable dwelling sufficiently light to be handled by swimmers in the water, yet comprehensive enough to be completely independent of the surface with no umbilical cable. Supplies would be brought in by divers and the only boat available an inflatable dingy.

There have been previous inflatable habitats, notably that of Edwin A. Link (MacInnis, 1966) and more recently one used by the Moscow "Dolphins" Club (Barton, 1969). Although little is known about the latter, the former, being a particularly deep dive, required considerable surface support and relied on power, gas and communication cables to the surface.

2. The Design

A two-man size was adopted as this represented the requirement of the smallest safe working team. After some tests which involved

* Presently with the Marine Unit Technology, Fort Bovisand, Plymouth, England.

students living in a sealed polythene enclosure, a size of 8 ft × 6 ft × 6 ft high (2.44m × 1.83 m × 1.83 m) appeared adequate. Allowing one foot clearance between the floor and the water surface raises the overall height to seven feet (2.1 m), and gives a submerged buoyancy of around eight tons. After a survey of some possible flexible materials, a nylon-neoprene fabric, manufactured by the Avon Rubber Company for the construction of inflatable boats, was selected.

Fig. 1. The underwater laboratory in use, Malta, 1969.

Although it was always realized that, if the fabric stress was to be minimized, the final shape of the inflated house should resemble an upturned water-drop, initially it was by no means clear how one could achieve this. A family of theoretical cross-sections was produced by an iterative process involving the relations linking the curvature of a membrane and the pressure differential across it. The cross-section that appeared to offer the most suitable height-to-width ratio was chosen. The next stage was to construct a straight sided approximation to this cross-section curve. When this process was extended to three dimensions a 48 faced (plus the base) figure evolved.

The reason for translating what should be a smooth shape into one with a number of flat faces, was that all the seams in the fabric could now be cemented along straight wooden jigs. Previously it had been found that a straight seam could be made with a strength nearly equal to that of the virgin material, whereas it was found difficult to produce a satisfactory curved seam. Before cutting the material and commencing construction, one final requirement had to be fulfilled. Under the calculated working stress the fabric would stretch by between 5 and 15%. Furthermore, the stress-strain relationship for this fabric varied markedly with the angle to the weave. It is difficult to foresee any accurate method of allowing for this anisotropic behaviour. The technique adopted in this design was to reduce each of the flat panels by an estimated fraction.

The base frame was a ring of 3 in steam pipe with 12 attachment points for moorings. The fabric was taken round the pipe and cemented back on itself. In use the floor area was 8 ft × 4 ft (2.4 m × 1.2 m) and the walls leant out to provide 9 ft × 4 ft 6 ins (2.74 m × 1.37 m) at working height. Estimates indicated that the maximum stress would not exceed 1/5th the breaking stress of the fabric.

It should be realized that although the structure was constructed as a series of flat panels, in operation the forces involved would distort the fabric into a smooth shape. The degree to which this has happened can be seen in Fig. 1.

3. Mooring

Available ballast, in the form of natural rocks, was used to provide an anchorage for the habitat. The initial preparation included surveying the site and attaching four wire ropes to the chosen rock. A series of wires were attached to the 12 points on the house ring and taken through four large shackles. In operation, these shackles lay about four feet below the ring (see Fig. 1).

The habitat was lashed as a flat package and manhandled into the water. At this stage the air that was trapped under the fabric tended to keep it afloat. After being towed into position above the ballast, the package was untied and a vent valve opened to allow it to sink. Four divers guided it onto the selected rock. Once the four mooring wires had been attached to the shackles, a small volume of air was introduced into the house. At this stage the length of the mooring wires could be adjusted by hand.

Once all appeared to be in order, the structure was partially inflated and the final levelling operation started. This was accomplished by attaching a Tirfor pulling machine in parallel with each of the mooring wires in turn. The tension was taken up by the Tirfor and it was then a relatively easy matter to adjust the length of the cable and release the machine. After some practice this whole levelling operation, involving all four wires, could be accomplished by three divers in just over an hour. Finally the house was fully inflated with an air line from the surface.

The wooden floor, supported on a Dexion frame, was bolted to the main ring as one complete sub-assembly. A total of about 20 man-hours were spent by personnel, working inside the inflated house, constructing all interior fittings. These included the bunks, table, and storage lockers and were made from wood and Dexion with the aid of a hacksaw, drill, screwdriver and spanners.

4. The Modular Life Support System

Once the habitat was inflated and level, the first item to be installed was the life support module, LSM. These units, of which three were constructed consisted of a case 17 in $\times$ 13 in $\times$ 19 in (43 cm $\times$ 33 cm $\times$ 48 cm) high with a sealed lid and a small air cylinder with valves to automatically pressurize the LSM to ambient sea pressure. Lead acid cells provided a 24 h supply of power and their weight gave the unit an overall negative buoyancy (10 lbs) and insured that the box had a sufficiently low centre of gravity to be stable when carried by a swimmer. Mounted above the batteries, a single integral unit housed the blower, absorbent chemicals and electrical control panel. The inlet air to the blower was drawn from the inside of the case so as to prevent the possibility of a build-up of hydrogen or acidic vapours. The waste heat from the motor and the work done compressing the air by the single stage centrifugal blower, produced a slight temperature rise and ensured that the air meeting the absorber was not quite saturated with water vapour. As is common European practice, a soda-lime charcoal absorber was used. This material has the advantage that it is most effective at the high levels of humidity that were bound to be present. The exhaust from the absorber was not ducted away, instead it was released through an orifice at a relatively high velocity. This "jet" of gas was deflected around the habitat by the curved roof. This method appears highly successful as no pockets of carbon dioxide could be detected.

Splashproof 12 V outlets were provided for lighting a miniature immersion heater (for hot drinks) and communication equipment. An inlet was available also for coupling to an external power source. All outlets were switched and protected by circuit breakers and the battery voltage was monitored. The main illumination was with low voltage fluorescent lights.

Although a conventional telephone was used in the habitat for some of the time, including the use for one call from the house to London, a "wireless" telephone was available for use in emergency. This communicator used the ionic conduction field generated by two spaced electrodes placed beneath the house. Although the communicator was normally powered from the LSM, it was arranged to switch over automatically to an internal battery in the event of a power failure. A pilot light in the communicator circuit indicated this condition and provided some illumination as it was assumed that the main lighting would have been inoperative in this situation.

The low pressure oxygen line in the habitat was fed from a cylinder and reducing valve situated below the house. A gauge was used to set and measure the flow into the atmosphere.

The level of carbon dioxide was checked with a "Ringrose" meter and oxygen with a paramagnetic meter, every four hours. Once a day these were in turn checked against chemical indicator tubes. This type of tube was also used for a daily check of carbon monoxide, hydrogen, stibine, and arsnine. Although the former was present in a concentration of 5 parts 10^6, no trace of the latter impurities was detected.

5. The Fuel Cell

At a late stage in the development of this habitat, a submersible fuel battery manufactured by the Electric Power Storage Co. Ltd., became available. This was a self contained unit using compressed hydrogen and oxygen. This 16 cell battery was housed in a light metal cylinder and was automatically pressurized to ambient seabed pressure with nitrogen. The battery, which could be handled by two divers, was placed on the seabed below the habitat (see Fig. 1) and was capable of supplying electrical power up to 100 watts. The gas cylinders were replaced weekly. Because of the ease of using this battery, the LSM was normally operated from the fuel cell in preference to its internal lead battery.

6. Operation

The site chosen was Marfa Point in the north of Malta. The shore facility was divided into the main camp, situated about 100 yards from the sea, and the landing place where the divers entered the water. Table 1 shows the main items of specialist equipment that were employed in this operation. The first habitat was moored at 30 ft in

Shore Equipment

Main Camp

Two vehicles.
Tent accommodation for up to 12 personnel.
Telephone connected to St. Pauls Bay G.P.O.
40 W fluorescent floodlight.
6 lead acid accumulators.
240 V converter for fluorescent light.
Hydrogen, oxygen and nitrogen store.
Soda lime store (5 cwts).
First aid kit.

Landing Place

Compressor.
800 cu ft 3300 psi air bank.
Normal diving equipment—aqualungs etc.
Floodlighting including fluorescent and quartz iodine spot lamps.
Derrick—5 cwt lifting capacity.
Diving ladder.
Avon Redshank and 5 Hp outboard.
Underwater tug.
Various watertight cases for transport of items to the house.

House Equipment

Domestic

Two bunks, top bunk converts to table.
Life support unit.
Fuel cell (on sea bed).
Servomex OA150 oxygen meter.
Ringrose carbon dioxide meter.
Draeger detector tubes for gas analysis.
Hot coffee heater.
Food store.
Low voltage fluorescent light.
Porta-shower fresh water shower.
Gapmeter for metering oxygen input.
Oxygen reducing valve and cylinder.
Telephone.
Conduction communicator.
First aid kit.

Scientific

Clevite CH13 hydrophone.
Bruel & Kjaer precision sound level meter and octave filter bank.
1 in. condenser microphone.
Submersible audiometer.
Avominor test meter.
2 Calypsophot cameras, one with flash.
2 × 20 W fluorescent lights for photography.

70 ft (9.1 m and 21.3 m) of water on the fifth of August. However, when preparations were in hand for the first diver to be put under saturation conditions a sudden storm destroyed the habitat complete with fittings and life support module and damaged the shore camp.

It was not until August 31st that the first pair of divers were ready to take up occupation in a second habitat.

After the first storm it was realized that 30 ft (9.1 m) was an awkward depth to face unsettled weather; it was not deep enough to ride out a storm, yet the length of decompression required made a rapid escape impossible. The choice now open was to operate much deeper at 60 ft (18.3 m) (the LSM was designed for a maximum of 70 ft (21.3 m) with nitrogen) where the effect of wave motion would be much less or to reduce the operating depth to 20 ft (6.1 m) where a rapid abort would be possible, leaving the support party time to deflate the habitat and stow it safely on the seabed. The lack of information on decompression after saturation with nitrogen mixtures determined that the latter course be adopted.

Somewhat to the surprise of the occupants, and the author who spent four days living in the habitat, it was comparatively easy to work and sleep underwater. The knowledge that the breathing atmosphere and safety was dependent on oneself and not on a third party who may not have been seen since leaving the surface several days previous, appeared to instil a sense of security.

The inside of the fabric was generally wet with condensation but the scientific apparatus and life support equipment did not collect moisture. After a freshwater shower the human body could be dried with a towel and would remain dry. No trouble was experienced when using writing paper or books. However, as has been reported on previous experiments of this nature, cold was the main problem. Although the water temperature was 25°C and air temperature 26°C, the divers found that it could take several hours to warm up after a relatively short dive. This was somewhat puzzling as on shore, where the air temperature was the same on cloudy days, there was no similar problem.

Through the experiment the partial pressure of carbon dioxide was around 0.5% and that of oxygen 28% of one atmosphere. All divers spent one hour free swimming at 10 ft (3.0 m) for decompression after one or more days in the habitat.

Acknowledgements

The setting up and operation of this habitat was the main programme of the joint Imperial College/Enfield College Expedition of 1969. We would like to thank the Royal Geographical Society and all our sponsors. We would also like to thank the organizations who helped us in Malta, including the Office of the Prime Minister, The Royal University of Malta, Malta Dry Docks, the Royal Navy Diving Team, and the Departments of Fisheries, Post and Telephones and the Land Commission, together with many other individuals.

References

MacInnis, J. B. (1966). Living under the sea. *Scientific American* **214** No. 3, 24.
Barton, R. (1969). International Oceanics. *Hydrospace* **2** No. 4, 10.

The Use of an Underwater Habitat as a Quiet Laboratory for Tests on Diver Hearing

B. RAY

*Kingston Polytechnic, London.**

1. Introduction	261
2. Experimental Procedure	261
3. Results	264
4. Discussion	265
Acknowledgements	266
References	266

1. Introduction

During the course of some tests concerning directional hearing underwater (Ray, 1969), the author was surprised when several subjects reported that they considered that a sound source appeared louder when they were facing it than when their backs were turned to it. This investigation set out to examine these reports.

2. Experimental Procedure

The arrangement of persons and equipment is shown in diagrammatic form in Fig. 1. The submerged audiometer is an instrument designed to produce a pure tone the amplitude of which can be accurately adjusted in 2 decibel (db) steps by means of an attenuator.

The transducer which converts the electrical signal to an acoustic one, is separated from the main instrument so as to prevent the operator's body from shading the sound. It would have been far more convenient if the electrical part of the audiometer had been situated inside the underwater laboratory. However at the time when this part of the apparatus was built the laboratory was still in a planning stage. The subject diver sat in 30 ft (9.1 m) of water with his head 15 ft (4.6 m) from the audiometer. The underwater laboratory was moored at 20 ft (6.1 m) slightly to the side of where the subject was seated. The acoustic signal at the subject's head was monitored with a hydrophone which

* Presently with the Marine Unit Technology, Fort Bovisand, Plymouth, England.

was connected to an octave filter bank and precision sound level meter in the laboratory.

The laboratory was run under quiet conditions with all life support functions switched off and the audiometer operator synchronized his breathing with that of the subject. This latter was essential as it was not until the larger exhaled bubbles had reached the surface that conditions were quiet enough for this work.

The audiometer was set so that the subject could clearly hear the tone and the subject was asked to show that he could do so by means of

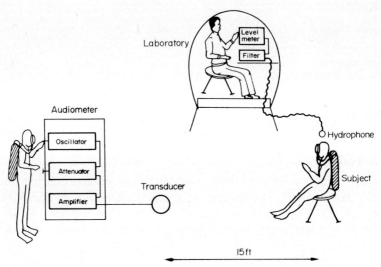

Fig. 1. Diagrammatic representation of the audiometer experiments.

simple hand signals. Once the subject had signalled, the tone level was reduced by 2 db during the next period when the subject was breathing. This was to make sure that any noise generated by operating the switch would be masked. When a level was reached where the subject indicated that he could no longer hear the tone, the setting of the attenuator was noted and the level increased in steps until the response was positive again. This setting was noted. Reference to Fig. 2 should clarify this procedure. The setting of the attenuator is shown against an arbitrary time scale measured by one division for every occasion that the subject signals a reply. The nature of the reply is also shown. In normal operation it was usual to only record the maxima and minima of the attenuator settings as it is the mean of these that determines the threshold.

Within the length of one dive it was normally possible to first record two or three excursions around the threshold with the subject facing the audiometer. The subject was then asked to turn around. It was important that only the subject moved and that the position of the audiometer was not disturbed as a small change here may have produced

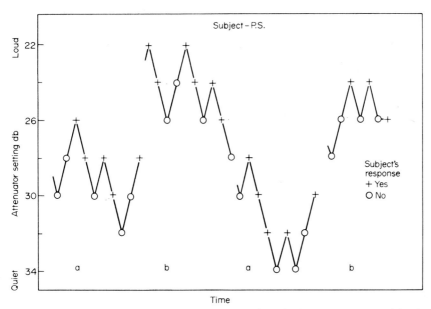

FIG. 2. Audiometer test with subject facing source (A), and with back to source (B). An attenuator setting of 30 db produced an absolute sound pressure of 54 db rel. 0.0002 dynes/cm² at the subject's head.

amplitude differences through changes in the standing wave pattern of the tone.

After several excursions with the subject facing away from the instrument, the whole sequence was repeated a second time. In this way it was hoped to reduce any slow changes through effects such as learning, boredom or cold.

Three of the subjects were tested with the pure tone audiometer at 1000 Hz. Before and after the tests the absolute acoustic pressure was measured from the underwater laboratory to provide a calibration. Four of the subjects were tested on an earlier occasion when no calibration facilities were available. In these cases it is only the differences in sensitivity that are meaningful. This earlier design of audiometer produced a wide-band noise which was predominantly a 600 Hz pulse train.

3. Results

Of the seven subjects who helped in these tests, six showed a marked improvement in hearing sensitivity when the sound source was in front of the body (see Table 1). In the case of the one subject who showed no

TABLE 1.

Subject	Difference front/back	Calibration and Tone	Conditions
R.L. (8)	5.4 db	Wide-band no calibration	Removed aqualung and allowed it to float about 2 m above his head.
P.N. (8)	4.8 db		
M.L. (8)	4.5 db		
J.W. (8)	5.1 db		
P.S. (12)	5.2 db	MFT 54 db at 1 KHz	Held aqualung in hands.
N.S. (6)	4.7 db	MFT for both runs 56 db at 1 KHz	Blacked-out facemask — Both tests on same dive
(7)	2.6 db		Without any facemask
M.B. (23)	−0.8 db	MFT 66 db at 1 KHz	Test performed rapidly so as to obtain a large number of "excursions". Subject had forced choice every breathing pause.

MFT = Mean front threshold (reference 0.0002 dynes/cm^2). Figure in brackets after the subject indicates the total number of threshold "excursions" involved (i.e. the total number of pairs of readings).

significant difference, the test was conducted faster than previous ones and the subject was asked to reach a decision in every breathing pause. It may be significant that the absolute value for the threshold is also somewhat higher in this case. On one other occasion when the front-to-back difference was small, the subject had removed his facemask and performed part of the test with no protection over his eyes or nose. With the exception of this test, the subjects could generally see the audiometer when it was in front of them, but they were instructed to fix their gaze on some object such as a rock and not to study the instrument. In any case they would not have been able to gain additional information from the position of the controls as these were not visible to the subject. All subjects were provided with a seat, an extra weight-belt for the lap and none wore a rubber hood.

In addition to these tests, the sound pressure field in the sea and in the laboratory were analyzed using the octave filter bank. These measurements are summarized in Fig. 3.

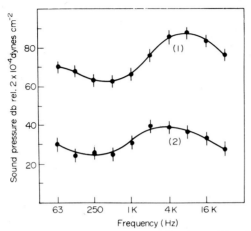

FIG. 3. Sound pressure level of ambient noise measured in the sea (1) and in the "air" inside the habitat (2).

4. Discussion

It is possibly worth illustrating just how sensitive a diver's ears can be if he is given the task of listening carefully. Any air leak from either his own or the audiometer operator's aqualung was sufficient to disturb the test.

As it appears that a considerable number of breathing sets do have small air leaks that are normally not noticed, it was sometimes necessary to take several sets underwater so as to try each in turn and select the most suitable. The running of a compressor on shore about 100 yards (91.4 m) away had to be halted and tests were not possible if any power boats were being used in the same bay.

It may seem surprising that the absolute values for the thresholds lie below the ambient noise in octave analysis. This is perfectly reasonable as the human ear is an excellent tool for detecting a distinctive sound below a broad band masking noise. This noise shows a distinct hump at about 8 KHz and appeared to be almost entirely biological and is normally credited to the "snapping shrimp". In the case of three subjects who were tested with calibrated equipment, the absolute threshold at 1 KHz has a mean of 59 db ref. 0.0002 dynes/cm². Although this is in reasonable agreement with the figures published by Hamilton (1957) and Brandt and Hollien (1967), the author feels that because of the relatively high ambient noise this must be regarded only as an upper limit to the true threshold.

Although further work is required before any firm suggestions as to the mechanism of this directional sensitivity can be put forward, a tentative model will be proposed. In the past there has been frequent discussion in the published literature as to whether hearing underwater is by bone conduction or by the sound being transmitted along the auditory canal in much the same way as in air. The author is of the opinion that it is not meaningful to consider underwater hearing being either bone conduction or tympanic. These ideas apply to hearing in air where the acoustic impedance difference between flesh-and-bone and air is so large (10^3–10^4) that one can consider a sound wave as travelling in one or the other. Flesh has a similar acoustic impedance to water and it may be more reasonable to consider that underwater the two hearing organs exist in an infinite fluid. This ignores the water-skin boundary completely! The middle ear, sinus cavities and the facemask along with other air spaces are impedance discontinuities in the neighbourhood of the receptors and must be taken into account. The effect of bubbles of air in the immediate vicinity of an acoustic transducer has been studied (Hunter, 1967). These can have a dramatic effect on the performance of the transducer and produce directionality in an otherwise omnidirectional device.

So far only one subject has been tested without a facemask. It is hoped that further tests with and without a facemask will shed additional light on this hypothesis.

The role that this directional sensitivity plays in underwater hearing is uncertain. A difference of about 5 db is relatively small in human terms and it is felt that this is unlikely to be significant in directional hearing.

Finally, mention must be made of the excellent quiet platform provided by the underwater laboratory. No noise could be detected as arising from this structure and the airborne sound pressure level inside (Fig. 3) appears to be due entirely to the "snapping shrimps" on the outside. Subjectively it appeared that the biological noise increased sharply just after dusk when a considerable number of small creatures between 1 mm and 3 mm in length, chose to climb up the inside wall of the house. No measurements are available because of humidity problems with the condenser microphone that was used for air measurements.

Acknowledgements

I would like to thank the members of the joint Imperial College/Enfield College Expedition of 1969 who acted as subjects in these experiments, and also Dr. R. W. B. Stephens for his help in planning the acoustic programme.

References

Brandt, J. F. and Hollien, H. (1967). Underwater hearing thresholds in man. *J. Acoust. Soc. Am.* **42,** 966.

Hamilton, P. M. (1957). Underwater hearing thresholds. *J. Acoust. Soc. Amer.* **29,** 792.

Hunter, W. (1967). The Influence of Gas Bubbles on the Generation of Underwater Sound. PhD Thesis, University of London.

Ray, B. (1969). Directional hearing by divers. U/W Assoc. Report.

Ray, B. (1971). Audio Communication Between Free Divers. *Underwater Acoustics* (Ed. R. W. B. Stephens). Wiley, London.

Towards the Development of a Practical Underwater Theodolite

R. FARRINGTON-WHARTON

Comex Diving Ltd., Bunns Lane, Southtown, Great Yarmouth, Norfolk

1. Introduction	267
2. The Instrument	269
3. Sea Trials of the Prototype Instrument	270
4. Results of the Sea Trials	272
5. Target Design	273
6. Ergonomic Considerations and Suggested Design Improvements	274
7. Practical Applications	275
8. Further Development	275
Acknowledgements	275

1. Introduction

This chaper traces the early development of the underwater theodolite from its inception to the successful field use of a prototype instrument. This must be seen in context as the first step in a development programme to produce a commercially viable instrument, robust and easy to operate.

It should be clearly understood that this initial development was to prove that accurate surveying was possible underwater and to show the way forward.

The author knew from the outset that the prototype instrument would bear no resemblance to the final development. While a laser is the ideal tool to mount as a light source an optical system was chosen for the prototype for cheapness and to cut down development time of the first prototype.

At the present time underwater surveying tends to be an inaccurate science. For large scale work hydrographic surveys and underwater contouring from aerial photography are sufficiently accurate for the preparation of marine charts, but not nearly accurate enough for detailed underwater construction work.

For small areas photogrametry, grid, tape and string line surveys are used. These methods are quite sufficient for charting wreck sites or setting out mass concrete foundations but they lack the precision needed for detailed work.

Partially for this reason, the scope of underwater construction work is very limited, and where it is carried out tolerances tend to be very large, frequently of the order of several inches. The physical difficulties of underwater setting out have led to structures being designed to accommodate appreciable errors in setting out and construction.

The next decade should see a steady increase in the volume and complexity of underwater engineering projects undertaken. With this growth of underwater activity will come the need for more accurate and reliable methods of underwater surveying and setting out, and it is to this end that the underwater theodolite has been developed.

With the sudden upsurge in offshore petroleum exploration and production there has been a parallel advance in underwater activity. The offshore oil industry uses large tolerances for underwater construction work but could use greater accuracy for level surveys, defect measurements and measurements of lack of fit.

Some two years ago a steel oil storage tank was installed in the Persian Gulf and when installed was found to be 5 degrees out of level due to inclination of the seabed. This inclination was not measurable by conventional means and meant imposing a reduction in storage capacity in the tank.

Two current requirements that cannot be met with existing equipment and techniques are, (1) to set up an accurate horizontal reference line, and (2) to set up an accurate vertical reference line.

Before going any further, the construction and use of a conventional theodolite will be described briefly. The theodolite and the level together constitute the basic tools of the land surveyor. Both the level and the theodolite are equipped with internally focusing telescopes with cross hairs etched on a graticule in the focal plane of the eyepiece. This graticule is adjustable so that the cross hairs may be brought accurately onto the axis of collimation, this operation being known as collimating the instrument.

The level is a simple instrument that allows the telescope to rotate through 360 degrees in a horizontal plane, and in conjunction with a graduated staff is used to measure differences in ground level.

The theodolite on the other hand is a more sophisticated instrument that allows the telescope to be rotated through 360 degrees in a horizontal and a vertical plane, and permits these rotations to be measured to within 20 seconds of arc. (To give some idea of this degree of accuracy, 20 seconds subtends one sixteenth of an inch at 50 ft or 0.16 cm in 15.2 m.) By setting the instrument up at either end of a measured base line the position of any point in visual range can be determined by simple trigonometry. Thus, the theodolite is a very powerful tool for the land surveyor, being cheap and easy to use and giving a direct visual readout of results.

2. The Instrument

The prototype underwater theodolite consists of a conventional 20 second vernier theodolite with a redesigned telescope. In order to overcome problems of sealing, the instrument was designed to work

FIG. 1. The underwater theodolite in use.

fully flooded with sea water at ambient pressure. It is used in conjunction with a wooden tripod and a conventional plumb bob to set the instrument up plumb over a survey station (Fig. 1).

Most forms of working diving dress use a face plate or mask set some 1 to 2 in (2.5–5.1 cm) from the diver's eyes. This prevents the diver from getting closer to a lens than say $1\frac{1}{2}$ in (3.5 cm). It therefore follows that, for an optical instrument to be used successfully by a diver, the eyepiece must have sufficient eye relief to overcome this difficulty. It is a further advantage if the eyepiece is fitted with a hood of convoluted rubber to cut down the scattered light reaching the diver's eye from the surrounding water, and to absorb jolts if accidentally touched by the diver's mask.

Because of the difference in refractive index of air and water, when a lens is immersed in water its power is greatly reduced. In order to keep the optical system to a manageable length, so that it will rotate within the telescope mountings of a conventional theodolite it is necessary to resort to more powerful lenses. A compromise may be achieved if a pair of lenses is used with an air space between them, the unit then being known as an air cell.

Air cells were used for both the eyepiece and the objective of the instrument and both were fitted with independent focusing controls. The magnification of the completed telescope was approximately two times. Because of the low magnification of the eyepiece it was not possible to use a conventional etched graticule. In its place cross hairs of fine tungsten wire were used, mounted in a graticule frame to allow adjustment for collimating the instrument.

Several other designs were considered for the optical system. An ideal solution would be to have a complete instrument enclosed in a watertight case, which might be pressurized to the ambient pressure to reduce sealing problems. While this technique works very well for camera cases it is doubtful whether it could be applied successfully to encasing a complete underwater theodolite. The difficulties involved in extending the numerous controls through a watertight case are enormous, and the very presence of such a case would preclude the close inspection of the verniers.

Perhaps a more simple solution would be to encase just the telescope in a watertight case. This, however, still presents problems of access to the eyepiece and collimation adjustment. The possibility of solely sealing the air space inside the telescope was also considered, but the focusing of the eyepiece alters the internal volume of the telescope and introduces considerable sealing difficulties. A fully flooded optical system was eventually chosen for simplicity.

3. Sea Trials of the Prototype Instrument

The sea trials of the instrument were carried out in the summer of 1969 off the north coast of Malta. The work formed part of the Imperial College/Enfield College Underwater Living Experiment. Malta was chosen as a test site because of the excellent visibility and the generally calm seas of the Mediterranean. In fact, the sea conditions were very rough for the time of year but it was found that the instrument could be operated without undue difficulty at 30 ft (9.1 m) while a ground swell of 8 ft (2.4 m) was running above.

An operating depth of 30–40 ft (9.1–12.2 m) was chosen, being sufficiently deep to avoid serious wave disturbance, and yet shallow enough to permit long dive durations and avoid the complications of decompression. There is no depth restriction to the instrument over the normal safe diving range. It is likely that at depths in excess of 150 ft (45.7 m) the accuracy and judgement of a diver breathing air would be impaired sufficiently by nitrogen narcosis to render his work unreliable.

To operate a theodolite underwater the diver must have accurate variable buoyancy control so that he may be positively buoyant to bring the instrument to the surface and transport it on the surface, but sufficiently negative to stand firmly for long periods beside the tripod when the instrument is set up. Varying degrees of buoyancy are required between these two limits for operations such as moving the instrument between stations. In order to achieve this range of buoyancy the diver must enter the water considerably overweighted and employ a constant volume buoyancy aid of the Fenzy type.

Apart from moving the instrument between stations there is little hard work involved in underwater surveying. This conserves the diver's air supply so that he can stay submerged at 30–40 ft (9.1 m–12.2 m) for $1\frac{1}{2}$ h using a 120 ft^3 (3.4 m^3) twin set.

In order to thoroughly evaluate the instrument a triangle with sides of 40 to 50 ft (12.2 m–15.2 m), which could be surveyed several times in the time available, was selected. This allowed a statistical assessment of the experimental error to be made. Three large rocks were chosen as survey points and were cleared of marine growth. The stations were marked with Oboe nails driven into the sandstone rock. In hard rock marker nails may be driven into lead plugs hammered into holes chiselled into the rock, and on a sandy or muddy bottom nails may be hammered into wooded stakes driven well in, as on land.

The nail heads, which were impossible to see from 50 ft away, were marked with upturned soft drink bottles filled with air and moored to the nails with fine nylon line. The stations were labelled with painted neoprene tags wired to the nails. With this arrangement it was a simple matter for the operator to carry the instrument to the station and set it up roughly, then undo the marker bottle, fit the plumb bob to the instrument and set it up accurately over the nail head.

The operation of the instrument was then as it would be on the surface, horizontal and vertical angles being recorded from each station. With the optical system used, parallex was very noticeable but it could be eliminated with careful focusing. The most tedious operation was reading the verniers. All the vernier magnifiers were removed from the instrument as they did not have sufficient eye relief. In place of these, a powerful double convex lens was mounted over the diver's face mask in such a way that it could be stored out of the way and swung down into place when required. This system had the advantage over a hand held lens of leaving both hands free to control the instrument. It was found quite possible to read the verniers to their accuracy of 20 s although this operation was very time consuming and was dependent upon the ambient lighting.

The readings were recorded in pencil on sheets of Ozalid polyester drawing film. The sheets were ruled out on the surface before each dive so that the writing time underwater was reduced to a minimum. This form of data recording is very cheap and effective, the results being transferred to permanent records after the dive.

Early on in the trials it was found that the bottle markers were hard to see in dull lighting and the lack of definition of the nail heads was giving rise to sighting errors. In order to overcome this the bottles were wrapped with red fluorescent adhesive film and moored tightly down so that their necks hung right on the nail heads. This arrangement gave rise to better results.

The theodolite was fitted with an adjustable bubble on the telescope tube to enable the instrument to be used as a level. A folding graduated wooden staff was used for the levelling work. This was weighted with lead to give it a negative buoyancy of about 2 pounds. The staff so weighted was very easy to use as it tended to assume an upright position at all times. The results of the levelling work were quite satisfactory, the accuracy achieved being that of the instrument. The levelling staff was found to be unworkable at shallow depths when a heavy sea or ground swell was running.

4. Results of the Sea Trials

The following data give an indication of the accuracy achieved by the prototype instrument.

An approximately equilateral triangle with sides of about 50 ft (15.2 m) was chosen for the trial surveys. The horizontal angles were each recorded ten times and the standard deviations were approximately one minute of arc. At 50 ft one minute of arc subtends three sixteenths of an inch (0.48 cm). This shows the small spread of results, as it may be assumed that in an unimodal distribution of results two thirds of the results will lie within one standard deviation of the mean. After ten surveys the sum of the means of the included horizontal angles was 180–00–52.

The survey triangle was on rocky ground and the survey points were all at different levels, so that the triangle did not lie in a horizontal plane. However, with the theodolite set up over one station point, knowing the height of the collimation axis above the station nail and the slope distance to another station nail, it was possible to calculate the true horizontal distance between the nails and the true difference in level between the nails simply by recording the angle of dip or rise to the second nail.

The slope distance between the three nails were measured using a linen tape that was later compared against a steel standard at the Royal University of Malta. These distances were measured several times using different operators to avoid bias error. The instrument height, or height from the station nail to the axis of collimation, was measured by tape at each set up of the instrument. All the true horizontal distances were calculated out for each of the ten surveys and the standard deviations were in the order of one sixteenth of an inch (0.16 cm).

These results are remarkably accurate for a prototype instrument, but it should be remembered that they are calculated means of several surveys and in practical use the instrument must be sufficiently accurate to give the desired results for one survey, not an average of several. The accuracy of the instrument could be greatly improved by consideration of the following points:

(a) using an eyepiece of greater magnification so that an etched graticule could be used, the tungsten wire cross hairs proved too thick for the optical system used.
(b) providing a glass sided tank in which to collimate the instrument under laboratory conditions as this operation is inaccurate when carried out underwater.
(c) using an internally focusing optical system to overcome the parallex problems encountered with the prototype instrument.
(d) use of improved targets.

5. Target Design

It is now clear that not enough forethought was given to the problem of target design. This is an important undertaking where visibility is limiting as it is underwater and it must be adapted to suit the conditions under which the target is used.

The physical parameters governing radiance transfer, visibility and image perception are discussed in detail by Cocking in this volume. Targets should always present a high inherent contrast and white upon black is usually best. Good colour contrasts can be obtained by using fluorescent colours but these tend to lose their effectiveness at extreme visual range. Perhaps the most visible objects are those which actually emit light and these should be considered in future work.

6. Ergonomic Considerations and Suggested Design Improvements

Surface time, compared with diving time, is very cheap indeed and for this reason alone underwater operations are only undertaken when there is no alternative method, and even then underwater time is kept to a minimum. There are numerous ways in which dive duration may be reduced by thorough pre-dive planning, and not the least of these is to reduce the task complexity to a minimum. Unfortunately, the actual operation of a theodolite below water is just as complex as its operation above. Emphasis for future development must, therefore, be placed on simplifying both the method of operation and type of construction of the instrument.

The most difficult task with the prototype was to read the vernier scales. These were read using a mask mounted magnifying lens. While this is feasible, the eye is still not close enough to the scales and there is some risk of jolting the instrument with the face mask. In order to overcome this problem, the author was fitted with experimental contact lenses. Experience gained with these indicated that they have a very worthwhile future for underwater operations, especially as they can be made to accommodate any prescription that the wearer might have.

Unfortunately, underwater contact lenses are unlikely to be in widespread use for several years yet and a method of readout is required that is both reliable and easy to use. A digital readout would be the most convenient for a diver to use, or possibly a conventional scale with a micrometer vernier. Operation of the instrument should not involve the operator in any calculation, unless strictly necessary. He must be simply an underwater theodolite operator who records only angles. All calculation from these angles should be carried out on the surface where time is cheap and thought is clear.

The suggested modifications of construction and operational procedure should produce an underwater theodolite much easier and faster to use than the prototype, which, with the exception of the telescope, was a land theodolite and in no way designed with the diver's particular problems in mind.

The ideal long term development for this instrument should be to mount a laser in place of the telescope. It would not take too much ingenuity to develop a self deploying tripod and a self levelling instrument head. An excellent solution for levelling would be to have the laser head rotating in a horizontal plane emitting pulse light. The diver could then stand anywhere within range of the instrument holding a levelling staff and note where the light beam intercepted it.

7. Practical Applications

The sea trials of this instrument were carried out in conjunction with the Imperial College/Enfield College Underwater Living Experiment in which an inflatable underwater house was moored to the seabed and inhabited by divers for a period of several days. In its initial trials this house was moored at shallow depths and was affected by the surface weather conditions. To assess the safety of the structure, and thereby its occupants, it was essential to measure its movements under different sea conditions. These measurements were made using the theodolite, set up a known distance from the house, to measure the angle described by a fixed point on the structure. With this information it was a simple matter to calculate the amplitude of the oscillations, their period being timed directly. This same principle could be applied to measuring deflections of offshore structures.

With this instrument it is also possible to survey and level a section of seabed to a degree of accuracy hitherto unattainable. This facility should enable more accurate underwater construction work to be undertaken, as the same instrument can be used to set out the works.

It must be appreciated that the use of the prototype underwater theodolite is limited by the range of visibility, the working depth, and to a certain extent surface conditions. Improved technology will make it possible to mount a laser on the trunnion axis of a theodolite and operate it over long distances underwater in conditions of low visibility.

8. Further Development

Accent for the development of the underwater theodolite should be placed on automating the system and reducing the diver development to the minimum. Since the field trials of the prototype instrument two significant developments have been carried out. The first being the development of an underwater plane table by Mr. T. E. McKay and Dr. P. H. Milne of the University of Strathclyde and the second being development work carried out in 1971–72 by Mr. Brian Ray on the use of lasers underwater.

Acknowledgements

I am indebted to the organizers and supporters of the Imperial College/Enfield College Underwater Living Experiment, Malta, 1969, for the facilities extended to me whilst carrying out the sea trials of the instrument. In particular to Mr. Brian Ray, the project co-ordinator, for his invaluable technical advice and to Mr. B. New for his excellent photographic coverage.

The Design and Applications of Free-flooding Diver Transport Vehicles

G. COOKE

Marine Unit, Fort Bovisand, Plymouth

1. Drag	277
2. Power and Propulsion	280
3. Structural Problems	280
4. Stability	280
5. Steering	281
6. Crew Support	282
7. Advantages	282
References	283

A free-flooding Diver Transport Vehicle is a submarine in which the entire hull fills with water when it is submerged. Vital components such as batteries, motor, switches and electronic equipment are individually protected from the water. The crew use self-contained diving gear or a built-in breathing supply and are, at all times, subject to the usual rules of diving.

1. Drag

The first problem that the designer of such a vehicle must consider is that of the hull shape. The size will depend on the number of crew members and the purpose for which the machine will be used. As with any other vehicle, the design must be a compromise. The most important factors in this case are the size and weight of the power supply and the speed and range required.

Both are governed by the amount of thrust necessary to propel the craft under water and this, in turn, is governed by the forces opposing the forward movement of the vessel, namely, drag.

As an example of the considerable resistance of water to the movement of an object through it, consider the formula for dynamic pressure, $\frac{1}{2}\rho V^2$, where ρ = mass density of seawater, V = speed of the current in feet per second.

In sea-water, as an approximation, $\rho = 2$, so $\frac{1}{2}\rho$ cancels out leaving the dynamic pressure equal to V^2.

From this it can be seen that the force on a one-foot-square plane normal to the direction of travel will be 100 lb (45.3 kg) at a speed of only 10 ft (305 cm) per s, about 6 knots.

It follows from this that some less abrupt introduction of the body shape to the flow of the water is essential. This may be achieved by streamlining the body so that there is a gradual lead-in to the flow and, what is more important, a very gradual tapering off towards the tail. The latter is in order to reduce the drag caused by the wake turbulence. The now familiar streamlined shape is the tear-drop, aerial bomb or airship shape (Prandtl and Tietjens, 1934).

The significant factor in the calculation of power and propulsion requirements is the drag coefficient for the particular size and shape of hull being considered.

The factors influencing this coefficient are the speed and shape of the vessel, its cross-sectional area and the density of the medium. The drag coefficient, C_d, is obtained by experimental means using models scaled down in accordance with the Reynolds number. This is a dimensionless number (Daugherty and Franzini, 1965).

Reynolds number $R_l = \dfrac{lV}{\nu}$

Where l = length of hull,
V = speed in fps,
ν = kinematic viscosity of the medium.

(ν for sea-water at 10°C equals 1.46×10^{-5} ft²/s or 1.36×10^{-6} m²/s.)

If the same medium is used for the model tests as will be used for the full-sized vessel then the speed of the model must be increased in order to keep the Reynolds number the same. It has been found that, for a shape similar to that of the American Akron airship or the British R101, the drag coefficient in water can be as low as 0.04 for the hull alone. Appendages, such as stabilizing fins and hydroplanes, will increase the drag coefficient considerably. The drag of each individual appendage can be calculated but an allowance should be made for interference effects caused by the proximity of the appendage with the hull.

These hulls are, of course, bodies of revolution.

The total drag D on a submerged body is given by

$$D = C_d \frac{A\rho V^2}{2} \text{ (lbf)}$$

where C_d = drag coefficient,
A = maximum cross-sectional area, ft²,
ρ = mass density of sea-water (1.99 slugs/ft³),
V = speed through water, fps.

This formula applies to a reasonably smooth streamlined body since surface roughness will increase the drag force (Hoerner, 1958).

From this information we can plot a graph to show thrust horse-power required at any given speed. A typical result is shown in Fig. 1.

The calculations assume that $C_d = 0.05$
$$D = 0.54 \ V^2 \ (\text{lbf})$$
Propeller efficiency $= 0.65$
Motor efficiency $= 0.65$

Also shown is the plot of input horse-power.

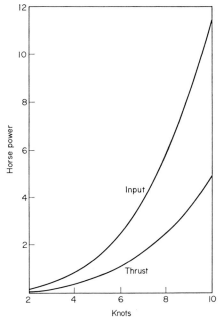

FIG. 1. A graph of input horse-power and thrust horse-power plotted against speed for a typical, streamlined, fully-submerged hull.

It may be seen that the graph is of the form $y = ax^2$ and that, as one approaches the steep part of the curve, it becomes obvious that a large increase in horse-power will result in only a small increase in speed.

Although the application of these data is still something of an art, it is possible to calculate with some accuracy. As an example, the calculated average drag coefficient of the French TotalSub, a large, four-man free-flooding Diver Transport Vehicle, was 0.055. As a result of model tests the coefficient was given as 0.0565.

2. Power and Propulsion

At present, there does not appear to be a power source available which can improve on the advantages of electric batteries. The lead-acid battery is the most economic type. Its only disadvantages are that it is heavy and that it gases during discharge, giving off hydrogen which, when mixed with oxygen, can be highly explosive.

Fuel cells, hydrogen peroxide and liquid monofuel are some of the power sources that have been experimented with. Let us consider the battery-powered electric motor. Speed variation may be obtained by grouping cells, simple rheostat control (wasteful of current), solid state control devices which draw negligible current and variable pitch propellers. The latter enable a speed range of from full forward to full reverse to be achieved with no switching of the batteries.

The position of the propeller is important if the efficiency of the DTV is to be kept high. Ideally, the propeller should be mounted a definite distance behind the tail of the hull so that minimum thrust deduction may be achieved (Beveridge, 1963).

The efficiency of a propeller increases with diameter but the rotational speed must be reduced when using larger propellers in order to maintain the correct load on the motor (Taylor, 1943).

The overall efficiency of the vessel consists of the resultant efficiency of the batteries, the motor, the transmission (gearing for example), and the propeller system (including the duct or shroud, if any).

3. Structural Problems

Because the DTV is free flooding there are relatively few structural problems. The absence of pressure differentials on the hull greatly simplifies construction. Some difficulties may arise with the waterproofing of the batteries, motor and switches, but established methods of sealing these items are quite satisfactory.

Hulls may be easily fabricated from glass fibre. Moulded plastics, although more expensive, would be more durable.

4. Stability

Again, I should like to use the analogy of the airship to show its similarities with underwater craft. The airship does not roll, neither should the enclosed type of DTV. In a fast turn, it may lean into the turn because of the centrifugal force acting on the centre of gravity.

In pitch, some control is needed when under way. Roll stability is simply a matter of ensuring a large righting moment by placing the centre of gravity below the centre of buoyancy and at a specific distance from it. This distance is critical when considering the dynamic behaviour of the craft. Too small a distance will result in instability and too great a distance could result in unfavourable reaction to the controls in the pitch plane.

Dynamic stability is also much affected by the static margin—the distance between the centre of pressure of the hull and the centre of gravity. The larger this is, the greater will be the tendency of the craft to "weather-cock" and follow a true course.

The stability requirements of the vessel when it is on the surface are different from those when it is underwater. The DTV floats dry on the surface and can be driven or towed to the diving site thus keeping the crew warm. A low C of G is needed in this condition but this is not necessarily in the same position as it is when submerged.

The criterion for stability on the surface is that the C of G be below the metacentre which, for a body of revolution, is at the centre of revolution.

Ballast tanks, full of air when on the surface, are vented in order to submerge. Once below, the contents of a trim tank are adjusted until the craft is neutrally buoyant. In this condition, the weight of the vessel is exactly counterbalanced by its buoyancy. The only factors that will alter this condition are the compression of the divers' suits with depth and the reduction in the weight of the air or gas carried as it is used up by the crew. These variables will necessitate occasional corrections of trim.

5. Steering

Steering in yaw is most simply achieved by using a conventional rudder behind the propeller. If precise manoeuvring is essential, then consideration must be given to bow thrusters or other side thrust units. The rudder, of course, is only effective when under way and, for most purposes, is satisfactory.

Ground speed may be zero or even negative when in a current but rudder control is still possible.

Since the DTV is roll stable, there are no great problems concerning steering in yaw. Steering in pitch poses more difficult problems. The effectiveness of hydroplanes is closely linked with the C of G to C of B distance, the lift of the hull itself when the angle of incidence is not zero, pitch moments about the control shafts and, finally, the aspect ratio of the control surfaces and their lift-drag ratios.

If rear elevators are fitted there may be difficulties in lifting the sub off the bottom without blowing tanks because the dynamic pressure may force the tail down. Forward-mounted hydroplanes overcome this difficulty but they are more vulnerable to damage. From the pilot's viewpoint it is desirable to have all steering controls on one handle, leaving his other hand free to operate buoyancy controls, communications apparatus and so on (Cooke, 1969).

6. Crew Support

A large quantity of breathing gas may be carried in a DTV giving the crew long endurance and added safety. A built-in breathing supply is used while the crew are in the craft but when they leave it underwater they can either change to a self-contained breathing apparatus or they can be supplied with gas through umbilicals from the BIBS.

Keeping divers warm is a continuing problem but in the DTV a built-in heating supply is quite feasible. Stored hot water would be one possibility. Electrical heating is also possible but it would impose an undesirably heavy load on the main batteries. There is ample space in a well-designed craft for auxiliary batteries of the nickel-cadmium type, for example.

7. Advantages

There is no depth limit on the craft itself. DTV's are relatively lighter in weight, smaller, less expensive and more easily handled than dry submersibles. The crew can leave the craft at any time and can be provided with navigation, homing, communications and inter-communication equipment. They can take power tools with them.

The crew are physically at rest while travelling and use less breathing gas. The DTV offers shark protection in addition to providing those things that help to keep the diver comfortable and safe.

With the increasing interest in the continental shelves, mapping the seabed will take on a new importance. Mapping may be done by photography in addition to side-scan sonar and magnetometer methods. The resolution obtained by photography is of the highest order but good visibility is essential, whereas the other techniques are independent of visibility.

Mapping the seabed by photogrammetry is likely to result in the greatest amount of information being acquired. A DTV, specially equipped, can traverse the seabed taking pictures automatically in sequence so that each successive pair of pictures represents a photogrammetric pair. The success of this system depends on accurate navigation.

The DTV could be used for photographic reconnaissance of wrecks and archaeological sites after it had located them by other means.

Pipe-lines could be surveyed in a very short time with minimum discomfort to the divers. Magnetometer searches could be carried out for geological purposes or for wreck finding by a DTV specially equipped and able to travel close to the seabed.

Sporting applications could include fish-hunting or simply photography and other attractions for the diving tourist.

A DTV could really make an impact on underwater technology as a mobile workshop since it is able to carry power tools, electrically, hydraulically or pneumatically operated.

When, eventually, the seabed habitat system is in common use, the logical form of transport between the units of the system is the free-flooding DTV.

There is much scope for imagination in the field of military applications but high-speed versions of the DTV would have obvious advantages.

Fish studies could be done by operating from a boat in shoals with an observer reporting directly to the surface by sonio.

Open water fish farming could be speeded up by the use of DTV's.

Finally, for observation and photography of relatively high-speed objects underwater, towed or otherwise, the DTV would appear to be the best choice of vehicle in view of its ease of handling, its low running costs and, last but not least, its crew safety aspects.

References

Beveridge, J. L. (1963). Effect of Axial Position of Propeller on the Propulsion Characteristics of a Submerged Body of Revolution. David Taylor Model Basin, Report 1456.

Cooke, G. W. R. (1969). Underwater Swimming, an Advanced Handbook. Chap. 8—Diver Transport Vehicles. Kaye and Ward, London.

Daugherty, R. L. and Franzini, J. B. (1965). Fluid Mechanics with Engineering Applications. McGraw-Hill Book Co., New York.

Hoerner, S. F. (1958). Fluid Dynamic Drag. Published by Author, New Jersey.

Prandtl, L. and Tietjens, O. G. (1934). Applied Hydro and Aero Mechanics. McGraw-Hill Book Co., New York.

Taylor, D. W. (1943). The Speed and Power of Ships. U.S. Government Printing Office, Washington.

Practical Considerations for Quantitative Estimation of Benthos from a Submersible

J. F. CADDY

Fisheries Research Board of Canada, Biological Station, St. Andrews, N.B., Canada

1. Introduction 285
   A. Area of study 285
2. Methods 286
   A. Dive procedure 286
   B. Navigation 286
   C. Scientific instrumentation 288
   D. Viewport calibration 291
3. Results 291
   A. Correction for changes in attitude 291
   B. Effect of changes in attitude on width of visual field . . . 295
   C. Scallop population parameters 296
4. Discussion 297
Acknowledgements 298
References 298

1. Introduction

In 1968, Caddy (1970) attempted to estimate the population density of scallops, *Placopecten magellanicus*, from a 2-man submersible (the Perry "Cubmarine") over a shallow-water scallop bed in the Gulf of St. Lawrence. Notable deficiencies of this early attempt were the absence of accurate information on changes in attitude of "Cubmarine", and instrumentation for measuring them. While working with "Shelf Diver" (Sept.–Oct., 1969) over a deep-water population of scallops in the Bay of Fundy, measurements of a number of variables affecting the field of view of the observer were made simultaneously with the population counts. The effects of these parameters were allowed for in calculating population density of scallops for a sample dive (dive 2).

A. Area of Study

Dives were made off Digby, Nova Scotia, in the Bay of Fundy (Lat. 44°40′, Long. 56°00′), an area notable for its strong currents and great tidal range: (the greatest tidal amplitude in the world (16 m) has been recorded in Minas Basin at the head of the bay). Currents of up to 3.7 km/h were recorded in the study area off Digby Gut by Forrester (1959). Tidal speeds of approximately the same magnitude were met, both at the surface, and close to bottom. Because of strong upwelling in the area (Lauzier, 1967), dense plankton was encountered throughout the water column, restricting visibility to less than 20 ft (6.1 m).

2. Methods
A. Dive Procedure

The relatively low maximum submerged speed of "Shelf diver" (4.6–5.6 km/h) meant that dives were scheduled for the period of slack water, although occasionally work continued into the ebb and flood periods. When operating in strong currents and low visibility, safety considerations dictate that bottom transects should be conducted heading up-current to avoid the possibility of collision. This consideration greatly reduced the mobility of "Shelf Diver" and the distance covered during each dive, as well as increasing power consumption. Because of heavy battery drain, dive duration rarely exceeded an hour.

B. Navigation

The gyro compass of "Shelf Diver" was calibrated from the ship's compass of the CCGS "C. D. Howe" before launching. After launching, the submersible moved approximately 1 km away from the mothership under her own power before diving. A fix on the submersible's position was provided by a buoy towed behind the submersible on a thin line whose length was approximately three times the depth of the water. A support launch equipped with a radar reflector mounted on a pole, stayed in the immediate vicinity of the buoy, and was tracked by radar from the "C. D. Howe". This introduced an error in positioning of not more than ± 350 m in 107 m depth of water. The distance travelled by "Shelf Diver" from its position of first arrival on bottom was measured by an odometer wheel on a flexible tow cable, towed along bottom behind the submersible (Fig. 1). Each revolution covered 1.91 m.

Submersible speed could be calculated by comparing the distance between successive odometer spikes and successive minute time pulses on the chart paper (Fig. 2). Submersible speed rarely exceeded 1 km/h. Because bottom current speed was of the same order as submersible speed, the true course often differed widely from that registered on the gyro compass. Yaw angles of up to 75° were recorded.

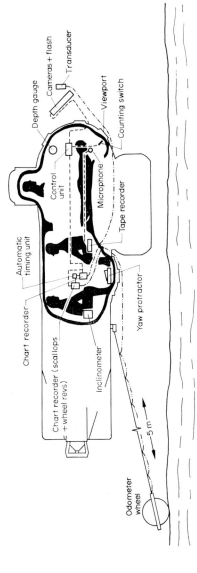

Fig. 1. Instrumentation installed in "Shelf Diver" for benthic surveys and measurements of vehicle performance.

During the 1968 charter, an approximate correction for yaw was provided by a continuously recording current meter moored in the survey area. This was not entirely satisfactory because variations in current speed and direction could have occurred between the positions of the submersible and the current meter.

In 1969, tubmersible heading was given by a gyro compass calibrated against the ship's compass of the "C. D. Howe", and yaw angles were measured directly by means of a protractor mounted in the lower

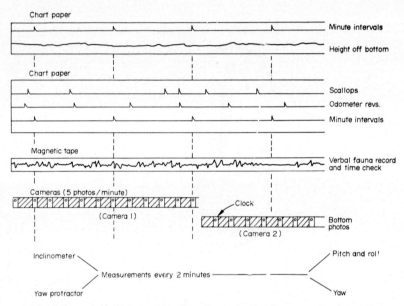

Fig. 2. Schematic representation of method of data recording over a common time base.

viewport. The two methods are illustrated in the vector diagrams of Fig. 3.

From the submersible heading, yaw angle, and distance travelled, a corrected course was reconstructed after the dive. When the distance travelled during dive 2 was plotted both along the compass heading and along the corrected course, the end points of the two courses were separated by 426 m after 1550 m of travel across bottom (Fig. 4).

C. Scientific Instrumentation

"Shelf Diver" (Fig. 1) has two personnel compartments: a forward compartment for pilot and scientific observer at atmospheric pressure, and a rear lock-out chamber, which can be isolated from the forward compartment and pressurized to ambient water pressure. A diver can then leave through a hatch in the floor of this chamber (Schafer, 1969). No lock-out dives were made during this operation; two scientific support personnel and the recording equipment occupied the rear chamber. The hatch between the two compartments was kept open.

Data recorded by the observer in the forward compartment were

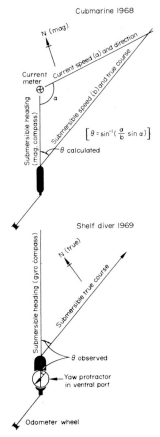

FIG. 3. Methods of calculating course corrections during 1968 "Cubmarine" and 1969 "Shelf Diver" charters.

relayed to recording instrumentation in the rear chamber. Scallops seen through the forward viewpoint were recorded by a manual switch onto an electrically driven chart recorder. Two other channels on the same recorder registered each revolution of the odometer wheel (Caddy, 1970) and a 1 min interval pulse from an automatic timer (Fig. 2). A second paper chart recorder registered submersible height off bottom, as measured by a transducer mounted on the bow framework just above the viewpoint, and 1 min time checks, registered as 5 s interruptions of the (1 s) transducer pulses (Fig. 2). Height off bottom was measured to an accuracy of ±15 cm. Other records such as depth, associated fauna, bottom type, compass heading and time checks were read into a microphone by the observer and recorded on a portable tape recorder. All electrical instrumentation was run off batteries independently of the submersible power supply.

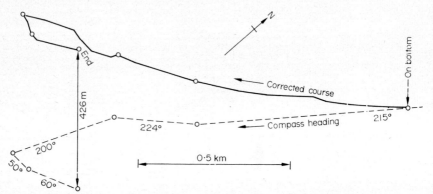

Fig. 4. A comparison of dive tracks before and after correcting the compass heading for yaw (dive 2).

Bottom photographs were taken at 12 s intervals throughout the dive by two EG and G cameras mounted on the bow plane framework. A control box in the forward chamber recorded the number of exposures, and allowed switching from one camera to the other. A total of 1000 exposures could be taken over a period of 3 h.

Support personnel in the rear chamber supervised recording instrumentation, and made measurements of submersible performance at 2 min intervals, recording their observations in a notebook.

A pendulum inclinometer placed on a flat surface in the rear compartment was orientated successively fore and aft, and side to side, to measure angles of pitch and roll, respectively, to an accuracy of ±0.5°.

An estimate of yaw (deviation from compass heading) was provided by a perspex "protractor", mounted in the 18 cm diameter window of the lock-out hatch. Parallel lines etched on the surface of a transparent plate rotating on a circular scale, were lined up with the apparent direction of movement of bottom features across the viewport. The yaw angle was then read off the 360° scale on the lower part of the instrument with an estimated accuracy of ±2°. (This instrument can also be used to measure bottom current direction: vehicle heading is adjusted to give a zero yaw angle, and current direction read from the submersible compass. Bottom current speed could then be estimated from the amperage drain necessary to remain stationary with respect to the bottom on this heading.)

D. Viewport Calibration

Because of the depth of water (105 m), viewport calibration had to be carried out in air, as "Shelf Diver" lay in its cradle with the battery pod 13 cm above the deck of the CCGS "C. D. Howe", and the midpoint of the viewport 102 cm above deck. The observer assumed a prone position in the front chamber of "Shelf Diver" as when observing under water, with his eyes approximately 30 cm. from the centre of the viewport. An assistant then drew a line on the deck of "C. D. Howe" from vertically below the observer viewport, extending forward parallel to the longitudinal axis of "Shelf Diver". A 3 m pole, graduated in 10 cm units, was then laid on deck directly below the viewport, and transversely to the longitudinal axis of the submersible. The pole was moved away from the viewport in 15 cm increments until visible to the observer inside the submersible. The observer then recorded the number of graduations visible on the pole at each position until it passed out of the top of his field of view. A plan of the field of view was then drawn (Fig. 5).

The visual field of the viewport in water is restricted by refraction of light at the water-glass, glass-air interface, causing the so-called "tunnel effect". An approximate correction for the effects of refraction was obtained by adopting a value for u (the refractive index from air to water) of 1.34 and recalculating the dimensions (Fig. 5).

3. Results

A. Correction for Changes in Attitude

From a knowledge of the viewport angle to vertical (45°), its internal diameter (18 cm), and the limits of the area visible from the viewport 102 cm off deck, the cone of vision of the observer porthole was reconstructed (Fig. 5). This gave an indication of the width of visual field under conditions of zero pitch and roll, with the submersible 13 cm above bottom. Several factors influence the width of the strip of bottom scanned by an observer from a moving submersible. The independent effects of height, pitch, roll and yaw on the width of visual field BC, were estimated from Fig. 5 by geometric projection. Variations of less than 5% in the length of BC caused by the above parameters were considered to be within the limits of experimental method and were ignored.

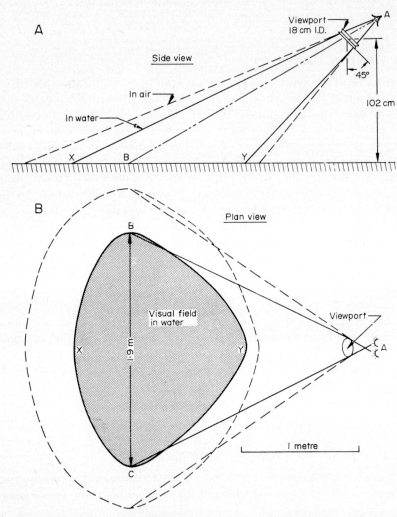

FIG. 5. Calibration of visual field of "Shelf Diver" in air, and calculated visual field in water A. side view; B. plan view.

Height off bottom

Values for BC at different heights off bottom were generated graphically from Fig. 5A by drawing horizontal sections across the visual cone at different vertical distances below the viewport midpoint. The corresponding values for BC could then be obtained for Fig. 5B at the point where the line AB intercepted the bottom at B. The relationship between width of the visual field (BC) and height off bottom was linear, of the form:

$$BC = 1.72 + 1.65 \text{ height (Fig. 6)}$$

A 5% change in BC occurred with a 3.5 cm change in height.

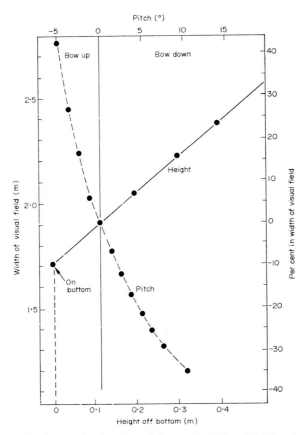

FIG. 6. Calculated corrections for the effect of changes in pitch and height off bottom on the width of sea floor visible from the observer viewport of "Shelf Diver".

Pitch

Pitch was defined as the inclination between the submersible's longitudinal axis and the horizontal, resulting from rotation of the vehicle about a transverse axis through the midpoint of the viewport. The effect of pitch was simulated by rotating the visual cone (Fig. 5) in the vertical plane parallel to the longitudinal axis of the vehicle, while maintaining the midpoint of the viewport at a constant height off bottom. BC was then determined graphically for a series of pitch angles. Pitch apparently exerts a considerable effect on the width of visual field; a 5% change in the length of BC results from 1°15′ of pitch (Fig. 6).

Roll

The effect of roll on BC was simulated by rotating the visual cone about the longitudinal axis of the vehicle. Apparently roll angles of up to 5° have relatively little effect on the width of visual field, although subsequently the width of visual field increases sharply (Fig. 7). A 5°30′ roll angle was necessary to produce a 5% change in the length of BC.

Yaw

If the visual field is not circular, the effective width of visual field scanned while travelling across bottom will differ from BC if the submersible's true course deviates from the compass heading.

Compression of the field of view along the longitudinal axis as demonstrated by deck calibration (Fig. 5) is probably a consequence of binocular vision: the observer was able to see further from side to side than from top to bottom of the field of view.

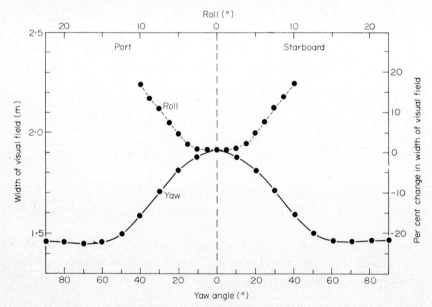

FIG. 7. Calculated corrections for the effect of changes in roll and yaw on the width of sea floor visible from the observer viewport of "Shelf Diver".

An estimate of the effect of yaw on the width of visual field was obtained from Fig. 5. The distance between two lines drawn tangentially to either side of the visual field, and parallel to the true course, was measured for a series of yaw angles (Fig. 7). Yaw exerted a relatively minor effect on the width of field; a yaw angle of 20° is needed to effect a 5% change in width of the visual field.

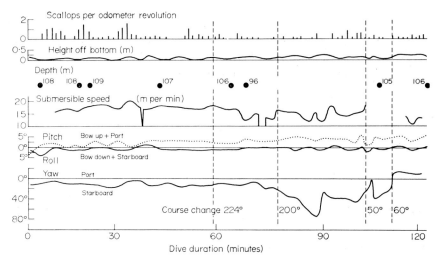

FIG. 8. Scallop counts and measurements of "Shelf Diver" performance parameters during dive 2.

B. Effect of Changes in Attitude on Width of Visual Field

Height off bottom

Submersible height remained fairly constant during dive 2 and other dives over level bottom (Fig. 8). The mean height off bottom of 13 cm corresponded almost exactly to the height at calibration so that no correction for this parameter is required.

Pitch

Pitch angles rarely exceeded 5° over level bottom. During dive 2, a fairly constant bow-up attitude was maintained (mean pitch angle: $-2°48'$) (Fig. 8). A 19% increase in the length of BC could be expected as a result of a $-2°48'$ pitch angle, corresponding to a corrected value for BC of 2.27 m during dive 2.

Roll

Probably because of the battery pod slung under "Shelf Diver" (540 kg weight in water), there is a wide separation between the centres of gravity and buoyancy. As a result, "Shelf Diver" shows little tendency to roll (Fig. 8). A mean roll angle of 0°36′ was calculated for dive 2. This is considerably less than the 5°30′ angle of roll needed to effect a 5% change in width of visual field; so that roll angles noted during dive 2 can be considered to have had negligible influence on BC.

Yaw

Yaw angles varied widely throughout dive 2; the maximum of 75° was reached immediately after the second course change (Fig. 8). A mean yaw angle of 22° was calculated, which from the calibration curve corresponds to a width of visual field of 1.79 m, a 6.3% decrease in BC.

C. Scallop Population Parameters

Area surveyed

The area of bottom surveyed is a product of the distance travelled and the width of visual field at right angles to the direction of travel. "Shelf Diver" travelled 1553 m (813 odometer revolutions) during dive 2.

Population density calculation

In calculating the total area scanned during dive 2, and hence the population density of scallops, corrections to the width of visual field need only be made for pitch and yaw. Simultaneous corrections for both of these variables are beyond the scope of the present study, since changes in pitch and roll alter the shape of the visual field, and hence invalidate the correction for yaw. It was calculated however, that the $-2°48′$ pitch angle adopted during dive 2 only resulted in an increase in the length:breadth ratio of the visual field (XY/BC) from 1.32 to 1.35, so that it is reasonable to apply pitch and yaw corrections simultaneously.

A $-2°48′$ pitch angle corresponds to a 2.27 m width of field at 13 cm off bottom. The effect of yaw is to reduce this width by approximately 6% to 2.14 m. The total area scanned from "Shelf Diver" during dive 2 was therefore approximately 2.14×1553 m^2 = 3323 m^2. Three hundred and ninety-five scallops were seen during the dive which corresponds to a population density of 0.12 scallop/m^2. Population densities during the four other dives made in the area were below this figure. Dickie (1955) calculated from underwater photographs that the population density of scallops in the same area was 0.62 scallop/m^2 in 1950. The present low abundance of scallops off Digby was confirmed by the low catches of the commercial fishery in 1969.

4. Discussion

The successful use of submersibles for quantitative surveys of benthic animals obviously dictates performance characteristics of the vehicle. A submersible must maintain a constant altitude and attitude, in order to ensure that the area scanned remains constant. To eliminate the effect of yaw on the width of visual field, both viewport configuration and observer positioning should be such as to ensure that the width of field remains constant, irrespective of yaw (i.e., a circular field of view).

Evidently the true course is not given by the compass heading because of the low speed of most submersibles with respect to bottom current speeds. *In situ* methods of measuring yaw offer the possibility of navigating without external navigational aids (e.g., sonic beacons). More accurate navigational instrumentation e.g., Doppler Sonar (Goulet, 1969), or an Inertial Navigation System (Lee, 1968)) are expensive and bulky, and have not been installed in most small submersibles.

Simulation studies on the combined effects of pitch, roll, yaw and underwater visibility on population counts will be needed before precise population density estimates can be made from a moving submersible. Some account must also be taken of the effect of bottom contour and visual acuity on the area scanned from a viewport: considerations which were beyond the scope of the present study.

Facilities should be available on submersibles used for scientific purposes for automatically recording vehicle performance (pitch, roll, height off bottom, depth, compass heading, etc.) on magnetic tape with a common time base, so that the observer is entirely free to concentrate on his scientific mission (Saila and Flowers, 1968). Subsequent analysis of this record to determine the area scanned and hence population densities can then be carried out by computer.

Over undulating bottom, the odometer introduces major inaccuracies in measuring straight-line distances across bottom, but has some advantage for survey of benthic species, since bottom-dwelling organisms are usually distributed in relation to the area of sea floor available for colonization, rather than to a vertical projection downwards from a known area of the air-water interface. For this reason, an odometer wheel presents certain advantages over the more sophisticated methods of navigation mentioned earlier. Because the area of bottom scanned is significantly altered by changes in height off bottom smaller than can be adequately measured by echo sounder, it may prove more practical in future to employ wheeled vehicles for benthic survey purposes (e.g., the Cammell Laird Sea Bed Vehicle, Hydrospace, June, 1969). This would substantially reduce the problem of correcting for changes in vehicle attitude and result in a more accurate measure of population density.

Acknowledgements

The author wishes to express his thanks to the many people who assisted during the Shelf Diver charter; in particular, S. M. Polar who was responsible for instrumentation design, Jim Dudley and the support crew of "Shelf Diver", and Captain Ouillet and the crew of CCGS "C. D. Howe", also R. A. Chandler, S. C. Rand and P. W. G. McMullon who acted as support personnel and D. J. Scarratt who made valuable suggestions about the manuscript.

References

Anonymous (1969). British underwater vehicle under active construction. *Hydrospace* **2**(2), 18–19.

Caddy, J. F. (1968). Underwater observations on scallop (*Placopecten magellanicus*) behaviour and drag efficiency. *J. Fish. Res. Bd. Canada* **25**(10), 2123–2141.

Caddy, J. F. (1970). A method of surveying scallop populations from a submersible. *J. Fish. Res. Bd. Canada* **27**(3), 535–549.

Dickie, L. M. (1955). Fluctuations in abundance of the giant scallop, *Placopecten magellanicus* (Gmelin), in the Digby area of the Bay of Fundy. *J. Fish. Res. Bd. Canada* **12**(6), 797–857.

Forrester, W. D. (1959). Current measurements in Passamaquoddy Bay and the Bay of Fundy 1957 and 1958. Passamaquoddy Fisheries Investigations. A report of the International Joint Commission. *App. I Oceanography*.

Goulet, T. A. (1969). The use of pulsed doppler sonar for navigation of manned deep submergence vehicles. Proceedings of the Institute of Navigation, National Marine Navigation Meeting. Second symposium on manned deep submergence vehicles, San Diego, California, Nov. 1969, pp. 60–67.

Lauzier, L. M. (1967). Bottom residual drift on the continental shelf area of the Canadian Atlantic coast. *J. Fish. Res. Bd. Canada* **24**(9), 1845–1859.

Lee, T. R. (1968). Submarine navigation. *The Journal of the Institute of Navigation* **2**(4), 480–489.

Saila, S. B. and Flowers, J. M. (1968). Some aspects of the visual detection of targets from research submarines. Transactions of the National Symposium on Ocean Sciences and Engineering of the Atlantic Shelf (Philadelphia), 257–263.

Schafer, C. (1969). Lock-out diving in Gulf of St. Lawrence for geological samples, *Sea Harvest and Ocean Science* December, 26–28.

A Stereophotographic Method for Quantitative Studies on Rocky-bottom Biocoenoses[*]

TOMAS L. LUNDÄLV

Kristineberg Zoological Station, S-450 34 Fiskebäckskil, Sweden

Conventional quantitative rocky-bottom sampling techniques generally involve manual cleaning of test areas by a diver. The procedure is time-consuming, and only a very limited bottom area may be investigated in a diving day. Furthermore, many organisms are difficult to detach from rocks, or are easily damaged or lost during the sampling. The heterogenity of rocky bottoms creates intricate sampling problems. Rocky surfaces in the same locality and depth often accommodate different biocoenoses, according to differences in slope, lighting, turbulence and sedimentation. Therefore, small test squares only provide a fragmentary picture of the biocoenoses composition in a certain locality. Furthermore, they constitute a poor basis for comparative investigations on dynamic processes within the biocoenoses. The effort needed to make repeated adequate random samplings for comparative purposes is usually too great.

To overcome some of the problems, a new technique for quantitative studies on sublittoral rocky-bottom biocoenoses has been developed at Kristineberg Zoological Station with support from the Swedish Environment Protection Board and the Swedish Natural Science Research Council. The basic principle is time-lapse stereophotographic recordings of permanently-marked test areas, followed by photogrammctrical analysis of the photographs.

Test areas are marked on steep rocky bottoms to depths of 25 m by means of underwater rock-drilling with a light-weight, self-contained pneumatic drilling-machine (Atlas Copco Wasp). Each test area is marked with two holes, 3 m horizontally apart. The holes are plugged with plastic dowels. At a test area to be photographed, a rigid rod is

---

[*] This is a short account of two papers which are published elsewhere: 1. Lundälv, T. (1971). Quantitative studies on rocky-bottom biocoenoses by underwater photogrammetry. A methodological study. *Thalassia Jugoslavica* **7**, 205—213. 2. Torlegård, A. K. I. and Lundälv, T. (1974). Under-water analytical system. *Photogrammetric Engineering*, 287–293.

fastened between the plastic dowels, and a photography frame is hung on the rod (Fig. 1). The frame is fitted with a camera support, which permits the camera to be consecutively placed at two defined positions, separated by a horizontal distance of 20 cm (the stereo base) approximately 60 cm from the rock face. In this way a stereoscopic pair of

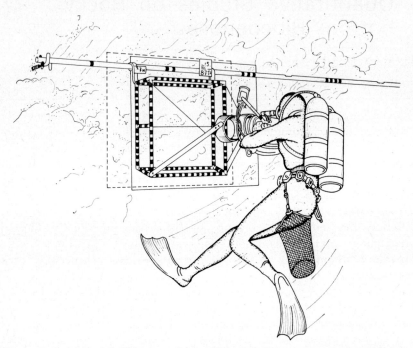

Fig. 1. A diver at work with stereophotographic recording of permanently-marked test areas. The areas covered by the photographs are indicated in the figure. From Lundälv, 1971.

pictures can be obtained by the use of a single camera. The photography frame, constructed as a parallel-epiped measuring 0.50 m × 0.50 m × 0.10 m, is used for calibration purposes at the picture analysis. The frame can be moved along the rod, so that six squares are photographed at each test area. An example of the test square photographs is shown in Fig. 2.

The camera is a Hasselblad SWC fitted with a corrective front port and an electronic flash unit. Positive 70 mm colour film is used.

The stereophotographs are analyzed in a Wild microstereo-comparator, MSTK, with respect to species composition, density, cover, size distribution, age distribution, spatial pattern and biomass (determined indirectly from correlations size-biomass). Population dynamics, growth rates and productivity can be studied by the comparison of photographs from the same test area obtained at intervals. The calculations needed for photogrammetrical determinations of size, area, volume, etc., are performed in a medium-sized programmable electronic calculator. All parts of the photogrammetrical analysis are simple enough to be made by the biologist in his own laboratory.

The accuracy of the photogrammetrical size-determinations was found by planting objects of known size on test squares. The test showed

FIG. 2. A stereoscopic pair of pictures from a test square at 20 m depth in the Gullmar Fjord on the west coast of Sweden. The pictures can be viewed stereoscopically with parallel eye-axes at a distance of approximately 30 cm. The test square is dominated by anthozoans, sedentary polychaets and ascidians. Since the pictures are black and white reproductions from the original colour transparencies, much detail is unfortunately lost.

that the errors were small and uniform enough to be neglected for most practical purposes.

The efficiency of density determinations from photograph analysis was tested by the comparison of results obtained from photograph analysis and careful manual cleaning of identical test squares. The test showed that the efficiency of photograph analysis was comparable to or superior to manual sampling for many species of the sessile macrofauna. Photograph analysis, however, is selective, and certain components of the biocoenoses can not be studied quantitatively. This is true for certain mobile animals like errant polychaets, nemerteans, amphipods and isopods. It is also true for very small specimens from all taxonomic groups (the lower limit being different for different species) and for species which can not be identified with certainty from photographs, e.g. certain hydrozoans, bryozoans and algae.

The efficiency of the method is also dependent on factors such as visibility, configuration of the rock surface and composition and nature of the biocoenoses.

Some of the principal advantages of the method are: (1) That quantitative data can easily be obtained from large rocky-bottom areas within a short time. (2) Practically simultaneous observations can be made along ecological gradients. (3) Dynamic processes within the biocoenoses, and their dependence on environmental factors can be studied, since test areas and individual specimens can be observed over long periods of time without interfering with the organisms. (4) In many respects, the analysis of photographs yields more information than manual sampling techniques, e.g. in problems concerning cover, spatial patterns and interspecific relationships.

The technique should be a useful tool for pollution impact studies.

Some Underwater Techniques for Estimating Echinoderm Populations

J. K. G. DART

and

P. S. RAINBOW*

Cambridge Coral Starfish Research Group, c/o Eastern Telegraph Company, P.O. Box 99, Port Sudan, Sudan

1. Introduction 303
2. Techniques 304
    A. Quadrats 304
    B. Transects 304
    C. Colony samplng 305
    D. Open search 305
    E. Patch counts 306
    F. Towing 307
    G. Marking and recapture 308
3. Discussion 309
    A. Errors in estimating echinoderm populations 309
    B. General considerations in the use of the techniques . . . 311
References 311

1. Introduction

The techniques described are those used in the work of the Cambridge Coral Starfish Research Group on reefs in the Red Sea near Port Sudan. The main object of the group has been to study the ecology, behaviour and population dynamics of the Crown of Thorns starfish *Acanthaster planci*. This asteroid is a significant predator of corals and its destructive capacity has been well documented. In the course of these studies the ecology of other echinoderms (Dart, 1972) in the Port Sudan region was also investigated and in consequence a number of techniques have been employed to estimate population levels and distributions of different echinoderms.

In selecting a technique for estimating the population of a particular species its size, abundance and behaviour must be considered as these

*Present address: Department of Zoology and Comparative Physiology, Queen Mary College, University of London.

parameters affect the ease with which that species may be found underwater. Also no single technique is equally efficient for estimating population densities of different species because of differences in these parameters. The diving procedure has also to be simplified as much as possible to increase the speed of the search.

Initially random searches of an area may indicate approximate distribution limits with respect to ecologically defined reef zones, or the most common habitat of a particular species. This allows subsequent searches to be concentrated on the main areas of distribution. Additionally the buoying of underwater reference points recognizable on a map enables the geographical localization of results and facilitates follow up studies.

2. Techniques

A. Quadrats

The use of quadrats is a standard ecological technique and was found useful when investigating inconspicuous species requiring careful searching. In each case a square of one metre was laid down and the area within it searched meticulously. Scaling up provides a reliable estimation of numbers over a small area. This technique was mainly used to estimate numbers of brittle stars living deep in the coral which must be broken up by the investigator. Similarly the echinoid *Eucidaris metularia* occupies dead coral rubble and the same method of estimation was employed. Limitations of this method lie in the fact that it is extremely laborious and time consuming, and large numbers of quadrats must be laid down to obtain population density estimates for even small areas of reef. Additionally varying relief of terrain leads to considerable differences in the areas covered by quadrats but the method is accurate over small areas.

B. Transects

Transects are found useful when an animal can be observed without disturbing the substratum. In the same manner as quadrats, results may be scaled-up to provide approximate population density values for larger areas, transects covering a greater area than quadrats. A line weighted at each end is laid over the reef so that it rests on the coral. A pole (usually metal) is then taken down one side of the setline such that one end of the pole remains on the line and the pole is at right angles to the line. The pole is moved slowly along the line by the worker who carefully searches the area covered. The same procedure is then carried out as the pole is brought back along the other side of the line. For our purposes a 12.5 m line was laid down and a 2 m pole used, giving a total coverage of 50 square metres. This method may be used in fairly shallow regions by a snorkel diver since the pole may be left in position as the worker surfaces for air.

Transects were employed particularly to estimate populations of the sea urchins *Echinometra mathaei* and the boring *Echinostrephus molare*. These echinoids often inhabit similar reef areas and the same transect usually provided data for both species. 50 square metre transect areas were chosen in view of the relatively small size of reefs being investigated but the line may be extended and the area so increased if the environment is more uniform, and Forster (1959) gives an example of this use of transects. Transects do give reliable estimates especially when populations are evenly distributed in small areas but as with quadrats, this method may not be used for species which are colonial and spread out such that colonies may be missed.

Ladder transects may be useful on steeply sloping reef faces where they can be used in conjunction with depth measurements to provide reef profiles and comparisons of species population densities.

C. Colony Sampling

Colony sampling is really the arithmetical extension of transects and quadrats. Colonies may be sampled when their limits are clearly defined and such samples are taken within a colony either by quadrats or by a transect. The density within each colony is therefore known and the total population may be estimated. The method also depends on the populations being reasonably widespread and has been used for *Echinometra mathaei* and *Echinostrephus molare* mentioned above, and also for *Tripneustes gratilla*, another echinoid, found in eel grass (*Thalassia*) beds in coastal lagoons.

D. Open Search

Open searching was used for conspicuous species which could be discovered by cursory observation and which live in widely scattered colonies easily missed by transect techniques. This simple method entails a diver swimming along a natural line such as the reef crest which has been divided into known lengths by buoys. The width of the search is estimated and the numbers of echinoderms seen may be referred to a scale map of the area, also showing the position of the buoys so that regional variations in density may be plotted. This method was used principally for estimating populations of the larger echinoids such as *Heterocentrotus mammilatus* and *Diadema setosum*, the spines of these species being visible as they protrude from crevices used as hiding places during the day. The same technique of open searching was also used in the initial stages of our work to estimate *Acanthaster planci* populations but other techniques were later devised for this starfish.

Advantages of this method lie in the fact that it is relatively quick and a whole reef may be surveyed, rather than selected parts as in the case of transects. Searching may be executed from the surface or with an aqualung but the main limitation of the method is the low level of accuracy in that a significant proportion of the population is missed.

E. Patch Counts

The use of patch counts is a specialized technique developed for *Acanthaster planci* and probably not of any use for the majority of other echinoderms. *A. planci* predates coral polyps and after feeding leaves an

Fig. 1. Underwater photograph showing white area grazed by *Acanthaster*. One starfish can be seen partially hidden by the coral.

approximately circular white patch of dead coral about 15 cm in diameter, as shown in Fig. 1. These patches are easily recognized by a trained observer and a group of patches indicates the presence of a starfish. Each patch is distinguishable for about ten days before becoming obscured by algal growth and the number of patches can be related to the number of starfish. A reef is divided into areas by buoying and the number of patches counted in each area. A very careful search is then conducted of a marked area, preferably at night when the starfish emerge from hiding and can be counted. This number is compared to the number of patches present in that area and the ratio of the number of patches to the number of starfish is thereby known.

An advantage of this technique is that once the ratio is known (approximately one starfish per six patches), it is not necessary to find individual starfish which are usually well concealed in the coral crevices. Patches could be seen from about 10 m underwater and this method allows large areas of reef to be covered extremely rapidly. The accuracy of this technique is increased if the same well trained observer is used for searches of areas where comparisons of population densities are to be made. The starfish:feeding patch ratio should be rechecked at intervals to pick up any change in feeding behaviour which might supply erroneous results. Further advantages of this method are that it allows simple detection and following of starfish movement over the season and we have found that it provides an accurate measure of relative densities of starfish populations.

F. Towing

Towing of a diver behind a boat is the speediest method of estimation used in our studies and is only useful when an animal can be detected easily. Thus it may be used as a rapid method of patch counting. An

Fig. 2. Observer towed with SCUBA and MANTA board.

observer with SCUBA and planing board (MANTA board) is towed as illustrated in Fig. 2. Since he is being towed the diver is unable to record observations on a formica board as used in other methods and so results are relayed by a one-way intercom system to an associate recording information in the boat. The diver uses a "Normalair" mask with microphone, and a 25 m rope and a planing board 1 m × ¾ m were found ideal. 60 cubic foot capacity cylinders giving one hour endurance at 10 m depth were used and a 200 m diameter patch reef could be surveyed in about 30 min as opposed to 6 h using open search. This survey is extremely rapid and although it does not provide a high degree of accuracy it does give a rough estimation of fairly sparsely distributed species and pinpoints higher concentrations for later investigation.

G. Marking and Recapture

A marking and recapture method was used to estimate the levels of the large populations of *Acanthaster planci* found on two of the patch reefs, see Table 1. The marking process assigned individual codes to the

TABLE 1

Size groups Diam.: cm	19–23	24–28	29–32	33–37	38–42	43+
Total No. recovered	12	56	70	109	45	9
No. recovered that were marked	1	18	16	30	11	1
% marked in recovered sample, i.e. recovery (%)	8.5	32	23	27.5	24.5	11
No. originally marked	16	43	82	92	37	1
Total population	188	138	358	334	151	9

TOTAL POPULATION = 1178
(above 19 cm diameter)

starfish and since the reef was well buoyed the marking procedure also allowed us to follow individual movement patterns.

Our group had previously investigated various methods of marking starfish in an attempt to find that most suitable for use on *Acanthaster planci*, the main subject of our study. Various tags (including coloured P.V.C.) had been attached either to the epidermis or to the spines of starfish, as also were cable clips and coloured threads. None of these methods however were found suitable for *Acanthaster* for the starfish is able to reject offending objects from its body wall which, as in other spinulosans, contains relatively few ossicles (Hyman, 1955) and the spines are also able to rub against each other to remove any tag, the most permanent lasting only a week. Any plastic threads connected to the animal were passed through the epidermis and were therefore of no use in a long term marking programme. It was finally necessary to make use of the fact that the starfish has many arms and is covered in spines, and spines were trimmed off certain arms producing unique codes. Our evidence suggested that the shock of this traumatic procedure affects the behaviour of the animal for up to a week but that no increase in mortality occurs. Furthermore the code is still recognizable for up to nine months before the spines have regrown.

On one reef 271 starfish were originally marked and 77 of these were recaptured giving a recovery percentage of about 28%. The total population was calculated by means of the Lincoln Index (Jackson, 1939; Leslie and Chitty, 1951; Bailey, 1952) and as can be seen in Table 1 an estimate of nearly 1200 starfish was obtained.

The main difficulty of using this method was to find a really suitable method of marking and the final method used is still too physically drastic to be completely satisfactory. As can be seen from the table the recovery percentages of the different size groups varied, being considerably smaller for the lowest size range. Estimates therefore tend to be carried out for each size group as opposed to one for the whole population, and a total figure can only be applied to the whole population if a lower size is stipulated (19 cm in this case). This point probably applies to the majority of techniques involving the finding of animals but it must be borne in mind in the consideration of final total populations.

3. Discussion
A. Errors in Estimating Echinoderm Populations

It must be recognized that a proportion of animals is overlooked when using any of these techniques for estimating population levels. In comparing distributions of one animal in different areas, the error in estimation of absolute population levels is of considerably less importance than in strictly quantitative work, but it is certainly necessary to estimate differences in errors incurred, especially if different techniques are used to obtain the values being compared.

The three most important variables producing errors are the use of different techniques as stated, the character of the terrain and the behaviour of the animals. For example one survey of the population of

Heterocentrotus mammilatus showed that if these animals were living in a certain coral (*Acropora*) community, they could not be counted using normal searching techniques because they were so well concealed, whereas such techniques were adequate when investigating a population of the same echinoid if situated in a different coral (*Porites*) community. Another echinoid *Echinothrix diadema* could not be found whilst concealed during the day and a night search was needed before an estimation could be made. However by careful selection of the correct technique initially and by cross checking different techniques, the errors involved can be estimated.

Cross checking by use of different techniques is the main method used in estimating any error. Transect and quadrat techniques are

TABLE 2. Cross checking of methods by night searching.

Species	Open Search of Buoyed Areas	
	Day	Night
Diadema setosum	17	18
	11	26
	1	11
	1	1
	1	0
	0	3
Heterocentrotus mammilatus	14	29
	10	16
	4	6
	1	2
	0	3
	0	1
	Transects (50 m²)	
	Day	Night
Diadema setosum	9	10
	3	2
Heterocentrotus mammilatus	12	17
	0	2

accurate when carried out carefully and therefore make a good check. The open search technique can therefore be cross checked by one of these two methods, a small area being first covered by an open search and then by a transect. Comparison of the results gives a value for the percentage error inherent in an open search technique and this can then be used to adjust other estimates obtained by open searching. The patch count technique can similarly be checked by marking and recapture. Errors may change with terrain and so when terrain is variable the habitats should be divided into groups and checks carried out in each group. Improvements of estimates may thus be carried out by comparing non-absolute methods, one method being used to cross check another.

Lastly a change in behaviour such as might occur between day and night, may be used to estimate errors. Many echinoderms are nocturnal, hiding during the day, and so night searches will reveal the population more completely, see Table 2. If marked areas are searched at night and during the day the percentage error of daywork may then be calculated. Once errors have been calculated by whatever method chosen, correction factors may be introduced into the results where necessary.

B. General Considerations in the Use of the Techniques

The use of the techniques described here, except for quadrats and transects where the work is carried out in a small well-marked area, is dependent on good visibility and so not all techniques may be of use in temperate waters. Careful choice of a technique suited to a particular animal and terrain is most important. Once results have been obtained they must be tested to ensure that the number of samples has been sufficient and that differences are significant. A discussion of this lies outside the scope of this paper but is integral to the uses of the techniques, and such statistical methods are described in most standard works on ecology such as Southwood (1966).

References

Bailey, N. T. J. (1952). Improvements in the interpretation of recapture data. *J. Anim. Ecol.* **21**, 120–7.
Dart, J. K. G. (1972). Echinoids, algal lawn and coral recolonisation. *Nature (Lond.)* **239**, 50–51.
Forster, G. R. (1959). The ecology of *Echinus esculentis* L. Quantitative distribution and rate of feeding. *J. Mar. Biol. Ass. U.K.* **38**, 361–367.
Hyman, L. H. (1955). Echinodermata. The Invertebrates, Vol. IV. McGraw Hill, New York.
Jackson, C. H. N. (1939). The analysis of an animal population. *J. Anim. Ecol.* **8**, 238–46.
Leslie, P. H. and Chitty, D. (1951). The estimation of population parameters from data obtained by means of the capture-recapture method I, *Biometrika* **38**, 269–92.
Southwood, T. R. E. (1966). Ecological Methods. Methuen, London.

Nocturnal Behaviour in Aggregations of *Acanthaster planci* in the Sudanese Red Sea

R. G. CRUMP

Orielton Field Centre, Pembroke, Wales

1. Introduction	313
2. Methods	314
3. Results	314
4. Discussion	317
Acknowledgements	318
References	318

1. Introduction

One of the major aims of the 1970 Cambridge Coral Starfish Research Group's Expedition to the Sudanese Red Sea was to establish the diurnal pattern of behaviour of *Acanthaster planci* (Linnaeus). Goreau (1964) found that for solitary *A. planci* in the southern Red Sea "feeding took place mostly at night, but sometimes continued to 0900 hours. By day the *Acanthaster* retreated into a dark crevice where it stayed immobile for between twelve and seventy-two hours before coming out to feed again". More recent observations off Port Sudan (Campbell and Ormond, 1970) confirm that under "normal conditions" of low density, *Acanthaster* is nocturnal, although occasional individuals may be found feeding during the day. By contrast, both Chesher (1969) and Endean (1969) have suggested that under infestation conditions in the Pacific, *A. planci* feeds both by day and by night.

The 1969 Red Sea Expedition found very few *Acanthaster* in the Port Sudan area (Campbell and Ormond, 1970) but a more extensive survey in July 1970 revealed a large aggregation (approx. 500 sea stars) on a patch reef at the northern end of Towartit. Shortly after the discovery of this aggregation of *Acanthaster* on Red Beacon Reef it was decided to examine the pattern of feeding activity to determine whether the normal pattern of nocturnal behaviour was disrupted in aggregated individuals. Initial observations from 3rd–8th August suggested that the majority of sea stars were clumped together inactive under *Acropora* tables by day but moved onto these tables to feed at night. A continuous dusk watch from 16.00–21.15 hours on 13th August showed that four out of six *Acanthaster* under a single table moved up the stalk and onto the table shortly before and after dusk. The sea stars did not become inactive and start feeding, however, until 21.00 hours. In order to more precisely determine the pattern of activity in the absence of CCTV or underwater TV cameras it was decided to mount a continuous 24 hour watch over a single table using all the expedition personnel.

2. Methods

On the 18th of August, ten *Acanthaster* were discovered clustered around the stalk of a partially predated *Acropora* table in 35 ft (11 m) of water, which was labelled S_2 for use on the following day. At 12.00 hours on the 19th of August two divers mapped table S_2 and its environs, counted thirteen *Acanthaster* around the stalk, and tagged ten of these individually with colour coded plastic tags slipped over the spines. The initial position of each starfish was recorded on a simplified map of the table drawn in pencil on a formica board, which was subsequently handed on to the next pair of divers.

Using compass and tape measure each subsequent pair of divers mapped the position, direction and duration of movement of the sea stars on or under the table during their watch. A total complement of 14 divers worked five 50 min shifts in pairs from noon 19th August to noon 20th August. Information from the formica boards was transposed to a log book together with dive details to calculate decompression times.

Observation during darkness was achieved by using torches covered with red filters and this red light appeared to have no obvious effect on the behaviour of *Acanthaster* as has been found for other echinoderms (Yoshida and Millott, 1960; Crump, 1965). An inflatable boat was moored over the site as a diving platform and a similar craft was employed for the transfer of personnel to and from a larger vessel, the "Hafiz". The "Hafiz", moored some distance away, was used for rest, recuperation and charging of bottles. This rather complex operation ran surprisingly smoothly but made extensive demands on the stamina and diving skill of all concerned.

3. Results

A clearly defined pattern of nocturnal activity was shown by all of the sea stars observed and the results are summarized in Fig. 1. After a short period of initial disturbance, probably caused by tagging, all

13 *A. planci* remained inactive and clustered around the stem of the table in heavy shade during the afternoon. Shortly before 16.00 hours the first individuals began to move up the stalk onto the underside of the table. These *Acanthaster* were joined by others at intervals so that at 19.00 hours six animals were on the table and seven remained on the stalk. It is of interest that only one *Acanthaster* moved onto the top of the table before dusk (19.30 hours) whereas shortly after dusk all thirteen of the original sea stars were moving actively on both sides of the table. A fourteenth individual moved onto the table from a coral head several

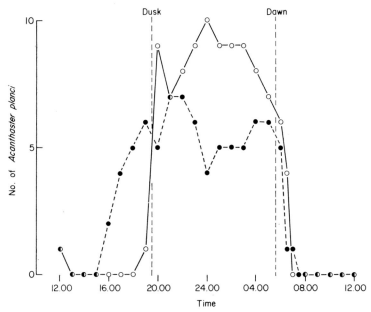

FIG. 1. Graph showing the distribution of *Acanthaster planci* at different times during the nocturnal behaviour study. Animals on underside of table indicated by closed circles and dashed lines and those on top by open circles and solid lines.

metres away at 20.00 hours and an intense period of activity ensued with a seething mass of *Acanthaster* jostling for position on the living part of the table. At 23.30 hours 15 *A. planci* were present lined up at the edge of the living coral on both sides of the table. From midnight until 05.00 hours most of the *Acanthaster* remained stationary and were seen to be feeding on the table with the exception of one individual which had moved off the table and started feeding on an adjacent coral head. Shortly before and after dawn (05.30 hours) a further period of intense activity was noted as the majority of the sea stars moved over the table top onto the underside and thence down the stalk where they again settled into hiding. By 07.30 hours all individuals were at least partially hidden beneath the table (around the stalk or in adjacent crevices) and no further activity was noted before observations terminated at 12.00 hours.

Apart from the position of the *Acanthaster* in relation to the table as a group, it was possible to map the exact movements of 7 individuals (3 tags were lost during the night) throughout the 24-hour period. The data for time, distance and direction of movement for one characteristic individual is shown in Fig. 2. It will be seen that the animal did not start

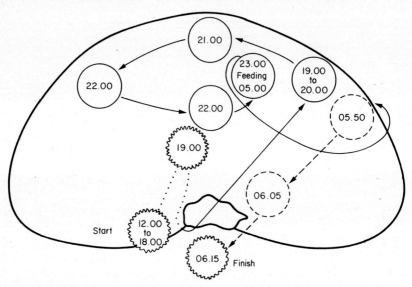

Fig. 2. Recorded movement of a typical starfish during one 24-hour period. Solid circles and lines indicate positions and movements on top of *Acropora* table, the broken ones those on the underside.

moving until just before dusk and moved up the stalk directly onto the top of the table. Feeding occurred on the top of the table but just after dawn the sea star moved onto the underside of the table and down into hiding.

Subsequent visits to table S_2 defined the history of predation upon it. Thus when first discovered sheltering 10 *Acanthaster* on 18th August approximately two thirds of the table had been killed in the characteristic pattern from the stalk outwards leaving a third unpredated at the edge. At the end of observations (12.00 hours) on 20th August the number of *A. planci* had risen to 14 and only a thin strip of living coral remained. By noon on the following day the number of sea stars present in the area had dropped to eight. Forty-eight hours later, on the 23rd of August only 3 *A. planci* were found within 10 m of the table and this had been completely denuded of living coral. It would appear that the period of observation had coincided with the zenith of predation of the table and in the next few days the *Acanthaster* moved on to adjacent unpredated *Acropora* tables where the food supply was more plentiful.

4. Discussion

It is clear from the above results that the basic pattern of nocturnal activity in *A. planci* was not materially altered by the existence of a high density aggregation in a specialized feeding habitat. None of the 15

TABLE 1. Activity phases in nocturnal behaviour pattern of *Acanthaster planci* feeding on an *Acropora* table.

Time	Activity
12.00 – 16.00	NO ACTIVITY (HIDING)
16.00 – 19.00	SOME ACTIVITY — Movement some individuals onto underside of table
19.30	DUSK
19.00 – 24.00	INTENSE ACTIVITY — Movement of majority of *A. planci* onto table followed by "food-searching" on both sides of table
24.00 – 05.00	NO ACTIVITY (FEEDING)
05.30	DAWN
05.00 – 08.00	INTENSE ACTIVITY — Movement off table down stalk into hiding
08.00 – 12.00	NO ACTIVITY (HIDING)

individuals observed were found to feed during the daylight period. It is suggested that the pattern of behaviour observed may be summarized into a distinct set of activity phases as shown in Table 1.

The evening emergence from hiding in *Acanthaster* was closely related to the dusk period and it seems likely that the phenomenon is linked to reduction in light intensity. It is particularly significant that while some individuals start moving 2–3 h before dusk this movement was confined to the underside of the table where conditions of deep shade exist.

Movement onto the top of the table took place immediately after dusk. It is also significant that most individuals moved off the table in the period immediately after dawn and those remaining at 06.30 had moved onto the underside of the table. Other observations show that *Acanthaster* may be found hanging upside down underneath *Acropora* tables at all times of the day. All the available evidence confirms the suggestion that *A. planci* is negatively phototactic and moves towards dark objects in strong sunlight.

The period of intense activity between dusk and midnight, however, may be peculiar to large aggregations of *A. planci* feeding in a confined space. There was considerable competition for feeding space between individuals and the resultant melee may have prevented starfish from settling down to feed before midnight. By contrast on the dusk watch 4 individuals out of a total of 6 were feeding on a partially predated table S_2 at 21.00 hours, and solitary individuals may be found feeding throughout the hours of darkness (Ormond; personal communication).

It has already been suggested that each *Acropora* table has a definite history of predation with a gradual build-up in numbers of *A. planci* over a period of a few days followed by a decline in numbers as the food supply is exhausted. The disturbance factor mentioned above could account for the fact that some tables are not completely predated with patches of living coral remaining after the starfish have passed on to fresh tables where competition for food is less intense.

Acknowledgements

I am extremely grateful to Dr. C. H. Roads for giving me the opportunity to join the 1970 Cambridge Coral Starfish Research Group's Expedition to the Red Sea and for providing such excellent diving and other facilities. The work of the expedition was only made possible by the generous financial support given by the Overseas Development Ministry and other sponsors listed in the Expedition Report.

References

Campbell, A. C. and Ormond, R. F. G. (1970). The threat of the "Crown-of-Thorns" Starfish (*Acanthaster planci*) to Coral Reefs in the Indo-Pacific Area. *Biological Conservation* **2**, No. 4, 246–251.

Chesher, R. (1969). Destruction of Pacific corals by the Sea Star *Acanthaster planci*. *Science, N.Y.* **165**, p. 280.

Crump, R. G. (1965). The diurnal activity of Holothurians. Symposium Underwater Assn., U.K., 1965, pp. 43–45.

Endean, R. (1969). Report on investigations made into aspects of the current *Acanthaster planci* infestations of certain reefs of the Great Barrier Reef. *Fisheries Branch, Queensland Dept. Primary Industries*, Brisbane, 37 pp. (Mimeographed).

Goreau, T. F. (1964). On the predation of coral by the spiny starfish *A. planci* L. in the southern Red Sea, *Bull. Sea Fisheries Res. Stn. Israel* **35**, 23–26.

Yoshida, M. and Millott, N. (1960). Shadow reaction of *Diadema antillarum* Phillipi, *J. Exp. Biol.* **37**, 390–397.

The Ecology of *Caryophyllia smithi* Strokes and Broderip on South-western Coasts of the British Isles

K. HISCOCK

Department of Zoology, Westfield College, University of London*

and

R. M. HOWLETT

Paediatric Research Unit, Guys Hospital, London†

1. Introduction	319
2. Methods	321
3. Sites	321
4. Results	324
A. Local distribution, density and morphology of corals	324
B. Reproduction, growth and mortality	327
C. Fauna and flora associated with *Caryophyllia smithi*	328
D. Food and feeding	332
5. Summary	333
Acknowledgements	334
References	334

1. Introduction

The Devonshire cup coral *Caryophyllia smithi* is a solitary scleractinian which grows to a height of about 12 mm. The polyps are variously coloured, common types having brown, white or green discs. *Caryophyllia* was once commonly found at low water in the south-west; P. H. Gosse, writing in 1853, described the rocks around Ilfracombe as ". . . studded at low water mark with this madrepore". Gosse is responsible for much of our present knowledge of the natural history of this animal but the collecting activities of the many Victorian naturalists who followed him must be considered a major contributory factor to the rarity with which *Caryophyllia* is now found on the shore. In areas

---

* Presently at the Department of Marine Biology, Marine Science Laboratories, Menai Bridge, Anglesey.

† Presently at the Department of Haematology, Royal Perth Hospital, Perth, Western Australia.

of the sublittoral however, *C. smithi* is an abundant and important part of the epilithic fauna.

Rees (1962, 1966) has described the distribution of the coral and it has been possible to add only a few records to his work (Fig. 1). However, it should be noted that only corals referred to *Caryophyllia smithi* have been included, although recent evidence indicates that corals referred to other species may be synonymous with it. These include the widespread *Caryophyllia clavus* Scacchi (Best, 1968 and others), *Paracyathus*

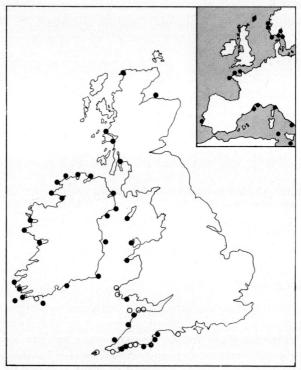

Fig. 1. Recorded distribution of *Caryophyllia smithi* showing the sites at which investigations have been carried out (open circles).

inornatus Duncan, *Coenocyathus dohrni* Doderlein (Rossi, 1957) and the North Pacific *Caryophyllia alaskensis* Vaughan (Rossi, 1960). In the past there has been considerable controversy over the possibility of *C. smithi* and *C. clavus* being mere growth forms of each other and recent evidence (Best, 1968) indicates that the two are varieties of *C. smithi*. Zibrowius (1970) reviews this problem.

ECOLOGY OF *Caryophyllia smithi*

The evidence suggests that *Caryophyllia smithi*, in its various forms, is more widespread than previously thought, although Ekman (1953) has already suggested that it occurs in "the arctic, boreal and antarctic seas and possibly is also found in the intermediate abyssal regions".

2. Methods

Except for a few observations on the shore in North Devon, all of the field work has been carried out using diving methods. During the course of the study over one thousand corals have been collected in order to analyze populations at various sites. The method employed at each site to obtain a representative sample entailed clearing an area of all corals to yield a sample of about one hundred specimens. Records were kept of (a) the size of each coral (taken as the area of the calice); (b) the presence and number of the barnacle *Pyrgoma anglicum* growing on the corallum; (d) the presence of boring organisms within the skeleton. For statistical purposes corals were separated into size groups 40 mm^2 apart. In some cases corals were measured underwater using plastic calipers.

Densities were measured using a sixteenth square metre quadrat, several quadrat densities being taken in the same area to give a mean density per square metre.

The relative expansion of coral polyps has been recorded by a simple scoring system and by the use of photographs where the same group of corals has been observed over a period. Corals collected for gut content analysis were preserved in 20% formalin at the end of dives.

3. Sites

Intensive investigations have been carried out in the Scilly Isles at Darritys Hole on the east coast of St Mary's and in North Devon at Cheyne Cove and Smallmouth near Ilfracombe. These provide two contrasting sites since Darritys Hole is situated in the clear oceanic waters of the Atlantic where the kelp forest extends to 15 m and deep water can be found close inshore with rock faces continuing to 35 m whilst in North Devon the turbid waters of the Bristol Channel restrict the kelp forest to a maximum depth of 4 m and rocks reach a sandy bottom at 9 m maximum depth.

Work has also been carried out in Cornwall (Penlee Point), Lundy, Pembrokeshire (Martins Haven and Skomer), and in south-west Eire (Lough Ine). Observations have been made at Great Britain Rock (Scilly Isles), Mounts Bay and Talland Bay (Cornwall), Ope Cove (Portland), Abereiddy (Pembrokeshire) and at various sites in North Devon. In the Falmouth Bay area (Cornwall) data has been collected by the St Mawgan Sub-Aqua Club.

TABLE 1. Summary of results for all areas where

Data	North Devon			Lundy
	Cheyne Cove north rocks	Cheyne Cove sewer pipe	Smallmouth Cove	Gannets Rock
Depth (m)	7–9	5–12	7–9	18
Density per m²	149	—	78	80
No. in sample	120		137	92
Mean size of corals (area of calice in mm²)	87	115	144	62
Size/frequency (size represented as the area of calice in mm² divided into groups 40 mm² apart)				
Pyrgoma anglicum, % infestation	53	81	56.5	49
Pyrgoma anglicum, size /mean number of barnacles per coral				
% of corals with epibiota (other than *P. anglicum*)	71	18	21	25
% of corals with boring worms	33	(very low)	20	30
% of polyps expanded/part. ex./closed	23/13.5/63.5 (Sept.) 28.1/18.5/53 (Dec.)	—	17/32/51	66/13/21

ll investigations have been carried out

	Scilly Isles		Cornwall	Pembrokeshire		Lough Ine	
					Skomer	(South-west Ireland)	
	(Darritys Hole)		Penlee	Martins	(North	Atlantic	South-west
10m	20m	Cable	Point	Haven	Haven)	Coast	Lough
10	26	18–25	8	10	18	21	16
32	126	—	48	128	120 (vert.)	198 & 482	15
			(local)		208 (horiz.)	(2 sites)	
285	105	46	65	90	72	79	60
55	61	75	53	60	97	40	90
26.5	61	43	20	20	43	33	0
54.5	40	5.7	68	43	40	80	56
(low)	(low)	(low)	48	32	40	54	(13)
—	72/9.5/18.5	—	100/0/0	58/18/24	93/4/3	77/9.5/13.5	80/9/11

4. Results

The results obtained at the main sites investigated are summarized in Table 1.

A. Local Distribution, Density and Morphology of Corals

The density of corals at various sites in the south-west is given in Table 1. It can be seen that there is considerable variation both locally and between different areas.

Variation with depth

It was found that at Darritys Hole at a depth of about 10 m on kelp-free rock faces the density of corals was only 32 per m^2 while at about 20 m the density was 125 per m^2. Shallow inshore areas in south Cornwall (Penlee Point, Talland Bay, Lamorna Cove, Mousehole, Mullion Cove, Prussia Cove) had extremely low coral densities compared to deep water parts of offshore rocks such as the Mancles, except in the case of the coral population at Penlee Point associated with caves and overhangs where algal growth is reduced.

In these areas, where water is generally very clear, differences in coral density with depth may be explained if we consider the much higher amount of epilithic biota in shallow water compared to deeper water below the kelp forest, especially *Laminaria* and other algae including the encrusting lithothamnions, animals such as *Corynactis viridis*, various Polyzoa and encrusting sponges, all of which enter into direct competition for bare rock. In some cases, these encrusting organisms will engulf the corals and skeletons of *Caryophyllia* have been found in the bases of large sponges or covered with lithothamnions. In North Devon where the water is turbid and the *Laminaria* forest extends to only 4 m, high densities of corals are found in shallow water.

Variation with water movement

Density also appears to be affected directly or indirectly by the degree of water movement to which the area is exposed. In areas sheltered from prevailing winds and strong tidal streams such as Darritys Hole in the Scilly Isles and Smallmouth in North Devon, coral numbers may be reduced by sedimentation. At depths of about 20 m on boulders and rock outcrops in Darritys Hole the density of *Caryophyllia* is greater on the sides of rocks (about 180 per m^2) than on the tops (about 80 per m^2) where sediment accumulates. In Smallmouth Cove the density of corals has been related to the different degree of silting that occurs at sites within the cove. Taking measurements at about the same depth, readings of 82 per m^2 at the mouth, 66 per m^2 just inside the cove and

29 per m² well within the cove, where silting is greatest, were obtained. At Abereiddy, *Caryophyllia* has been found in a flooded sea-water quarry which is extremely sheltered and in which, consequently, sedimentation is high. The corals occur on vertical rock faces where they are still liable to some silting. The greatest density of corals was found below the inlet where water movement is greatest and was reduced in areas exposed to little water movement and high rates of siltation. At Lough Ine, an extremely sheltered area, *Caryophyllia* has been found in generally low densities on vertical rock (about 15 per m²) although in one area a density of 225 per m² was measured. On the Atlantic coast outside of the Lough, at the same depth, a density of about 220 per m² was found generally.

At both of the extremely sheltered sites studied the coral is one of a very restricted fauna mainly of sponges tunicates and, at Abereiddy, Sabellid worms. It is interesting to note that *Caryophyllia* is able to establish and survive in conditions apparently unfavourable to many sessile animals. Silting may prevent the settlement of planulae or may suffocate corals already growing on the rock. Marshall and Orr (1931) have come to the conclusion that reef building corals can withstand large quantities of sediment falling from above, but mud coming from below kills in one or two days. In the laboratory, it has been found that *C. smithi* has an extremely efficient ciliary cleaning mechanism used to keep the polyp clear and that it can survive burial for at least five days.

On rocks exposed to strong currents where silting is prevented and the supply of oxygen and possibly food is increased, densities are generally high. Off Great Britain Rock, a short distance south of Darritys Hole at the same depth but in a strong tidal stream, coral densities of 368 per m² occur compared to 126 per m² within the shelter of Darritys Hole. Off Gannets Rock on Lundy and in parts of Falmouth Bay a similar situation has been observed. In extremely strong currents such as those encountered at the mouth of the Bristol Channel off Ilfracombe, where surface current velocities reach 2.5 knots, corals are not found on the tops of flat rock outcrops, and on the sewer pipe at Cheyne Cove, they are found mainly in the lee of joints.

It is interesting to note that the corals collected from the exposed parts of the Cheyne sewer pipe had exceptionally wide bases and in many cases were very squat, compared to the corals collected from sheltered areas such as Smallmouth, which, in a few specimens, had narrow bases. Lefargue (1969) has described a distinct form of *Caryophyllia smithi* which is found in the archipelago of Glenan off South Brittany; the form found on overhanging cliffs and in sheltered areas ("La forme des surplombs et des zones abritées") which is untypical in that the column has a narrow base and the skeleton is fragile. Our own observations have shown that such ecological variation does occur in North Devon and further that corals exposed to extremely strong currents show yet another form, but in both very sheltered and very exposed sites "typical" forms outnumber the distinct "ecological" forms. At Lough Ine none of the corals collected could be referred to distinct still-water forms.

Size distribution of coral populations

Table 1 shows size frequency histograms for corals cleared from sublittoral rocks in various parts of the south-west. It is immediately apparent that although some of the populations are very similar (Darritys Hole 10 m, Penlee Point, Gannets Rock, Martins Haven) others show obvious variation particularly in the maximum size reached by corals in different populations.

It may be proposed that variation in maximum size reached by corals from different sites is due to differences in growth rate or to increased or decreased longevity under the conditions in which they are living. Sufficient data is not available to compare the growth rates of animals in different areas and factors which may affect longevity are not fully elucidated, although it is known that boring organisms within the skeleton are liable to weaken the base and may result in the detachment of the coral and subsequent death.

Conditions seem to be particularly favourable for coral growth and maintenance in the shallow turbid waters of North Devon where corals reach a mean size of 108 mm^2 compared to about 60 mm^2 elsewhere. In sheltered areas such as Smallmouth where the largest corals reach a size of 425 mm^2, the reduction in wave action, which would normally displace corals damaged by boring organisms, may result in the maintenance of larger corals. The population at this site has a more normal size distribution than for other areas studied although conditions may not be considered optimum since density is low (78 per m^2) due largely to the effects of competition for space by other epilithic biota.

Work at Lough Ine provided an interesting comparison between corals collected on the open Atlantic coast immediately outside of the Lough and well within the Lough where water movement of any sort is slight. Corals collected from vertical rock at the same depth showed distinctly different size distributions. Corals collected from outside of the Lough had a mean size of 40 mm^2 whilst those collected inside were 90 mm^2 mean area. This may result from the lower density of competing forms and lower density of boring organisms in the Lough population.

The excess of small corals which are present in most areas studied except Smallmouth may be explained if there has been a high larval settlement in recent years, or more probably, that growth becomes extremely slow at about the mean size with consequent accumulation of individuals followed by very slow growth to a larger size during which mortality occurs.

B. Reproduction, Growth and Mortality

Reproduction

We have not observed the production of planulae. Laboratory experiments have caused corals to release sperm on transfer from 8.5°C to 15°C on March 22nd 1970. *Caryophyllia smithi* have been observed to produce sperm and ova in the laboratory at Plymouth for about 2 weeks during February for several years in succession (P. R. G. Tranter, personal communication).

Asexual division occurs fairly commonly in North Devon, double corals being found often, triple corals occasionally, and on two occasions quadruple corals have been collected. Sections of the skeletons reveal that the soft parts of the individual corallites are continuous with each other indicating that the multiples are not formed by the chance settlement of planulae on an established coral. Each polyp of a multiple is identical in colouration. No examples of multiple corals have been found at other sites. Division may result from distortion of the corallum caused by the proximity of other sessile animals and by the presence of epibiota on the coral including the barnacle *Pyrgoma anglicum*.

Growth

An indication of growth rate has been given by measurements taken of corals at four sites and remeasured one and two years later, the results are given in Table 2. Difficulties were experienced in remeasuring the corals due to the often irregular shape of the corallum and consequently results were very erratic, varying from apparent gains of upto 2 mm to apparent losses of 1.7 mm in diameter over one year. At Cheyne Cove 17 corals under 80 mm^2 and 25 corals over 80 mm^2 in size have been measured over one year. The small corals showed an apparent mean increase in diameter of 0.06 mm whilst the large corals showed an apparent loss in diameter of 0.08 mm. Taking into account the inaccuracies in measuring such low growth, it seems probable that the smallest corals grow at a faster rate than of larger individuals.

Attempts to obtain newly settled corals by the use of settling blocks put down in 1968 were unsuccessful and very little information has been obtained from the measurement of corals from dateable objects since most such structures are too old. In the northern Mediterranean Zibrowius (1970) has obtained records of growth in specimens of *Caryophyllia smithi* growing on panels suspended at 300 m. On panels suspended for four years, the largest coral measured 18 mm × 23 mm in calice diameter (325 mm^2); a very large coral compared to samples from Britain and an extremely fast growth.

TABLE 2. Records of growth in marked coral population.

Site	No. in sample	Measured	Remeasured	Mean difference in diameter of calice
Smallmouth	14	11.8.68	7.8.69 7.8.70	−0.405 mm
Cheyne Cove	42	30.7.69	1.8.70	−0.073 mm
Darritys Hole	4	27.8.68	17.8.69	+0.145 mm

The barnacle *Pyrgoma anglicum* which lives on the edge of the calice may inhibit growth of the coral (evidence is given for this later).

Mortality

Very few dead corals have been seen during the investigations. Of those that have, the majority were engulfed by encrusting organisms or were found under a layer of mud. One cause of death which appears to be fairly common in North Devon, and probably elsewhere, is the detachment of corals from the rock face after the skeleton has been weakened by the action of boring organisms.

The only observation of predation of a coral was observed in the aquarium when a top shell *Calliostoma zizyphinum* settled on a specimen of *Balanophyllia regia* and consumed much of the tissue overnight. Whilst diving, gastropods and echinoderms have often been removed from the rockface but in no case have they been found to be feeding on corals.

C. Fauna and Flora associated with Caryophyllia smithi

In *Caryophyllia smithi* the outside of the corallum is covered and protected by the expanded polyp which extends over the edge of the calice but as the coral grows in height the lower parts of the skeleton are unprotected and open to the settlement of other organisms. Epibiota growing on the corallum of *Caryophyllia* is usually epilithic on the surrounding substrate and includes sponges, hydroids, *Corynactis viridis*, *Verruca stroemia*, *Monia squama*, polyzoans, lithothamnions, unicellular green algae and the barnacle *Pyrgoma anglicum*. The number of corals with epibiota is greatest where the density of sessile fauna and flora is also at a high level. Most organisms are kept clear of the calice but lithothamnions, and massive sponges such as *Cliona*, *Pachymatisma*, and *Alcyonium digitatum* will engulf the coral and kill it. Distortion of the calice may occur due to the growth of organisms, encrusting or not, in close proximity to the coral and this may be observed on large corals in shallow water where the density of sessile flora and fauna is high. Burrowing organisms, mainly worms, have been found to occur in the skeleton of about 20%–30% of the corals collected from most sites. This is reduced to a very low level in the Scilly Isles where corals grow on granite and the rock population of boring forms is probably low, and on the sewer pipe at Cheyne Cove where corals grow on a bituminous layer on iron. The fan worm *Potamilla reniformis* commonly uses the base of the coral as a refuge, its tube extending for one to two centimetres through an opening on the lower part of the corallum. *Phoronis hippocrepia* burrows into the column but in no case have they been found to be in contact with the coral tissue. The bivalve *Hiatella arctica* is often found in the dead shells of large *P. anglicum* and in the base of the coral; it is probable that they enter worm holes in the base of the skeleton and as they grow enlarge the cavity by their own activities. The structural damage resulting from these activities has already been cited as a cause of mortality in the corals.

Pyrgoma anglicum

The barnacle *Pyrgoma anglicum* is an especially interesting member of the corals epibiota since it is found solely on corals and appears to restrict the growth of *Caryophyllia* in some cases. The nauplius is capable of gaining access to the skeleton of the coral through the coral tissue in order to attach at a position about 2.5 mm below the level of the calice (Moyse, 1971). Previously, *Pyrgoma anglicum* was recorded in the British Isles only from *C. smithi* but present observations have shown that it also occurs on *Leptopsammia pruvoti*, *Balanophyllia regia* and *Hoplangia durotrix* in the south-west. Individual corals have been found to carry from one to thirty-one barnacles. The barnacles live from three to four years (J. Moyse, personal communication). Rees (1962) states that *Pyrgoma* occurs on about one in six corals collected from the south-west. Recent investigations indicate that in most areas the figure is nearer thirty to fifty percent.

In the British Isles the distribution of *Pyrgoma anglicum* is restricted to the south-west, reaching Exmouth in the English Channel, Elwill Bay (North Devon), south-west Pembrokeshire, and south-west Ireland; thus its recorded distribution differs from that of its host but accords well with that of other warm water species at the northern limits of their distribution.

The barnacles are usually found at the edge of the calice oriented so that the operculum is about level with the top of the corallum and the cirri point outwards. There is a tendency to aggregation (Moyse, 1971) and young barnacles sometimes settle on older ones. In locations affected by strong tidal streams, *Pyrgoma* has been recorded from the column, base, and even on the substrate a short distance from the coral.

The number of barnacles associated with individual corals varies with location and probably with the year. It was found that in Darritys Hole the mean number of barnacles per coral and the percentage infestation was greater at 20 m (61%) than at 10 m (26.5%) (see Table 1). The barnacle population on the corals collected from the telephone cable also showed a lower infestation (43%) than that of the surrounding rocks and this may be accounted for if it is realized that the cable population only dates from 1947 or 1922 at the earliest and that heavy infestation is probably more pronounced in long established groups. At Lough Ine, samples from the Atlantic coast show a 33% infestation but of 119 corals collected from within the shelter of the Lough only one held a barnacle. Reduced water movement within the Lough may prevent the effective distribution of larvae and of food for adult barnacles.

Of the populations for which adequate records are available, corals from the sewer pipe at Cheyne Cove show the highest barnacle population. In this situation ten *Pyrgoma* per coral is a common figure and twenty-one were recorded from one coral, the infestation being 81% compared to 53% amongst the rocks of the cove (see Table 1 histograms). It is suggested that the larger volume of water passing over the pipe may bring a greater number of barnacle larvae and a greater quantity of food which maintains settled barnacles. The largest population of barnacles on one coral was recorded from a specimen collected on the wreck of the M. V. Hera in Falmouth Bay; it held 31 *Pyrgoma*. The wreck is exposed to strong tidal streams.

Any advantages which arise out of the association appear to be almost entirely on the side of the barnacle which obtains protection by the polyp nematocysts, elevation from the substrate, is kept free of encrusting organisms, and probably obtains some if the food captured by the coral. The coral possibly only obtains occasional nauplii. Although the histograms of number of barnacle plotted against coral size generally show a higher population of barnacles on the larger corals, some observations indicate that the largest corals have a lower population of barnacles than those of a medium size. Analysis of records from 137 corals collected at Smallmouth (Fig. 2) shows clearly that the largest population of twelve barnacles per coral occurred in corals with a mean circumference of 49 mm whilst in corals with a mean circumference of 62 mm the number of *Pyrgoma* per coral was seven. Furthermore, the five largest corals, all over 75 mm circumference, held 0, 1, 2, 3 and 3 *Pyrgoma* each. It would appear that *Pyrgoma* does not generally settle on large corals or that the growth of the coral may actually be restricted by the presence of barnacles. A large number of barnacles on one coral may steal some of its food but it seems much more likely that

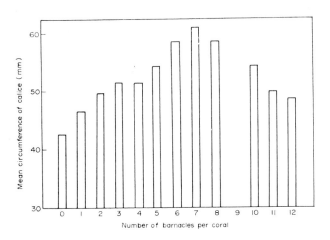

Fig. 2. Number of *Pyrgoma anglicum* per coral plotted with the mean coral circumference holding that number of barnacles. From a sample of 137 corals collected at Smallmouth.

the prevention of growth would be a mechanical effect; barnacles at the edge of the calice restricting the polyp. The effect of this can be seen clearly where the coral is affected by a number of large barnacles, these cause distortion of the corallum which grows only on each side of one barnacle or between two barnacles.

The relationship between the coral and barnacle cannot be described as commensial since the barnacle appears to affect the growth of the coral, neither can it be called parasitic since no part of the living barnacle is in contact with the tissue of the coral. The most accurate description of the relationship is probably that it is an inquilinism.

D. Food and Feeding

The gut contents of 140 corals collected at various sites in Darritys Hole have been examined. The corals were scraped of all epibiota and initially examined under low power microscopy for whole prey in the mouth or gastrovascular cavity, and finally gut scrapings from split corals were inspected with high power microscopy. The results of these investigations are given in Table 3.

TABLE 3. Material recovered from the gastrovascular cavity of *C. smithi*.

		Number of items recovered from 140 corals
Algae	Chlorophyceae (fragmented filaments)	29
Protozoa	Foraminifera	8
Cnidaria	Hydroids (fragments)	2
Polychaeta	(fragments)	8
Crustacea	(unidentified fragments)	53
	(identified, complete)	10
	? *Pyrgoma* (nauplii)	11
Mollusca	Gastropoda (Nassarids, Rissoids, ?veligers)	9
	Bivalvia	3
Polyzoa	(fragments)	10

All of the organisms recovered from corals except nauplii and veligers are to be found living on the rock faces or associated with epilithic growths. The presence of algal filaments is unusual and it is possible that they entered the coral by chance with the animal food. It can be seen that *Caryophyllia smithi* appears to take in anything organic which comes its way subsisting, in the area studied, mainly on small crustaceans. A great deal of the food is probably suspended in the water passing over the coral or attached to algae brushing against the polyps.

The proportion of polyps fully expanded, and presumably actively feeding, partly expanded, and closed (not actively feeding) has been recorded at most sites (see Table 1). The results vary from one site to another but in most areas a high proportion of polyps are expanded and it is only in the shallow turbid waters of North Devon that the majority of coral polyps are not fully expanded. Within limited areas, counts of polyp expansion have been made over two weeks during which turbidity steadily increased (Darritys Hole), over a five hour period during which surface tidal flow over the site varied from 0.18 m/s to 0.5 m/s (Cheyne Cove), at very exposed and extremely sheltered sites (Lough Ine), at different times of the year (Cheyne Cove, Martins Haven), and at different times of day and night (Darritys Hole); no significant difference has been observed in the proportion of polyps expanded in any one area during these observations.

Roushdy (1962) considers that polyp expansion in *Alcyonium digitatum* is primarily concerned with food capture. Although it may be suggested that polyp expansion is related to food supply and respiratory requirements, it is not clear what conditions elicit expansion and why a higher proportion of corals are expanded in clear-water conditions.

5. Summary

(1) At most sites studied in the south-west *Caryophyllia smithi* shows densities of 120 and more per m^2 on rocks below the *Laminaria* forest. In the clear waters of south Cornwall, the Isles of Scilly, Lundy, and south-west Ireland coral densities are low in shallow water because of the density of other epilithic organisms. In extremely sheltered areas where silting occurs *Caryophyllia* is found in reduced number, but in areas affected by strong currents corals show a recorded maximum density of 482 per m^2.

(2) Reproduction probably takes place in the spring. Asexual division and the production of multiple corals has been observed in many corals collected from North Devon.

(3) Growth is very slow. Corals reach a larger size in the turbid shallow waters of North Devon than in the clear oceanic or Channel water. Encrusting organisms cause distortion of the coral and may engulf corals.

(4) The barnacle *Pyrgoma anglicum* has been found to occur on about 50% of corals collected; this figure being higher in populations exposed to strong tidal streams and lower in very sheltered areas. Large barnacles cause distortion of the corallum and a high population of barnacles may restrict the coral growth.

(5) *Caryophyllia* feeds mainly on small crustaceans and other small animals which are found on epilithic growths and on the substrate. Feeding activity appears to be greatest in clear oceanic and Channel waters where the percentage of polyp expansion is high.

Acknowledgements

We are very grateful to the St Mawgan Sub-Aqua Club under Squadron-Leader A. Salter, who provided samples and much valuable information from the Falmouth Bay area; to G. Waite who assisted in work during 1968; Dr M. W. Robins, Dr W. A. M. Courtney and Mr M. Thurston for their help and advice throughout the study and the latter for the identification of crustacea. The work at Lough Ine was carried out during part of a programme of diving investigations supervised by Professor J. A. Kitching and Dr J. C. Gamble and we are very grateful for the opportunity to use this information. Our thanks are also due to Dr J. Moyse who supplied us with data on *Pyrgoma anglicum* before its publication and to Dr C. C. Emig who kindly identified the phoronid worms.

References

Best, M. B. (1968). Notes on three species of madreporarian corals known as: *Caryophyllia smithi, Caryophyllia clavus, Coenocyathus dohrni. Bijdr. Dierk.* **38**, 16–21.

Ekman, S. (1953). Zoogeography of the Sea. Sidgewick and Jackson, London.

Gosse, P. H. (1853). A Naturalists Rambles on the Devonshire Coast. John Van Voorst, London.

Gosse, P. H. (1860). *Actinologia Britannica*: A History of the British Sea Anemones and Corals. John Van Voorst, London.

Lefargue, F. (1969). Peuplements sessiles de l'Archipel du Glenan I: Inventaire Anthozoaires. *Vie Milieu* **20**, 415–437.

Marshall, S. M. and Orr, A. P. (1931). Sedimentation on Low Isles Reef and its relation to coral growth. *Sci. Rep. Gr. Barrier Reef Exped.* **1**, 93–133.

Moyse, J. (1971). Settlement and growth pattern in the parasitic barnacle *Pyrgoma anglicum. Proceedings of the 4th European Marine Biology Symposium* (Ed. D. J. Crisp), 125–141. Cambridge University Press.

Rees, W. J. (1962). The distribution of the coral *Caryophyllia smithii* and the barnacle *Pyrgoma anglicum* in British waters. *Bull. Brit. Mus. Zool.* **8**,(9), 401–418.

Rees, W. J. (1966). Further notes on the distribution on *Caryophyllia smithii* Stokes and Broderip and *Pyrgoma anglicum* Sowerby. *Ann. Mag. Nat. Hist.* sec. **13**,(9), 289–292.

Rossi, L. (1957). Revisione critica dei Madreporarii del Mar ligure I. *Doriana, Annali del Museo civico di Storia naturale "G. Doria"* II, nr. **86**, 1–9.

Rossi, L. (1960). Madreporaires. *Res. Sci. Comp. N.R.P. Faial. Portugal.* **3**, 1–13.

Roushdy, H. M. (1962). Expansion of *Alcyonium digitatum* L. (Octocorallia) and its significance for the uptake of food. *Vidensk. Meddr dansk naturh. Foren* **124**, 400–420.

Zibrowius, H. (1970). Étude qualitative et quantitative des salissures biologiques de plaques expérimentales immergées en pleine eau. 3. *Caryophyllia smithi* Stokes & Broderip et considerations sur d'autres espèces de madréporaires. *Tethys* **2**, (3), 615–632.

Light, Zonation and Biomass of Submerged Freshwater Macrophytes

D. H. N. SPENCE

Department of Botany, University of St. Andrews, Scotland

1. Introduction	335
2. Light, Specific Leaf Area and Zonation	336
3. Depth Limits of Colonization	340
4. Biomass Maxima and Depth Distribution	342
Acknowledgements	345
References	345

1. Introduction

This chapter is concerned with the freshwater analogues of attached marine angiosperms and algae, the fully submerged macrophytes, which mainly comprise flowering plants but include mosses and larger algae (Characeae or stoneworts). In it, a look is taken at the extent to which light may determine the vertical distribution of these plants and their biomass This is part of a broader study of factors controlling the general distribution and performance of freshwater macrophytes (Spence, 1964, 1967; Spence and Crystal, 1970a, b; Spence et al., 1973) that has included analysis of underwater spectral intensity (Spence et al., 1971).

Loch Uanagan in Inverness-shire provides an example of the vertical distribution of submerged vegetation in terms of biomass per unit area at various depths of water (Fig. 1). Data were collected in August 1967 by diving, mostly with snorkel. Figure 1(a) shows total biomass of all submerged species in relation to increasing rooting depth, or the depth of water above the soil surface in which each plant is rooted, along a transect while Fig. 1(b) indicates zonation with increasing rooting depth in terms of the relative maximal performance of each of a series of species. *Potamogeton praelongus* Wulff has a peak in deeper water than *Littorella uniflora* L. and it may be asked whether this is typical of these species and if so whether there exist habitual deep-water species; Fig 2 which is based on data from an extensive, earlier, survey of Scottish lochs (Spence, 1964) shows that there are. For example, *Potamogeton obtusifolius* Mert. and Koch is rooted in significantly deeper water than *Potamogeton polygonifolius* Pourr.

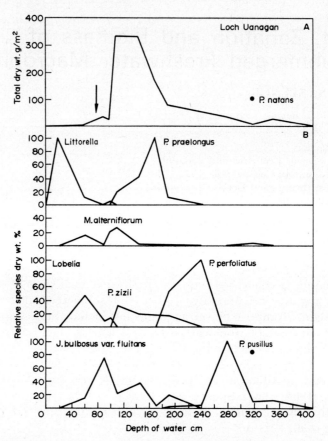

Fig. 1.(a) Biomass in g oven-dry weight per m² of all submerged species in relation to rooting depth, or depth of water (cm) above the surface of the soil where they were rooted; sample areas exclude reedswamp. Loch Uanagan 13 August 1967. Arrow indicates base of littoral shelf. (b) Biomass of named species at sampled rooting depths as percentage of total biomass of all species at that sample depth (All data collected by hand by P. Denny and author; redrawn from Spence 1972).

2. Light, Specific Leaf Area and Zonation

Flowering plants have the capacity in varying degree to adjust leaf morphology to sun and shade conditions by altering their thickness and their specific leaf areas (SLA: cm² leaf area/mg leaf dry weight) (Evans and Hughes, 1961). Sun leaves are thick with a low SLA while shade leaves are thin and have a high SLA. In addition to their mutually exclusive rooting depths in nature, Spence and Chrystal (1970) found that *P. obtusifolius* and *P. polygonifolius* produced mutually exclusive specific leaf areas when grown in sun and shade conditions in a glasshouse, and that the larger was the SLA the smaller was the dark respiration rate per unit area. As a result a shade leaf achieved a higher net photosynthetic rate per unit area at low irradiance, in the range 400–750 nm, than a sun leaf at that irradiance.

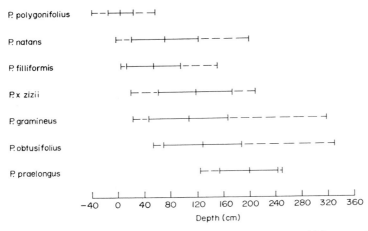

FIG. 2. Mean, standard deviation and range in depth of water at which a number of *Potamogeton* species are rooted in Scottish lochs (redrawn from Spence and Chrystal 1970a).

These observations suggested that range in SLA is an intrinsic characteristic of a species which can in nature relate directly to zonation (Spence and Chrystal, 1970b). Using SCUBA, whole shoots of selected species were subsequently collected from precisely measured depths in five lochs (Table 1) and specific leaf areas were estimated on 10 cm stem segments. Analysis and discussion of these data have been published (Spence *et al.*, 1973) and only some of the findings are outlined here. The relationship between SLA and depth of water is illustrated in Fig. 3 for pairs of species, on single sampling dates in two lochs, as a series of regression lines for each of which the correlation coefficient between SLA and depth is significant at the 5% level or less. The regression lines of *Potamogeton x zizii* Roth. and *P. perfoliatus* L. in Loch Baille na Ghobhainn show that the slope of SLA with depth is typical of the loch and not of the species.

There is a seasonal increase of SLA of *P. obtusifolius* over the water depth 210–270 cm in the brown-water Loch of Lowes (Fig. 4) but no such increase in SLA of *Potamogeton praelongus* in the much less brown-water Loch Lanlish (Fig. 5). Such attenuation data as there are indicate a seasonal increase of E_e, 400–750 nm, in Loch of Lowes over this period but little change in Loch Lanlish (by Loch Croispol, Table 1), which suggests that variation in shading rather than in other environmental or ontogenetic factors causes seasonal drift in SLA.

The SLA field range for *P. obtusifolius* (Fig. 4), is 1.43 to 1.98, compared with a laboratory range of 1.60 to 2.05. For *P. perfoliatus*,

TABLE 1. Lochs in which specific leaf areas have been measured, with map references and, where appropriate, attenuation coefficients (E_e), 400–750 nm, or E_B, E_G, E_R.

Loch	Map ref.	E_e[1]	E ln units[2] E_B E_G E_R	Date
Baille na Ghobhainn, Lismore	NM 860425	0.50		24.6.69
Croispol, Durness	NC 390680	0.56	0.29 0.18 0.43	19.6.70
		0.55		4.8.70
		0.57		16.9.70
Lanlish, Durness	NC 388680	—	0.77 0.36 0.56	August 1974
Lowes, Dunkeld	NO 050440	0.76		30.4.70
		1.20		14.5.70
		1.30	0.97 0.60 0.59	18.8.70
Uanagan, Fort Augustus	NH 370070	1.50	0.62 0.37 0.43	15.7.70

[1] Measured with spectroradiometer over 1 m (Spence *et al.*, 1971) or, [2], with blue(OB), green(OG) or red(OR) Chance filters, June–August (Spence 1975).

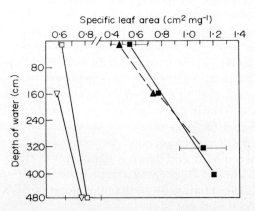

FIG. 3. Linear regressions of specific leaf area upon depth of water, of which the correlation coefficients are significant at the 5% level or less. Least and greatest error mean square are indicated. Loch Baille na Ghobhainn: *Potamogeton crispus* (▽), *P. perfoliatus* (□) 24 June 1969. Loch Uanagan: *P. × zizii* (▲), *P. perfoliatus* (□----□), 12 August 1969; *P. perfoliatus* (□——□), 14 July 1970. (Redrawn from Spence *et al.*, 1973.)

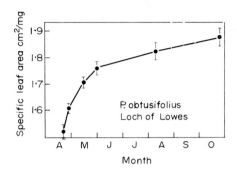

FIG. 4. Seasonal variation in mean SLA (with standard errors) of *Potamogeton obtusifolius* over the water depth 210–270 cm in Loch of Lowes, 1970, a brown-water loch. (Redrawn from Spence *et al.*, 1973.)

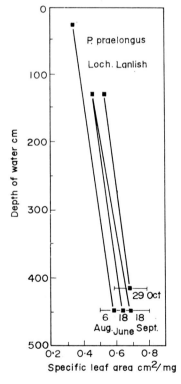

FIG. 5. Linear regressions of specific leaf area upon depth of water, of which correlation coefficients are significant at the 5% level or less, for *Potamogeton praelongus* in Loch Lanlish on various sampling dates in 1970. (Redrawn from Spence 1972.)

field values extend from 0.55 to 1.75 and, in the glasshouse, from 0.80 to 2.20. Thus the range in SLA of each species overlaps with the other and is matched in laboratory and field. Since any two such species have different capacities for shoot extension, they may have mutually exclusive rooting ranges in a particular site like *P. x zizii* and *P. perfoliatus* in Loch Uanagan (Fig. 3) or both their rooting ranges and their specific leaf areas may overlap, as with *P. crispus* L. and *P. perfoliatus* in Loch Baille na Ghobhainn or *P. obtusifolius* and *P. perfoliatus* in Loch of Lowes, when substratum rather than irradiance must determine their zonation. We still cannot directly test the hypothesis that, subject to modification on many shores by factors like turbulence or substratum or competition which are not strictly depth-controlled, light determines the potential depth range of a species (Spence, 1967), since we lack field measurements of any two species that do not have overlapping specific leaf areas in the laboratory. *P. polygonifolius* for instance was absent from all our sites. Moreover, presently available evidence concerns species from the same genus. However, using indirect evidence it would seem certain that if the specific leaf areas at least of two related species do not overlap then neither can their rooting ranges which, given the known relationship between SLA and function, would then be a direct cause of zonation.

3. Depth Limits of Colonization

The photic zone in freshwater is very much shallower than in the sea and it is seldom necessary in the United Kingdom to dive below 10 m to study attached plants which rarely penetrate anywhere nearly as deep as this. A brief outline follows of limits of colonization, and of biomass, in relation to depth of water. The limestone Loch Baille na Ghobhainn on Lismore was visited because West (1910) reported "dense beds" of the aquatic moss *Fontinalis antipyretica* L. at 13.1 m depth, suggesting that this loch was colonized to amongst the greatest depths of any in Scotland. Diving revealed however that while plants occurred to a depth of 13.1 m these were only fragments below 6 m, except at the very narrow ends of the loch. In June 1969, the loch was thermally stratified with a clear epilimnion down to 6 m and, below, both thermocline and hypolimnion contained large quantities of suspended matter which produced opaque layers with considerable attenuation but there was no stratification in August 1974. As determined by diver survey the shore again excepting its narrow ends has a gradient of 1 in 4 to a depth of 5 m; there, 1 m above the lower limit of continuous vegetation, it suddenly steepens to 1 in 1.05. Either the instability of this marl slope, or the existence at 6 m of the intermittent, opaque thermocline, causes an apparently premature limit to downward penetration of macrophytes in most of this loch.

Data are also available for two more limestone lakes, Lochs Borralie and Croispol at Durness in Sutherland which are unstratified and of the same water colour, clarity and homogeneity (Spence, unpublished data), resembling the epilimnion in Loch Baille na Ghobhainn and the clear coastal waters of the Moray Firth described by Hemmings (1966). Vascular plants do not grow in either loch below 6 m depth of

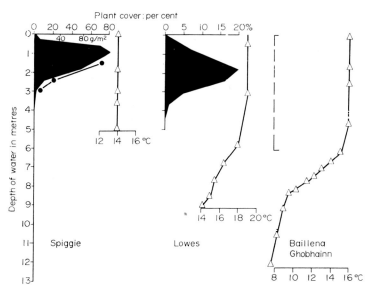

FIG. 6. Temperature profiles for Loch Spiggie, Shetland, (July), Loch of Lowes (August) and Loch Baille na Ghobhainn (June), together with available data on depth of colonization, percent plant cover, and oven dry weight of crop in g/m² (●—●) (all data obtained by diving).

water; as in Loch Baille na Ghobhainn this is also the downward limit for all attached macrophytes in Loch Croispol, whereas in Loch Borralie species of Charophyta, mainly *Nitella opaca*, which is absent from Loch Croispol, have been found by diving to form beds of vegetation down to 11 m depth of water. Our experimental evidence is insufficient as yet to explain the differences in depths of colonisation in these three limestone lochs or the fact that the angiosperms (*Potamogeton* species) cease to grow at the same depth in each.

Further examples are given in Fig. 6 of the commoner situation where there is no thermal stratification even in relatively calm weather or, if it is present, where the epilimnion extends well below the lower limits of attached vegetation. On another count this may be the commoner situation at least in the United Kingdom, because Loch Spiggie and Lowes represent the brown coloured water of low alkalinity (<0.3 m eq l^{-1} $HCO_3 + CO_3^= + OH^-$) on the greatest length of shoreline in Scottish freshwater lochs (Spence, 1964) and indicate, therefore, the sort of depths to which attached plants usually penetrate. It is not of course as simple as that since there are also nutrient-poor waters of great clarity like Ennerdale in the English Lake District where vegetation is reported, by using a grab, at 10 m (Pearsall, 1921) or Lake Grane Langsø in Denmark (Nygaard, 1958) where vegetation was found by diving at 11 m (Fig. 7).

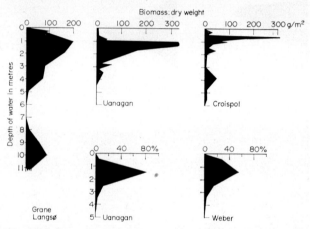

FIG. 7. Depth profiles of biomass in Lake Grane Langsø (Denmark), Loch Uanagan, (Inverness-shire) (August), Loch Croispol (Sutherland) (June) and Lake Weber (Wisconsin). (Data for first three obtained by diving; data for Lake Weber obtained with a grab and expressed as percent weight per 1 metre depth interval; lower graph for Loch Uanagan also on this basis for comparison.) Lake Grane Langsø data adapted from Nygaard (1958), Lake Weber data from Potzger and Van Engel (1942).

4. Biomass Maxima and Depth Distribution

It may be noted in Fig. 7 that biomass tends to reach a maximum in about 1.0 to 1.5 m depth of water while in the clear limestone Loch Borralie this maximum is reached between 3 and 4 m. Biomass in Loch Uanagan along with the intrinsic stature of the species (Fig. 1) decreases steadily with depth; biomass decrease is in fact logarithmic (Fig. 8) suggesting that, for the given level of nutrients in this moderately alkaline loch, biomass is limited by irradiance, which is also reduced or attenuated logarithmically with depth in this optically homogeneous medium (Spence *et al.*, 1971). Data are insufficient so far to confirm or refute the existence of such a biomass trend from maximal values as the depth increases in Loch Borralie or, indeed, in general.

The water depth at which maximal biomass occurs is worth a comment. In Loch Uanagan this lies at and below the base of the littoral shelf (Fig. 1, arrow) while in Loch Borralie it coincides with the slope of a less obvious, though still discernible, littoral shelf in deeper water. It seems likely that aspect expressed as excessive turbulence may directly,

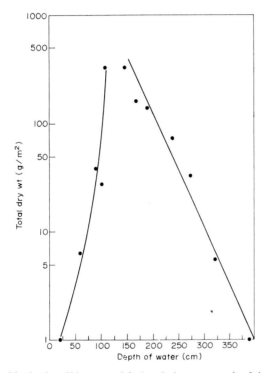

FIG. 8. Semi-logarithmic plot of biomass weight in relation to water depth in Loch Uanagan. (Data as in Fig. 1.)

or indirectly through soil particle size and substratum chemistry, affect summer growth of winter survival of macrophytes and thus cause limitations in shallow water.

Unpublished data indicate that the seasonal maximal biomass is reach in August. Such maxima, (Fig. 7) are only exceeded by values of Forsberg (1960) of 400 g over-dry weight per m² for nutrient-rich Lake Ösby in Sweden and approximately the same value for Loch Borralie. Even in water with the least attenuation, like that of Lochs Borralie and Croispol, biomass is low relative to that in optically comparable, clear coastal seawater (Fig. 9). This to an unknown extent reflects differences in growth-forms and, therefore, in winter biomass which, for some *Potamogeton* species that survive as overwintering buds, is as little as 1/20th the summer biomass. *In situ* production figures suggest a more meaningful comparison and are given in another paper (Campbell and Spence) in this volume.

I conclude with an attempt to summarize these inter-relationships, particularly with regard to the role of light. The most bulky submerged macrophytes like *Potamogeton praelongus* produce shoots up to 3.5 m tall in the United Kingdom so that in standing as opposed to running

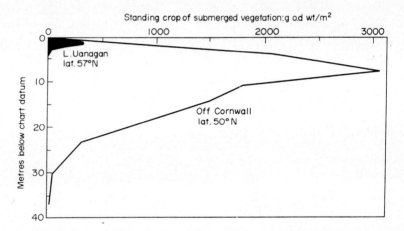

FIG. 9. Comparison of submerged macrophyte biomass in Loch Uanagan and a marine site off Cornwall (*Laminaria hyperborea* forest) (Loch Uanagan data as for Fig. 1. *Laminaria* data from Bellamy and Whittick, 1968.)

water the zone of maximum submerged biomass should occur in not less than 3.5 m depth of water. In shallower standing water, the submerged biomass is inevitably the product of smaller plants. Excessive turbulence will directly or indirectly limit this further while the downward extension of crop maxima and indeed the depths that any plant can grow depends primarily on the degree of light attenuation. This leads to the proposition that the resultant of the opposing factors of increasingly excessive turbulence and high light attenuation determines the rooting depth at which maximal submerged biomass occurs on any shore. The chemistry of the substratum and the extent of grazing, and of periphyton and sedimentary cover on leaves, will thereafter set the limits of the actual submerged maxima on a particular shore.

Acknowledgements

The work on which this paper is based is supported by grants from the Natural-Environment Research Council. Thanks are due to Dr. Patrick Denny for his help in collecting the data presented in Fig. 1.

References

Bellamy, D. and Whittick, A. (1968). Kelp forest ecosystems as a "phytometer" in marine pollution. *Underwater Ass. Rep.* 1967–68.

Evans, G. C. and Hughes, A. P. (1961). Plant growth and the aerial environment I. Effect of artificial shading on *Impatiens* parviflora. *New Phytol.* **60**, 150–180.

Forsberg, R. (1960). Subaquatic macrovegetation in Ösbyjön, Djursholm. *Oikos* **10**, 233.

Hemmings, C. C. (1966). Factors influencing the visibility of objects underwater. *Light as an Ecological Factor* (Eds. R. Bainbridge, G. C. Evans and O. Rackham), 359–374. Blackwell Scientific Publications, Oxford.

Nygaard, G. (1958). On the productivity of the bottom vegetation in Lake Grane Langsø. *Verh. int. Verein. theor. angew. Limnol* **13**, 144.

Pearsall, W. H. (1921). The development of vegetation in the English Lakes, considered in relation to the general evolution of glacial lakes and rock-basins. *Proc. Roy. Soc.* **B, 92**, 259–284.

Potzger, J. E. and Engel, W. A. Van (1942). Study of the rooted aquatic vegetation of Weber Lake, Vilas County, Wisconsin. *Trans. Wis. Acad. Sci. Arts Lett.* **34**, 149.

Spence, D. H. N. (1964). The macrophytic vegetation of freshwater lochs, swamps and associated fens. *The Vegetation of Scotland* (Ed. J. H. Burnett), 306–425. Oliver and Boyd, Edinburgh.

Spence, D. H. N. (1967). Factors controlling the distribution of freshwater macrophytes with particular reference to the lochs of Scotland. *J. Ecol.* **55**, 147–170.

Spence, D. H. N. (1972). Light on freshwater macrophytes. *Trans. Bot. Soc. Edinb.* **42**, 491–505.

Spence, D. H. N. (1975). Light and plant response in fresh water. *Light as an Ecological Factor II.* (Eds. R. Bainbridge, G. C. Evans and O. Rackham). Blackwell Scientific Publications, Oxford.

Spence, D. H. N. and Jean Chrystal (1970a). Photosynthesis and zonation of freshwater macrophytes I. Depth distribution and shade tolerance. *New Phytol.* **69**, 205–216.

Spence, D. H. N. and Jean Chrystal (1970b). Photosynthesis and zonation of freshwater macrophytes. II. Adaptability of species of deep and shallow water. *New Phytol.* **69**, 217–227.

Spence, D. H. N., Jean Chrystal and Campbell, R. M. (1973). Specific leaf area and zonation of freshwater macrophytes. *J. Ecol.* **61**, 317–328.

Spence, D. H. N., Campbell, R. M. and Jean Campbell (1971). Spectral intensity in some Scottish freshwater lochs. *Freshw. Biol.* **1**, 321–327.

West, Geo. (1910). A further contribution to the comparative study of the dominant phanerogamic and higher cryptogamic flora of aquatic habit in Scottish lakes. *Proc. Roy. Soc. Edinb.* **25**, 967–1023.

Preliminary Studies on the Primary Productivity of Macrophytes in Scottish Freshwater Lochs

R. M. CAMPBELL
and
D. H. N. SPENCE

Department of Botany, University of St. Andrews, Scotland

1. Introduction	347
2. Methods	347
3. Results and Discussion	349
A. Productivity	349
B. Light and productivity	350
C. Nutrients and productivity	351
Acknowledgements	355
References	355

1. Introduction

In the previous chapter in this volume it was shown that some freshwater angiosperm plants are morphologically adapted for growth in different light intensities by possessing leaves which exhibit sun and shade characteristics. The degree of development of this capacity by individual species is one of the factors which is linked with the zonation of vegetation from shallow to deep water (Spence *et al.*, 1973). Evidence was also produced that the standing crop of submerged vegetation in several Scottish lochs is much less than that reported from both terrestrial and marine communities. This paper reports preliminary findings from a study designed to measure the *in situ* production rates of several species of *Potamogeton*, *P. obtusifolius* Mert. and Koch, *P. praelongus* Wulft and *P. perfoliatus* L., and to investigate possible limiting factors.

2. Methods

Several techniques are available for measuring the productivity of aquatic plant communities (Vollenweider, 1969). However, many are unsuitable for use with angiosperms (Wetzel, 1964a). The carbon 14 technique first developed by Steeman-Nielsen (1952) is now standard for the measurement of planktonic algal productivity and has recently also been applied to macrophytic algae (Drew and Larkum, 1967). Wetzel (1964b), reports the first application of this technique to freshwater macrophytes, but his method using whole plants is not ideally suited either for large scale replication or deep water studies.

Rather than utilize whole plants the method employed in the present study used the main photosynthetic organs of the plant, the leaves. Leaves with a leaf area (one surface) of not more than 8 cm^2, were detached and placed in sealed 28 ml glass bottles containing loch water labelled with carbon 14 sodium bicarbonate. The method required a Scuba diver to collect the plant material and set up the experiments. Because of possible accidental leakage of isotope and the dangers of contamination in small bodies of water, carbon 14 was injected into the experimental bottles either in a boat or at a land station. During this procedure the bottles were kept in the dark and wrapped in aluminium foil. The diver clipped the bottles onto a frame placed at the depth where the plant material had been collected, removed the foil and left the bottles for 3–4 h, after which time they were again wrapped in foil and brought ashore. The leaves were washed, their area outlined on paper then placed on a planchet for measurement of radioactivity. From a knowledge of the specific activity of the carbon in the lochwater in the bottles, a figure was calculated for the rate of carbon fixation after Strickland (1960).

Previous laboratory manometric studies on the oxygen production rates of detached leaves of *Potamogeton* species in bicarbonate-carbonate buffer solutions have been carried out (Spence and Chrystal, 1970). The same buffer solution—Warburg no. 11—was used to replace lochwater in some of the present investigations in order not only to compare results with laboratory findings but also to have a base line for comparison of the production rates of species growing in lochs of different nutrient status.

Following the standard bioassay procedure of Goldman (1965) and Wetzel (1965) for phytoplankton, additions of phosphate were made in several experimental treatments to investigate its possible limiting effect on productivity.

To compare experiments done on different days and in different lochs the underwater irradiance during the course of an experiment was recorded using an I.S.C.O. Spectroradiometer and chart recorder with an underwater remote probe. Experiments, however, were always carried out at depths greater than those at which the irradiance could be measured directly, due to the limited length of the probe. Therefore the light energy transmitted to the experimental depth was calculated from a vertical attenuation coefficient (Westlake, 1965b) measured using the spectroradiometer as described by Spence et al. (1971). Production can thus be expressed in units of µg C/cal as well as in carbon fixed per unit leaf area or plant dry weight.

Underwater light quality has been measured by calculating the attenuation with depth of water for 25 nm wavebands of the spectrum as described by Spence et al. (1971).

3. Results and Discussion
A. Productivity

A series of 30 experiments were carried out in five lochs over the entire growing season. These gave a mean figure for the productivity of the three submerged *Potamogeton* species investigated of 3 µgC/cm²

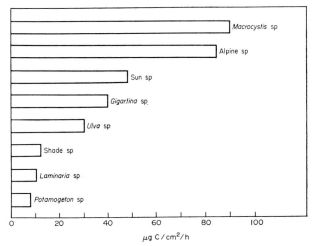

FIG. 1. Comparison of productivity (µg C/cm² tissue area/h) of different plant communities measured under natural conditions. Data from E. A. Drew (personal communication) and recalculated from Spector (1965) and Verduin (1953).

leaf area/hour, with a maximal rate of about 8 µgC/cm²/hour. Figure 1 presents this maximal figure in relation to similar reported measurements for other types of vegetation. For beds with a biomass in the region of 400 g oven-dry weight/m² (Spence, 1974), it is calculated that a daily production value for submerged macrophytes in Scottish lochs in unlikely to exceed 20 gC/m² loch floor/day. While this is smaller than figures published for most terrestrial and marine communities and freshwater phytoplankton (Westlake, 1963, 1965a), it should be noted that it is difficult to assess the accuracy of such an extrapolation from short-term measurements to a daily rate.

B. Light and Productivity

The decrease of light intensity with increasing depth obviously influences the productivity of aquatic vegetation but, as mentioned previously, some submerged plants are adapted to low light intensities. Figure 2 illustrates the rate of carbon fixation for two species of *Potamogeton* measured in Loch Uanagan. Terminal, mature leaves of *P.*

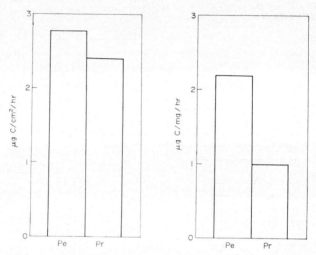

FIG. 2. Rates of carbon fixation for *Potamogeton perfoliatus* (Pe) and *P praelongus* (Pr) measured simultaneously in Loch Uanagan on 15 July 1970. Results are expressed as rates per unit leaf area and, from a knowledge of their specific leaf area (SLA: cm^2 leaf area/mg leaf dry weight), per unit leaf dry weight. SLA for *P. perfoliatus* is 0.77 and for *P. praelongus* is 0.34.

praelongus were taken from a water depth of 40 cm, and similar leaves from *P. perfoliatus* at 140 cm, and both sets were exposed at a depth of 140 cm which, in this loch, is about 100 cm above the lower rooting depth of *P. perfoliatus* (Spence, 1974, Fig. 1). Results expressed on a leaf basis are similar but when one takes into account the specific leaf area (SLA) or the ratio between leaf area and leaf dry weight of the two species, *P. perfoliatus* which is the deeper ranging species is in fact twice as effective at producing organic matter per unit leaf weight as *P. praelongus*.

Figure 3 illustrates the difference in percentage composition of light at subsurface and at 1 m depth in Loch Croispol, Sutherland, and Loch Uanagan, contrasting clear and brown water lochs. Although as yet there is no direct evidence, the effect these changes in light quality have on productivity is probably very slight when compared to the overall effect of light intensity (Spence et al., 1973).

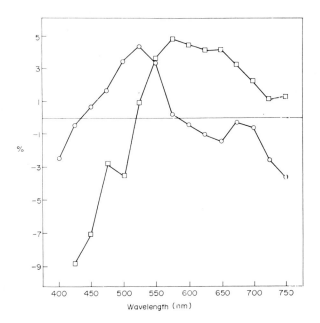

FIG. 3. Percentage composition of spectral intensity in 25 nm wavebands, 400–750 nm, at 1 m depth of water as more or less than the percentage composition of the same wavebands at subsurface (0 m). Loch Croispol on 4 August 1970 ○——○ and Loch Uanagan on 15 August 1970 □——□.

C. Nutrients and Productivity

Figure 4 presents the results from an *in situ* production experiment carried out in Loch Lanlish, Sutherland, using leaves of *P. praelongus* placed in three waters; the buffer solution, Loch Croispol water and Loch Lanlish water. It illustrates how carbon fixation rates can be influenced by nutrient concentrations in the bathing solution. Table 1 gives some chemical data for the two lochs which are situated only some 500 m apart on the Durness limestone.

Data presented in Fig. 5 illustrates the effect of the addition of phosphate on the rate of carbon fixation by detached leaves of *P. perfoliatus* in Loch Croispol. It would therefore appear that the productivity of the submerged macrophytic vegetation of limestone lochs can be limited by a lack of at least one nutrient.

A similar result is shown in Fig. 6 for two species found growing in Loch of Lowes, a brown-water loch in Perthshire with completely different water chemistry (Table 1). The effect of enrichment with phosphate is similar for both species but that in lochwater is twice that

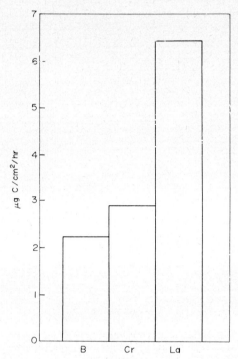

FIG. 4. Rates of carbon fixation (μg C/cm^2 leaf area/h), for *Potamogeton praelongus*. Measured in Loch Lanlish on 6.8.70 using three bathing solutions; a bicarbonate-carbonate buffer (B), Loch Croispol water (Cr) and Loch Lanlish water (La).

TABLE 1. Some chemical properties of three lochs in which productivity experiments have been carried out. Some of the data were kindly provided by Mr. A. V. Holden.

Loch	Croispol	Lanlish	Lowes
pH	8.8	8.7	6.8
alkalinity (parts/10^6 CaCO$_3$)	133.0	78.0	16.0
Ca parts/10^6	27.0	24.0	8.2
Mg parts/10^6	22.0	4.5	2.8
Na parts/10^6	21.0	29.0	5.6
K parts/10^6	1.4	0.8	0.8
P parts/10^6	0.008	0.017	0.018

in the buffer solution. The pH of Lowes is some two units lower than that of the limestone lochs and the much lower levels of calcium and magnesium produce a poorly buffered water. Addition of phosphate to this water produces a small drop in pH which changes the ratio of bicarbonate to free CO_2 and thus possibly alters the availability of carbon for photosynthesis. It is thus possible that in some waters

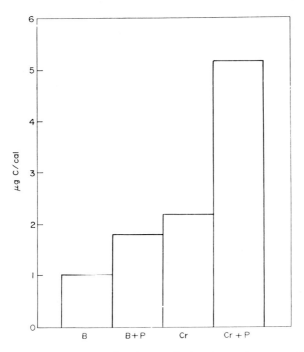

FIG. 5. Mean rates of carbon fixation (μg C/cal) for 10 experiments carried out with *Potamogeton perfoliatus* in Loch Croispol. Using a bicarbonate-carbonate buffer (B), the buffer with phosphate (B + P), Loch Croispol water (Cr), Loch Croispol water with phosphate (Cr + P). Phosphate was added in the form of K_2HPO_4 at the rate of 20 μg/25 ml.

there is not only a strict nutrient deficiency but also a carbon deficiency as well.

Apart from the effect that basic lack of nutrients has on productivity it is possible that the planktonic algae may also be competing with the higher plants for the nutrients which are available. Figure 7 presents data from a phosphate enrichment experiment carried out in Loch of Lowes with the leaves of *P. obtusifolius* Mert. and Koch, in three waters; the buffer solution, Loch of Lowes water, low in phytoplankton and Loch Leven water rich in phytoplankton. Results illustrate that enrichment of both Lowes water and buffer solution produce similar enhanced rates of carbon fixation, whereas enrichment of Leven water containing abundant phytoplankton has much less effect. In conclusion we should say that phosphate is not the only nutrient which could be limiting production and our future field programme is designed to

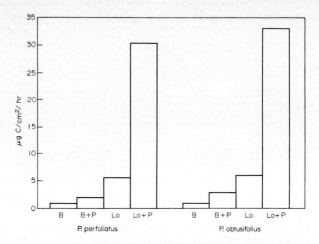

Fig. 6. Rates of carbon fixation (μg C/cm^2 leaf area/h) for *Potamogeton perfoliatus* and *P. obtusifolius*. Measured in Loch of Lowes on 12.8.70 using a bicarbonate-carbonate buffer (B), the buffer with phosphate (B + P), Loch of Lowes water (Lo) and Loch of Lowes water with phosphate (Lo + P). Phosphate was added in the form of K$_2$HPO$_4$ at the rate of 20 μg/25 ml.

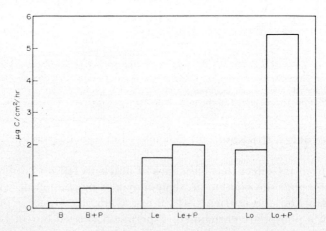

Fig. 7. Rates of carbon fixation (μg C/cm^2 leaf area/h) for *Potamogeton obtusifolius*. Measured in Loch of Lowes on 13.8.70 using a bicarbonate-carbonate buffer (B), the buffer with phosphate (B + P), Loch of Lowes water (Lo), Loch of Lowes water with phosphate (Lo + P), Loch Leven water (Le) and Loch Leven water with phosphate (Le + P). Phosphate was added in the form of K$_2$HPO$_4$ at the rate of 20 μg/25 ml.

investigate a wide range of nutrients and to try and gain some more information on algae-macrophyte inter-relationships.

Acknowledgements

This study forms part of a research project financed by the Natural Environment Research Council.

References

Drew, E. A. and Larkum, A. W. D. (1967). Growth and photosynthesis of Udotea—a deep water plant. *Underwater Assoc. Rep.*, 1966–1967, 65–71.

Goldman, C. R. (1965). Micronutrient limiting factors and their detection in natural phytoplankton populations. *Mem. Ist. Ital. Idrobiol.* **18**, Suppl., 121–135.

Spector, W. S. (1965). Handbook of Biological Data. W. B. Saunders Co., Philadelphia and London.

Spence, D. H. N. and Chrystal, J. (1970). Photosynthesis and zonation of freshwater macrophytes. *New Phytol.* **69**, 205–215, 217–227.

Spence, D. H. N., Campbell, R. M. and Jean Chrystal (1971). Spectral intensity in some Scottish freshwater lochs. *Freshwat. Biol.* **1**, 321–337.

Spence, D. H. N., Campbell, R. M. and Jean Chrystal (1973). Specific leaf area and zonation of freshwater macrophytes. *J. Ecol.* **61**, 317–328.

Steeman-Nielsen, E. (1952). The use of radioactive carbon (C-14) for measuring organic production in the sea. *J. Cons. int. Expl. Mer.* **18**, 117–140.

Strickland, J. D. A. (1960). Measuring the production of marine phytoplankton. *Bull. Fish Res. Bd. Canada*, 122.

Verduin, J. (1953). A table of photosynthetic rates under optimal near-natural conditions. *Amer. J. Bot.* **40**, 675–679.

Vollenweider, R. A. (1969). A Manual for Measuring Primary Production in Aquatic Environments. I.B.P. Handbook No. 12. Blackwell, Oxford and Edinburgh.

Westlake, D. F. (1963). Comparisons of plant productivity. *Biol. Rev.* **38**, 385–425.

Westlake, D. F. (1965a). Some basic data for investigations on the productivity of aquatic macrophytes. *Mem. Ist. Ital. Idrobiol.* **18**, Suppl., 229–248.

Westlake, D. F. (1965b). Some problems in the measurement of radiation underwater: a review. *Phytochem. Photobiol.* **4**, 849–868.

Wetzel, R. G. (1964a). A comparative study of the primary productivity of higher aquatic plants, periphyton and phytoplankton in a large shallow lake. *Int. Rev. ges. Hydrobiol.* **49**, 1–61.

Wetzel, R. G. (1964b). Primary productivity of aquatic macrophytes. *Verh. int. Ver. Limnol.* **15**, 426–436.

Wetzel, R. G. (1965). Nutritional aspects of algal productivity in marl lakes with particular reference to enrichment bioassays and their interpretation. *Mem. Ist. Ital. Idrobiol.* **18**, Suppl., 137–157.

Some Aspects of the Growth of *Posidonia oceanica* in Malta

E. A. DREW
and
B. P. JUPP

Gatty Marine Laboratory, St. Andrews, Scotland

1. Introduction	357
2. Methods	358
A. Standing crop	358
B. Photosynthetic rates	358
3. Results	360
A. Standing crop data	360
B. Growth rate of rhizomes	361
C. Photosynthetic rates	362
4. Discussion	363
Acknowledgements	367
References	367

1. Introduction

Posidonia oceanica is the largest of the marine angiosperms, with massive rhizomes and linear leaves which may be more than a metre long but only a centimetre wide. It is a member of the Potamogetonaceae, although sometimes considered in a family of its own, the Posidoniaceae, and it is endemic to the Mediterranean. The only other extant species of *Posidonia* is confined to Australian and S. E. Asian coasts.

Unlike most of the marine angiosperms, which inhabit the intertidal zone and seldom penetrate far below water mark, *Posidonia oceanica* is entirely sublittoral and has been recorded from less than a metre below low water to depths of 40 metres. It forms extensive meadows on suitable substrates all round the Mediterranean and has been the subject of many investigations, especially on the French coast. Most of these investigations have been of a qualitative ecological nature or have concentrated on the marine algae and great variety of animals which live within the meadows.

The results of quantitative investigations of the ecology of *Posidonia* at various depths around the coast of Malta are described in this chapter. In order to assess the overall productivity of this plant, both standing crop and photosynthetic ability have been measured.

2. Methods

A. Standing Crop

The entire vegetation contained within square quadrats with areas of $\frac{1}{2}$ or 1 m^2 was harvested by divers and transported to the surface in sacks. Care was taken to include all marine algae present below the leaf canopy and also all the non-mobile animals.

In most cases the basal rhizome "matte" was too deep for complete collection to be possible; for this reason the biomass of rhizome material was ignored in these samples. Most of the rhizome material below the matte surface was, in any case, dead whilst the living portions represented several years growth; both were irrelevant in standing crop measurements. Collection of leaves was carried out by uprooting handfuls of shoots to include the tops of the attached rhizomes, and with them the epiphytic flora and fauna. The structure of a *Posidonia oceanica* shoot is shown in Fig. 1.

On land the samples were sorted and treated to obtain the following data:

(i) number of individual shoots (these were separated manually from the rhizomes);
(ii) number of leaves on a representative 10% of the shoots;
(iii) length and breadth of leaves on the representative batch;
(iv) dry weight of the total leaf crop; encrusting calcareous algae mostly flaked off during drying and were not included;
(v) distance apart of leaf scars on rhizomes; rhizomes growing both vertically and horizontally were sampled;
(vi) numbers of important animals, especially echinoderms.

B. Photosynthetic Rates

The carbon 14 method described by Drew and Larkum (1967) was used to determine the photosynthetic potential of portions of *Posidonia* leaves at various depths. Certain modifications to the technique were incorporated as follows:

(i) the horizontal platforms on which plant material was exposed for photosynthesis had rigid legs which were pushed into convenient patches of soft substrate;

(ii) experimental vessels were 30 cm lengths of perspex tube (internal diameter 2.5 or 1.5 cm) closed with suba-seal serum caps;

(iii) the radioisotope was injected into the tubes underwater, using 1 ml disposable syringes filled on the surface and each containing 10 μCi of $NaH^{14}CO_3$. The tubes were shaken after injection to ensure adequate mixing of the contents.

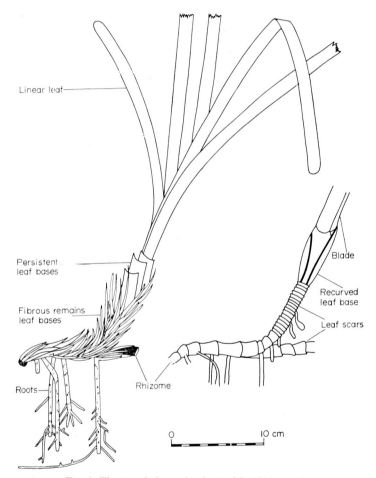

FIG. 1. The morphology of a shoot of *Posidonia oceanica*.

Plant material was collected from the depths of the experimental platforms and suitable portions of epiphyte-free leaves were selected. Experiments were of four hours duration and were carried out over noon.

The long perspex tubes were more appropriate for incubation of the narrow *Posidonia* leaves than the 500 ml Kilner jars previously used for marine algae, although in retrospect it seems likely that the radio-isotope was not thoroughly mixed along the length of the tubes. This should, however, have had little effect on the total amount of carbon 14 fixed since the leaf portions extended the entire length of the tubes and were therefore exposed to the average carbon 14 concentration *in toto*.

It is unlikely that the somewhat variable results obtained in replicate samples in these experiments with *Posidonia* were due to inconsistencies in the method, since parallel experiments with the green alga *Caulerpa prolifera* showed good replication in identical vessels. It is more likely that the *Posidonia* leaf material was itself variable, a feature which will be discussed later.

One experiment with another sublittoral marine angiosperm—*Cymodocea nodosa*—was also carried out and will be mentioned briefly.

At the end of the experiments the vessels were wrapped in aluminium foil by divers and brought to the surface where the tissues were killed and preserved in 80% ethanol for analysis in the laboratory. Water samples were taken from the dark treatment vessels (wrapped in foil throughout the experiments) for subsequent determination of the specific radioactivity of the inorganic carbon in the experimental vessels.

3. Results

A. Standing Crop Data

The data set out in Figs. 2 and 3 show that both the biomass of leaves and the number of individual shoots per unit area decreased with increasing depth of the meadows, although not dramatically in either case. The considerable variation between samples bears little relation to substrate differences, although there was a tendency for higher yields on rocky terrain.

The other features investigated showed no change correlated with increasing depth. The number of leaves per shoot was surprisingly constant at all depths (Fig. 4), whilst features such as the total area of leaves per unit area of seabed and the density of the leaf tissue were very variable at all depths, as indicated in Table 1, but showed no distinct trends.

The herbivorous sea-urchin *Paracentrotus lividus*, already shown by Neill and Larkum (1966) to consume marine algae in Malta, was found in shallow *Posidonia* meadows and had easily recognizable leaf fragments in the gut. The abundance of this herbivore and its restriction to shallow sites are indicated in Fig. 2.

B. Growth Rate of Rhizomes

Posidonia is known to shed its leaves regularly towards the end of summer; the spatial distribution of leaf scars on the rhizome should therefore offer some guide to the rate of growth of this organ. Data

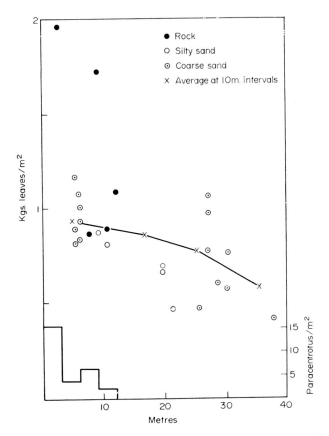

Fig. 2. Biomass of leaves of *Posidonia* per metre square at various depths.

set out in Table 2 show a clear difference between the growth rates of horizontal and vertical rhizomes. The annual growth rates indicated are necessarily approximate; a vertical increment for the rhizome matte as a whole of between 0.5 and 1.0 cm per year means that a matte 1 m deep would take 100 to 200 years to develop, so that the lower layers of mattes several metres deep are of considerable antiquity. Such mattes are of frequent occurrence, especially in sheltered bays.

TABLE 1. Quantitative features of *Posidonia* meadows (values are averages for all quadrats from 3 to 38 m depth), (fiducial limits in parentheses).

Leaves per shoot ($n = 22$)	Area of leaves (m^2) per m^2 of seabed* ($n = 21$)	Leaf tissue density (mg/cm^2) ($n = 20$)
4.5 (3.2–5.8)	7.3 (3.8–11.5)	11.0 (6.9–16.9)

* Very high values for shallow rocky substrates omitted: these were 19.6 and 24.4 m^2.

TABLE 2. Growth rate of *Posidonia* rhizomes.

Sample depth (m) / Rhizome type	* mm separation leaf scars			Average (mm)	Growth† rate (mm/year)
	9	12	20		
Vertical	2.55	1.00	0.92	1.50	6.0
Horizontal	4.68	5.00	3.30	4.33	17.3
H/V	1.85	5.00	3.50	2.85	

* Insertion of leaves on rhizome is alternate; total number recorded therefore twice that on single side.

† Assuming that only the four outer, partially dead and heavily encrusted leaves are shed each year.

C. Photosynthetic Rates

The data for photosynthetic rates at various depths, set out in Fig. 5, are from three separate experiments. Corrections for respiratory loss of fixed carbon 14 have been made according to data previously used for marine algae (Drew and Larkum, 1967), using respiratory rates determined for *Posidonia* leaves in the laboratory by Warburg manometry. No correction for dark fixation of carbon 14 has been made; this never exceeded 0.16 µg carbon/cm^2 per hour (5% of the lowest light fixation rate recorded) and averaged 0.08 µg carbon.

In all light treatments over 90% of fixed carbon 14 was recovered in the ethanol soluble fraction of the leaf tissue. Since sucrose was the only carbohydrate detected in significant amount in this fraction by gas-liquid chromatography, it is probable that this compound was the initial stable storage product formed in photosynthesis. Sucrose represented nearly 3% of the dry weight of the leaf tissue (Drew, unpublished data).

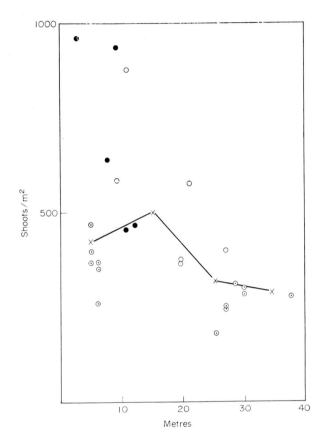

FIG. 3. Number of shoots of *Posidonia* per metre square at various depths.

4. Discussion

The results presented above represent a survey of the performance of *Posidonia oceanica* over its entire depth range in Malta. In very shallow water it is generally replaced by the much smaller marine angiosperm *Cymodocea nodosa*, and the interrelation between these two plants has been extensively studied (Molinier, 1960; Peres, 1967). In Malta, *Cymodocea* has also been found to replace *Posidonia* at its deepest penetration—36 m—and to occur as a wide fringe down to 39 m. As far as we are aware, *Cymodocea nodosa* has not before been reported from much below 10 m depth.

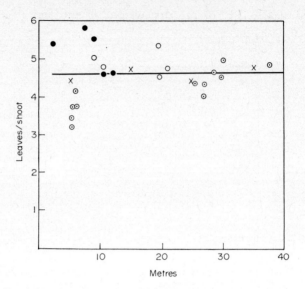

Fig. 4. Number of leaves per shoot of *Posidonia* at various depths.

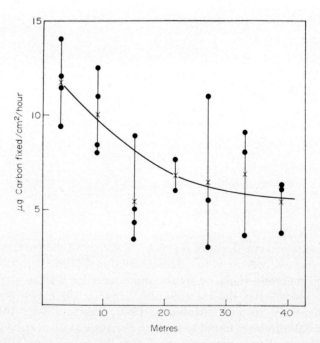

Fig. 5. Rate of photosynthetic carbon fixation by *Posidonia* leaves at various depths.

TABLE 3. Carbon balance sheet for *Posidonia* at various depths.

Depth (m)	Photosynth. (μgC/cm²/h)	Respiration (μgC/cm²/h)	Experimental leaf (μgC/cm²)			Meadow (gC/m²)*	
			Photo./12 h	Resp./24 h	Accretion/24 h	Photo./12 h	Resp./24 h
5	11.0	4.5	132	108	24	6.5	7.5
10	9.5	4.5	114	108	6	5.6	7.5
20	7.0	3.3	84	79	5	4.2	5.5
30	6.3	3.1	76	74	2	3.7	5.2
40	5.6	2.6	67	62	5	3.3	4.3

* Assuming 7 layers of leaves and overall photosynthesis to be 70% of sum of unshaded layers—see text.

TABLE 4. Comparison of photosynthesis and respiration rates in *Posidonia* and *Udotea*.

Plant	tissue density (mg/cm²)	A. Photosynthesis (μgC/cm²/h)			B. Respiration (μgC/cm²/h and μgC/mg/h)						
		Depth (m)		Temp (°C)							
		10	30		16		21		26		
					cm²	mg	cm²	mg	cm²	mg	
Posidonia	11.0	9.5	5.8		1.45	0.13	2.40	0.24	4.5	0.41	
Udotea	4.1	7.0	5.0		1.40	0.34	1.75	0.43	2.3	0.56	

N.B. ambient summer water temperatures—10 m, 26°C; 30 m, 23°C.

There is surprisingly little variation in the whole structure of the *Posidonia* meadows over their entire depth range, except a reduction in total leaf biomass, but then only by some 40%. This is despite a reduction in ambient light energy by about 90% between the surface and 40 m. It has been suggested by Drew (1971) that grazing of the shallower meadows by *Paracentrotus* may reduce their biomass below its potential level, whereas the deeper meadows may achieve their potential in the absence of this animal. *Paracentrotus* may also be responsible for the complete absence of epiphytic flora and fauna below the leaf canopy in the shallower meadows whereas epiphytic algae and sedentary animals such as bryozoans and sponges are relatively abundant in the deep samples.

Photosynthetic measurements indicate that the response of *Posidonia* to increased depth parallels that of green algae in the same waters (Table 4A). However, such data on photosynthetic rates means little unless considered in terms of the simultaneous respiratory loss of carbon and the resultant rate of organic accretion. The calculations set out in Table 3 indicate that under our experimental conditions, with leaf portions exposed horizontally in unshaded places, *Posidonia* leaves were able to maintain an overall daily accretion of carbon, albeit small below 10 m.

This was no longer the case when the meadows themselves were considered, for there were some seven layers of leaves respiring over every square metre of seabed, whilst photosynthesis in all but the uppermost layer must have been considerably reduced by shading. If it is assumed that a *Posidonia* leaf absorbs only about 33% of the light falling on it, transmitting the rest to the leaf below, then it can be estimated that *in toto* the seven layers would photosynthesize at a rate some 70% of that expected from seven completely unshaded layers. A value of 33% absorption or less must be assumed if enough light is to penetrate the canopy in deep meadows to sustain the *Udotea* plants known to grow there.

However, this level of photosynthesis would not be adequate to provide for even the respiratory needs of the canopy, discounting annual production of new rhizome and leaf material. Nevertheless, it seems unlikely that our estimates for the photosynthetic rate of *Posidonia* are too low since comparison of both photosynthetic and respiratory rates found for *Posidonia* with those determined in similar experiments for the green alga *Udotea* show a marked similarity between the two plants despite their completely different structures (see Table 4). The angiosperm has an excess of non-photosynthetic tissue and a more complex cellular organization than the alga; flowering plants are known to utilize blue light much less efficiently than algae (Gabrielson, 1940), and blue light is the major subsurface component in Maltese waters.

Thus one might not expect such high rates of carbon fixation for *Posidonia* in this environment as we have found, let alone postulate higher rates.

It should, however, be mentioned that in an exploratory experiment with *Cymodocea nodosa* between 27 and 40 m depth, very high photosynthetic rates were recorded. Under some conditions these plants must be able to utilize blue light very efficiently indeed.

At this point we can offer no definite explanation of the anomolous situation in which measured photosynthetic and respiratory rates cannot account for the ability of *Posidonia* meadows to grow at the considerable depths at which they are found. However, the determinations reported show considerable variation between replicate samples and it is possible that even greater variation exists between the epiphyte-free portions of leaves selected for experiments, and usually taken from within the canopy, and the rest of the leaf canopy. Indeed, the older leaves frequently appear somewhat moribund, although still green in colour over most of their length, whilst the youngest leaves are very bright green. These differences may reflect variation in respiratory rates as well as the more obvious changes in photosynthesis expected to be correlated with pigmentation. In addition, static conditions within the experimental vessels may have caused localized bicarbonate deficiency within the diffusion shell around the leaves, significantly reducing photosynthetic rates compared with natural conditions in turbulent water. Photosynthesis by marine algae can be increased up to twofold by agitation of the vessels although respiratory rates are hardly affected.

Experiments to ascertain the range of such variations are required before any further attempt can be made to establish an acceptable carbon balance sheet for *Posidonia* meadows.

Acknowledgements

The authors would like to acknowledge the assistance of all the persons involved with both the Leeds University Underwater Expedition to Malta (1967) and the St. Andrews University Underwater Expedition (1969) during which this work was carried out.

References

Drew, E. A. (1971). Botany. Underwater Science. (Ed. J. D. Woods and J. N. Lythgoe). Oxford University Press, London, N.Y., Toronto.

Drew, E. A. and Larkum, A. W. D. (1967). Photosynthesis and growth in *Udotea petiolata*, a green alga from deep water. *Underwater Ass. Rep.* 1966–67.

Gabrielson, E. K. (1940). Einfluss der Lichtfaktoren auf die Kohlensaure-assimilation der Laubblatter. *Dansk Botanisk Arkiv* **10**.

Molinier, R. (1960). Etude des biocenoses marines du Cap Corse. *Vegetatio* **IX**.

Neill, S. R. St. J. and Larkum, H. (1966). Ecology of some echinoderms in Maltese waters. *Proc. Symp. Underwater Ass. Malta* 1965.

Peres, J. M. (1967). The Mediterranean benthos. *Oceanogr. Mar. Biol. Ann. Rev.* **5**.

Photosynthesis and Growth of *Laminaria hyperborea* in British Waters

E. A. DREW,
B. P. JUPP
Gatty Marine Laboratory, St. Andrews, Scotland

and

W. A. A. ROBERTSON
Department of Botany, University of St. Andrews, Scotland

1. Introduction	369
2. Materials and Methods	372
3. Results and Discussion	374
4. Summary	378
References	378

1. Introduction

Laminaria hyperborea (Gunn.). Foslie is a brown alga that demonstrates the complexity and anatomical specialization characteristic of the order Laminariales. The species is essentially Northern in distribution, extending from Portugal in the South to the Norwegian coast and a short way into the U.S.S.R. at its Northern limits. Around British coasts it forms dense sublittoral forests from 1 m below Low Water Springs, with plants sometimes exposed at very low Spring Tides, down to 37 m in the clear water around Cornwall. The usual lower limit around Scottish coasts is about 15 m.

Figure 1 shows the form of the macroscopic sporophyte plant. It develops from a microscopic gametophyte which colonizes suitable rocky substrates and gives rise directly to the sporophyte, which may be up to 2.5 m in length from holdfast to frond tip. Growth soon becomes localized in the plant and several meristematic zones become differentiated. Haptera, which are small root-like appendages, develop to form a complex attachment organ, the holdfast, which allows strong adherence to the substrate. The stipe is rigid and keeps the plant upright to form tree-like forests; it is flexible enough to withstand swell but not appreciable wave action as indicated by cast-up weed on beaches after storms. Radial growth of the stipe is achieved by

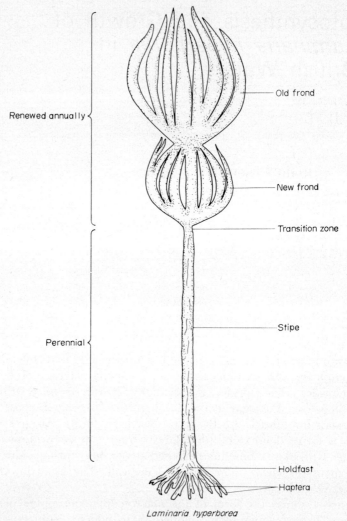

FIG. 1. Morphology of *Laminaria hyperborea*.

the activity of a meristem of dividing cells at the surface, and is rapid during the first half of the year and slow thereafter. It is thus possible to determine the age of a plant by taking a transverse section of the stipe near the holdfast and counting the annual rings in the same way as trees are aged. Plants up to 13 years have been found but the normal life span is 10 years.

The primary growing region is at the transition zone between the frond and the stipe. A new frond develops each year from this meristem and the old frond may remain attached by a narrow collar to the new frond up to May or June. The new frond grows rapidly from January

TABLE 1. Estimates of productivity of various communities.*

Ecosystem	Dominant species	Growing season productivity (g carbon/m^2/day)
1. Tropical agriculture	*Saccharum officinarum*	9.3
	Oryza sativa	8.9
2. Tropical rain forest	various trees	7.6
3. Tropical marine submerged macrophytes including coral reefs	green algae	5.4
4. Temperate agriculture	*Beta maritima*	5.0
5. Temperate marine submerged macrophytes	*Laminaria longicuris*	3.9
	Laminaria hyperborea (present study)	2.6
6. Temperate coniferous forest	*Pinus sylvestris*	2.0
7. Temperature freshwater submerged macrophytes	*Berula erecta* and *Ranunculus p'fluitans*	1.5
8. Marine phytoplankton	*Skeletonema costatum* (diatoms and flagellates)	0.32

* Data from Westlake, 1963.

to June, more slowly from July to October and then becomes reduced in size from October to the following May or June by physical attrition.

The lower limit of colonization by *L. hyperborea* is determined by the nature of the substrate, light intensity and the pressure of herbivorous grazing (Kain, 1963; Jones and Kain, 1967). Competition with other algae is unlikely to affect final growth since the species, once established, will dominate the less long-lived algae such as *Laminaria saccharina* and *Saccorhiza polyschides* (Kain, 1969).

The kelp forest is, in many ways, similar to a terrestrial forest and its production may exceed that of a coniferous forest.

In Table 1 it can be seen that production rates of *L. hyperborea* compare favourably with those of other communities. The most important processes in the plant are photosynthesis and respiration and of particular interest is the compensation point when, due to limiting factors such as light intensity, carbon supply and temperature, photosynthesis and respiration rates are equal and the plant will cease to grow, except by the use of stored materials. In this study photosynthesis and respiration of *L. hyperborea* both *in situ* and in the laboratory, and the growth of the plant in its environment have been measured.

2. Materials and Methods

Biomass data for *L. hyperborea* were obtained using plants collected from 0.5 m² quadrats from Arisaig, Inverness-shire, Scotland (56° 57′N; 05° 52′W). Ten quadrats, containing 218 plants, were collected from 3.1 m depth and 12 quadrats, containing 105 plants, were collected from 9.1 m depth over a period of 15 months. Ages of plants were determined by counting the annual growth rings in transverse

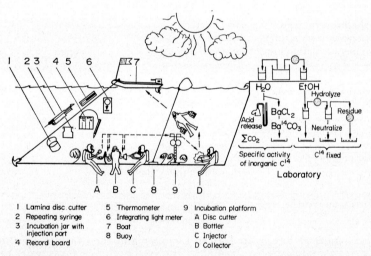

1 Lamina disc cutter
2 Repeating syringe
3 Incubation jar with injection port
4 Record board
5 Thermometer
6 Integrating light meter
7 Boat
8 Buoy
9 Incubation platform
A Disc cutter
B Bottler
C Injector
D Collector

FIG. 2. Synopsis of *in situ* ^{14}C technique.

sections of the stipes. Fresh and dry weights of holdfast, stipe and frond; stipe length; area and specific leaf area (cm² frond/mg dry weight frond) of old and new fronds were measured. The net annual primary productivity of the frond was estimated from seasonal maximum frond biomass, assuming little damage to the frond from wave action or grazing had occurred during the growing season. The method used by John (1968, 1969) and Whittick (1969) for estimating the net annual primary productivity of the perennial stipes and holdfasts was adopted in which age:biomass relationships are used as initially described by Bellamy and Holland (1966).

Photosynthesis was measured *in situ* using a ^{14}C technique (Drew, 1966, 1973). A synopsis of the technique is shown in Fig. 2. Tissues of the alga were incubated in sealed jars containing seawater, the carbon pool of which was labelled with ^{14}C sodium bicarbonate. At the diving site buoyant plantforms (9) were attached to anchors on the seabed and relocated after the experiments by means of a marker buoy (8). Divers took down 450 ml Kilner jars (3) filled with seawater at the surface;

TABLE 2. Biometric data for *L. hyperborea* at Arisaig. Data are the mean values for all quadrats.

Depth (m)	Biomass kg/m^2		Density plants/m^2	Net annual primary productivity; metric tonnes organic matter/hectare/year
	Fresh	Dry		
3.1	20.4	3.4	22	16.5
9.1	7.7	1.2	9	8.0

serum ports in the lids allowed subsequent injection of radioisotope. Discs of tissue were cut out from the frond with the cutter (1) and put in the jars which were then sealed. The injector (C) then delivered $10\,\mu$Ci of ^{14}C sodium bicarbonate from an automatic syringe (2). Dark controls were wrapped in aluminium foil throughout the experiment. The jars were inverted and held in position by clips on the platforms. Integrating light meters (6) were used to measure light intensity at the levels of the jars throughout the experiment and have been described in detail by Drew (1972). The incubation period was about 4 h over midday.

After incubation, each jar was wrapped in foil to terminate photosynthesis and brought to the surface where the tissue was killed and preserved in 80% ethanol. Water samples were taken from each jar to determine the specific radio-activity of the inorganic carbon pool. In the laboratory, complete ethanol extractions of tissue were made and ethanol-insoluble tissue was hydrolysed with 1N H_2SO_4 at 100°C for 3 hours. Radioactivity in the ethanol fractions, acid hydrolysates and acid-insoluble residues was determined using a gas-flow proportional counter. Total counts were corrected for dark fixation and converted to µg carbon fixed according to the specific radioactivity of the carbon in each jar. This gave a value for gross photosynthesis as laboratory experiments indicated little loss of fixed ^{14}C by subsequent respiratory processes during the time course of these *in situ* experiments.

The concentration of photosynthetic pigments in the alga was analyzed using techniques described in Goodwin (1965). Absorbance of extracts was measured on a Pye Unicam SP 1800 Spectrophotometer.

Translocation of ^{14}C labelled photosynthetic products within the plant was studied using apparatus similar to that of Lüning (1971), Lüning *et al.* (1971) and described more fully by Jupp (1972).

3. Results and Discussion

Biometric data for *L. hyperborea* are set out in Table 2.

It can be seen that biomass and productivity are reduced with depth. This is largely due to the reduced irradiance at depth and also to

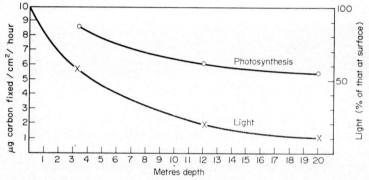

FIG. 3. Photosynthesis by *Laminaria* and light intensity at various depths.

grazing pressure affecting the settlement and development of the microscopic gametophyte in deeper waters, so that plant densities, as well as performance, are much reduced at depth (Kain, 1969, 1971).

Photosynthesis is largely limited by light intensity but it can be seen from Fig. 3 that the reduction in photosynthesis of *L. hyperborea* is not directly related to the attenuation of light intensity with depth. Data for photosynthetic efficiency, based on comparisons between calorific equivalents of carbon fixed and the recorded available light energy, showed values of 1.3% at 3.1 m and 6.8% at 18.3 m on a sunny day in Cornwall. On a dull day the value at 18.3 m rose to 13%. It is possible that the maintenance of high photosynthetic efficiency at depth is partly due to changes in frond thickness and pigment content with depth. Thus in Table 3 it can be seen that chlorophyll content on a weight basis increases with depth (this being due entirely to an increase in chlorophyll a content) and the specific leaf area of fronds also increases with depth.

TABLE 3. Chlorophyll content and specific leaf area for frond tissue of *L. hyperborea* at two depths. Material collected in January.

Depth (m)	Ether; $\mu g/g$ fr. wt. Total chlorophyll	90% acetone $\mu g/g$ fr. wt. chl. a	chl. c	Specific leaf area; cm^2 frond/mg dry wt. frond
3.1	150	115	39	0.05
9.1	200	139	31	0.10

Culture experiments indicated that light intensity plays some part in controlling frond thickness (Jupp, 1972), whereas Svendsen and Kain (1971) and Larkum (1972) have correlated changes in frond morphology of *L. hyperborea* with wave action. At present it is unlikely that these small changes in frond thickness and chlorophyll content indicate that deep growing plants are "shade adapted" but the increased specific leaf area and chlorophyll content may be of some benefit in maintaining a high light trapping ability and photosynthetic efficiency at depth, whilst at the same time reducing the respiratory loss per unit area of frond.

Kain (1963) found that the stipe length of young plants (1 to 3 years) at 5 m in the *L. hyperborea* forest were significantly smaller than stipe lengths of similarly aged plants at 10 m near the limit of the forest. This effect was explained in terms of the effectiveness of the dense forest canopy in shallow waters in reducing the light intensity available to young plants. An *in situ* photosynthesis experiment was carried out to investigate this effect and Fig. 4 indicates that older plants at the canopy level at 5 m have a high photosynthetic capacity whereas young plants under the canopy at 5 m have a lower photosynthetic capacity than similarly aged plants at 20 m in the more open community. Light intensity was about 1% of surface irradiance under the canopy at 5 m and 11% of surface irradiance at 20 m. An *in situ* growth experiment showed that removal of old plants can cause an increase in the area of new fronds developed by young plants during the growing season by up to 100% of those left under the canopy (Jupp, 1972).

Lüning (1969, 1970, 1971) has shown that a rapid expansion in the area of the new fronds of *L. hyperborea* occurs each year from January to June. This expansion occurs when, particularly from January to March,

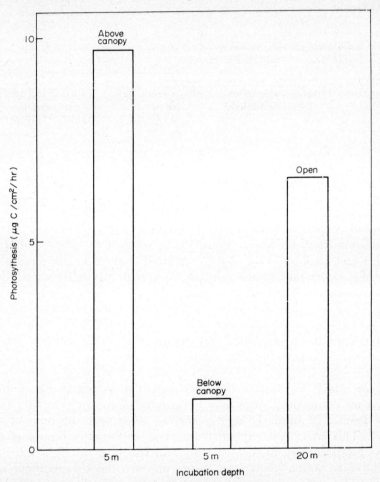

FIG. 4. Effect of forest canopy on carbon fixation of young plants.

underwater light intensity is low due to low surface irradiance, short daylength and high turbidity caused by frequent storms. The interesting question is how this growth is achieved under such poor illumination. Lüning (1971) has suggested that the new frond is below its compensation point up to March. An *in situ* experiment carried out in February also indicated that new frond tissue was below its compensation point (Jupp, 1972).

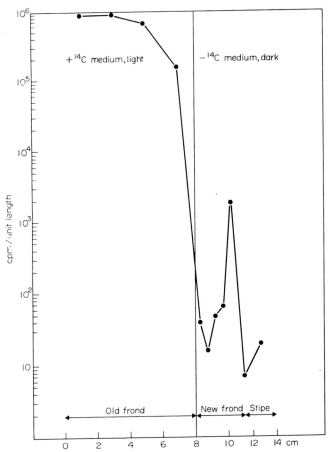

FIG. 5. Translocation of ^{14}C from old frond of *L. hyperborea* to new frond and stipe.

Lüning (1971) has argued from laboratory and field experiments using amputation techniques that the development of the new frond is dependent on the attachment of the old frond during winter and early spring. Lüning *et al.* (1971) have presented evidence from ^{14}C tracer experiments that a translocation of labelled compounds, mannitol in particular, occurs from the old frond to the new frond. Figure 5 shows the results of a similar translocation experiment on Scottish plants where ^{14}C labelled compounds formed during photosynthesis by the old frond moved to the new frond and accumulated at the junction between the new frond and stipe. The rate of movement of the tracer front was about 9 cm/h. Further experiments showed that the stipe can provide 0.3 mg mannitol/day to the new frond and that the old frond can provide 5 to 9 mg mannitol/day to the new frond. These experiments were carried out with young plants less than 6 years old. These values of mass transfer could account for the observed growth rates of new fronds of 4 to 5 year old plants of 16 mg dry weight/day up to March (Whittick, 1969). The data in general agree with that of Lüning et al. (1971). However *Laminaria hyperborea* seems to have a primitive form of translocation compared with that of the giant kelp, *Macrocystis*, which can translocate large amounts of labelled material at rates of up to 60 cm/h along the stipe (Parker, 1965). Nevertheless the translocation of metabolites in *Laminaria* could provide sufficient carbon for the development of a small new frond before light conditions improve and allow positive photosynthetic accretion by the new frond tissue.

4. Summary

Laminaria hyperborea has been shown to be a highly productive plant and can survive well at depth. This may in part be related to the efficient light trapping ability of its cells. No shade adapted ecotypes could be distinguished at depth. Translocation of reserves is important in establishing a new frond over the first few months of growth when environmental conditions reduce photosynthesis below the compensation point.

References

Bellamy, D. J. and Holland, P. J. (1966). Determination of the net annual serial production of *Calluna vulgaris* (L.) Hull, in northern England, *Oikos* **17**, 113–120.

Drew, E. A. (1966). A technique for the determination of photosynthetic ability of attached marine algae at various depths *in situ*. *Proc. Symp. Underwater Ass. Malta* '65, 65–67.

Drew, E. A. (1972). A simple integrating photometer. *New Phytol.* **71**, 407–413.

Drew, E. A. (1973). Primary production of large marine algae measured *in situ* using uptake of ^{14}C. A guide to the measurement of marine primary production under some special conditions. Monographs on oceanographic methodology, Unesco, Paris, 22–26.

Goodwin, T. W. (1965). Chemistry and Biochemistry of Plant Pigments. Academic Press, London and New York.
John, D. M. (1968). Studies on Littoral and Sublittoral Ecosystems. Ph.D. thesis, University of Durham, 168 pp.
John, D. M. (1969). An ecological study on *Laminaria ochroleuca*. *J. Mar. Biol. Ass. U.K.* **49**, 175–187.
Jones, N. S. and Kain, J. M. (1967). Subtidal algal colonization following the removal of *Echinus*. *Helgoländer Wiss. Meeresunters* **15**, 460–466.
Jupp, B. P. (1972). Studies on the growth and physiology of attached marine algae. Ph.D. thesis, University of St. Andrews, 206 pp.
Kain, J. M. (1963). Aspects of the biology of *Laminaria hyperborea*. II, Age, weight and length. *J. Mar. Biol. Ass. U.K.* **43**, 129–151.
Kain, J. M. (1969). The biology of *Laminaria hyperborea*. V. Comparisons with early stages of competitors. *J. Mar. Biol. Ass. U.K.* **49**, 455–473.
Kain, J. M. (1971). Continuous recording of underwater light in relation to *Laminaria* distribution. *Proceedings of the 4th European Marine Biology Symposium*, 335–346 (Ed. D. J. Crisp). Cambridge University Press.
Larkum, A. W. D. (1972). Frond structure and growth in *Laminaria hyperborea*. *J. Mar. Biol. Ass. U.K.* **52**, 405–418.
Lüning, K. (1969). Growth of amputated and dark-exposed individuals of the brown alga *Laminaria hyperborea*. *Mar. Biol.* **2**, 218–223.
Lüning, K. (1970). Cultivation of *Laminaria hyperborea in situ* and in continuous darkness under laboratory conditions. *Helgoländer Wiss. Meeresunters* **20**, 79–88.
Lüning, K. (1971). Seasonal growth of *Laminaria hyperborea* under recorded underwater light conditions near Helgoland. *Proceedings of the 4th European Marine Biology Symposium*, 347–61 (Ed. D. J. Crisp). Cambridge University Press.
Lüning, K., Schmitz, K. and Willenbrink, J. (1971). Translocation of ^{14}C-labelled assimilates in two *Laminaria* species. *Proc. VII Int. Seaweed Symp., Tokyo*.
Parker, B. C. (1965). Translocation in the giant kelp *Macrocystis*. I, Rates, direction, quantity of ^{14}C labelled products and fluorescein. *J. Phycol.* **1**, 41–46.
Svendsen, P. and Kain, J. M. (1971). The taxonomic status, distribution and morphology of *Laminaria cucullata sensu* Jorde and Klavestad. *Sarsia* **46**, 1–22.
Westlake, D. F. (1963). Comparisons of plant productivity. *Biol. Rev.* **38**, 385–425.
Whittick, A. (1969). The Kelp Forest Ecosystem at Petticoe Wick Bay, Lat. 55°N Long. 2°09′ W. An Ecological Study. M.Sc. thesis, University of Durham, 139 pp.

Deposition of Calcium Carbonate Skeletons by Corals: An Appraisal of Physiological and Ecological Evidence

R. K. TRENCH

*Department of Agricultural Science, University of Oxford**

1. Introduction	381
2. Morphology of Corals	382
A. The animal	382
B. The skeleton	384
3. The Distribution and Growth Form of Corals in Coral Reefs	384
4. Calcium Carbonate Deposition and the Control of Growth Rates	387
A. Effect of temperature on coral growth rates	388
B. Effect of adequate food supply	388
C. Effect of light on calcification rates	389
D. Zooxanthellae and coral nutrient supply	390
5. Synopsis	391
Acknowledgements	392
References	392

1. Introduction

The term "coral" is often loosely used to describe the stony or proteinaceous skeletal supports produced by some marine coelenterates. Thus, "soft coral" refers to the skeleton of alcyonarians which is composed of protein with spicules of calcium carbonate embedded in it, while "stony corals" refers to the aragonitic skeleton of the madreporarian or scleractinian corals. Mention should also be made of the hydrocorals or "stinging corals" (Milleporina) which secrete massive aragonitic skeletons and comprise a major source of calcium carbonate in Caribbean coral reefs, and the "blue corals" (Coenothecalia) which produce appreciable quantities of calcite in Indo-Pacific coral reefs.

Limestone producing corals may be divided into the two groups, hermatypic and ahermatypic. Hermatypic corals produce extensive shallow water reef communities in tropical seas, and they invariably

* Presently at the Biology Department, Yale University, New Haven, Conn., U.S.A.

contain symbiotic algae called zooxanthellae (belonging to the Dinophyceae) in their tissues. The presence of zooxanthellae is currently thought to be a direct reason for the ability of hermatypic corals to produce limestone at a rate rapid enough to overcome the destructive physical and biological forces at work in shallow water reef environments. Ahermatypic corals do not build shallow water reefs in tropical regions, but may form deep water reefs in temperate seas; they do not contain zooxanthellae.

The precise mechanism by which corals deposit limestone is not clearly understood, and the purpose of this chapter is to briefly review relevant physiological and ecological evidence. This paper will be concerned exclusively with hermatypic Scleractinia.

Elucidation of the mechanism by which corals produce calcium carbonate and of the factors controlling it will ultimately depend upon physiological experiments conducted under carefully controlled laboratory conditions. However, any information gained under these conditions will eventually have to be interpreted in the context of the coral as an integral part of the reef ecosystem. Thus, the solution of the coral reef problem today depends on both the experimental approach and on field observations.

In order to put the evidence into perspective, it is first necessary to describe general features of the morphology of corals and of reef coral communities. Most of the comments on coral ecology will be based on the reefs of Jamaica and British Honduras, with which the author is most familiar.

2. Morphology of Corals

A. The Animal

Detailed descriptions of the structure of corals may be found in various textbooks on invertebrate zoology (e.g. see Hyman, 1940 and Wells, 1956). Therefore, only a very brief account will be given here.

A coral is essentially a sea anemone that produces a skeleton. The coral polyp, like a sea anemone, is morphologically cylindrical with one end sealed off to form the "foot" or pedal disc, and the opposite end, the oral disc, bearing the mouth and tentacles. The animal is composed of two tissue layers; the epidermis or covering cells and the gastroderm, comprising digestive, secretive and absorptive cells. The gastrodermal cells contain the zooxanthellae. The two layers of tissue are separated by an essentially acellular material termed the mesoglea. The aboral epidermis, lying in close juxtaposition to the skeleton is termed the calicoblastic epidermis.

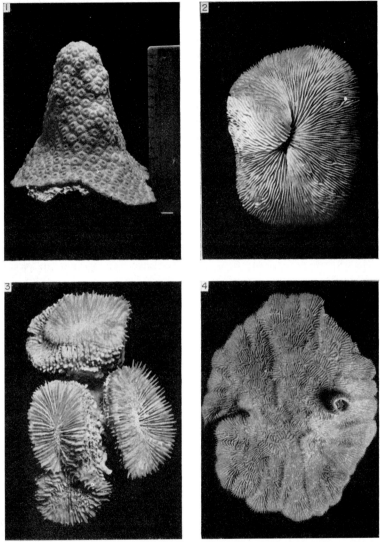

FIG. 1. A colony of Montastrea cavernosa (Linnaeus) collected from 20 metres, Pear Tree Bottom, Runaway Bay, Jamaica. Colonies may grow to several times the size of the one shown in the photograph.

FIG. 2. A solitary coral, *Fungia scutaria* (Lamark), was collected from very shallow water in Kaneohe Bay, Oahu, Hawaii. The skeleton shown (8 cm across) represents the work of a single polyp.

FIG. 3. The flower coral, *Mussa angulosa* (Pallas) was collected from about 25 metres off English Caye Reef, British Honduras (specimen 11 cm diameter). This coral has a rather wide range of distribution.

FIG. 4. *Mycetophyllia lamarkana* (Milne-Edwards and Haime), measuring 25 cm across the widest diameter. Two worm tubes can be seen projecting from the coral, the larger of a sabellid and the smaller of a serpulid worm. Note that the coral has grown around the worm tubes.

B. The Skeleton

The skeleton of scleractinian corals is composed of aragonite, which arise as "aggregates of sperulites" originating from "centres of calcification" (Bryan and Hill, 1941). Associated with the mineral phase of the skeleton is an "organic matrix" which may be chitinous (Wainwright, 1963) or proteinaceous (Young, 1968). The skeletons of four hermetypic corals are shown in Figs. 1–4.

The calcium in coral skeletons is derived directly from sea water, and not via the food chain. This was demonstrated by Thompson and Chow (1955), by showing that the atomic ratios of calcium to strontium in coral skeletons was similar to that in sea water. It is assumed that calcium is taken up by the coral by an active transport system (Goreau, 1959). The origin of carbonate in the coral skeleton is less clear, but because the $^{18}O:^{16}O$ and $^{13}C:^{12}C$ ratios in the skeleton is unlike that in sea water, it is reasonable to assume that the carbonate in the skeleton may be metabolic in origin, at least in part (see section on effect of adequate food supply, below).

3. The Distribution and Growth Form of Corals in Coral Reefs

The distribution of corals in coral reefs has been the subject of many articles and reviews. The reviews of Yonge (1963) and Stoddart (1969) are highly recommended. In this section, I shall restrict myself to West Indian coral reefs, and the paper by Goreau (1959a) on the ecology of Jamaican reef corals and of Stoddart (1962) on the distribution of reef corals off British Honduras, should be consulted.

A coral reef may be defined as "a balanced community of calcium carbonate secreting and frame-building organisms, associated biota and mainly biogenic sediments" (Cloud, 1959). The activities of the frame building corals and of the encrusting and cementing algae produce an "organized and coherent structure adapted for maximum attenuation of mechanical stresses set up by the constant battering of the seas" (Goreau, 1961). Thus, a coral reef is not just a random aggregate of corals, but is a well structured ecosystem.

The organization of a coral reef is such that several distinct zones, each with its characteristic coral community can be recognized. The underwater photographs in Figs. 5–8 represent a transect of a Jamaican reef from depths of 3 m to depths of 70 m. Most luxurious coral growth occurs between the surface and about 10–12 m. In this zone are found the surf zone communities, often composed almost exclusively of *Acropora palmata* (Fig. 5), (but this community may be

Fig. 5. Underwater photograph of the surf zone, looking landward, in the reef at Pear Tree Bottom, Runaway Bay, Jamaca. The dominant coral species seen is the elkhorn coral *A. palmata* which characteristically grows out to meet the onrushing surf. Because of the dominance of *A. palmata*, this zone has also been called the palmata zone. In other areas however, *A. palmata* may be replaced by a mixed community of corals dominated by the hydrocoral *Millepora complanata*.

Fig. 6. Underwater photograph of the Buttress and Canyon System at Pear Tree Bottom. The massive growth seen is that of *Montastrea annularis*. In this zone, the coral growth can rise up to 8 metres above the sea floor. Depth about 10 metres.

replaced in some areas by a more mixed community dominated by *Millepora complanata*, e.g. the reefs off Gallows Point, British Honduras), and the system of buttresses and canyons. The dominant species in this latter zone is *Montastrea annularis* (Fig. 6), where the coral growth may rise some 6 m above the ocean floor. The fore-reef slope is composed

Fig. 7. Underwater photograph of the "fore reef slope". The most obvious coral is again *M. annularis*, but note the flattened growth form, and the presence of gorgonians competing for bottom space. Depth about 40 metres.

Fig. 8. Underwater photograph of the deep fore reef slope off Discovery Bay, Jamaica. Note that coral cover has markedly decreased and sponges and gorgonians are abundant. Depth about 70 metres.

of very mixed coral communities, and with increasing depth, coral cover decreases, and sponges, gorgonians, antipatharians, etc. compete successfully for bottom space (Fig. 8).

Goreau (1959a) drew attention to the marked variation of coral growth forms in different areas of Jamaican coral reefs. In general, the fragile corals tend to grow in tranquil lagoon waters or below the wave base on the windward slopes of reefs. This is well exemplified by the staghorn coral *Acropora cervicornis*. Corals such as *M. annularis* which form massive colonies in the buttresses of Jamaican reefs, become flattened unstable colonies in deeper water (Goreau, 1963. Also see Figs. 6 and 7).

Goreau (1959a) and Stoddart (1962) drew attention to the highly oriented growth of *A. palmata*, the branches of which grow into the onrushing surf (Fig. 5). This pattern is characteristic of areas where the water current experienced by the corals tends to be unidirectional and relatively strong. In other areas such as reef lagoons where the rate of water flow is less rapid and unidirectionality is less obvious, growth of *A. palmata* becomes randomized (Trench, unpublished).

Nothing is known about the possible factors that may regulate and modify the pattern of coral growth. Goreau (1963) postulated that the flattening of *M. annularis* at greater depths was in response to decreasing light intensity. The animal grows in such a manner as to increase its surface area to expose the zooxanthellae to the maximum available light. A possible explanation for the growth response of *A. palmata* may be that by growing into the current, it offers a small cross sectional area as resistance to current. If the deposition of calcium carbonate at the growing tips of colonies of *A. palmata* were under the influence of pieso-electric effects as postulated for mammalian bone (Bassett, 1965) or molluscan shells (Digby, 1968), this mechanism could conceivably regulate the direction of colony growth, through the influence of the impinging current. D. J. Barnes (personal communication), holds the opinion that growth form of some coral colonies is a function of inter-polypery competition for space into which to grow. Clearly, more observations are necessary before these points can be resolved.

4. Calcium Carbonate Deposition and the Control of Growth Rates

Many investigators have measured growth in corals by such parameters as increment in length, diameter and weight (for comprehensive review see Stoddart, 1969). The general conclusions drawn were that branching forms of corals grew faster than massive forms. The investigations of Goreau (1959b) using ^{45}Ca have in general confirmed earlier findings, but have raised many compelling questions on the mechanism of calcium carbonate deposition and of the control of coral growth. In the remainder of this paper, evidence on factors affecting the rate of calcium deposition by hermatypic corals will be reviewed.

A. Effect of Temperature on Coral Growth Rates

Aragonite is an unstable form of calcium carbonate, particularly in cold water (Lowenstam, 1964). The fact that recent hermatypic corals only build reefs in shallow warm tropical regions of the world has been pointed out by Wells (1957). However, direct experiments to determine the effects of temperature on the rates of calcification in corals are few.

In field experiments, Shinn (1966) transplanted corals from their natural habitat to one of warmer water, and found an initial increase in growth rate followed by expulsion of the zooxanthellae and eventually death. Yamazato (1966) studied rates of deposition of ^{45}Ca in *Fungia scutaria* and found that the temperature regime for optimal growth was about 24°C. Similarly, Clausen (1968) found that temperatures of 23–25°C were best for calcium deposition in *Pocillopora damicornis*. It is nonetheless unresolved whether temperature effects observed are physical or biochemical, i.e. whether temperature effects physico-chemical aspects of calcium carbonate deposition or whether enzyme-mediated reactions are affected.

B. Effect of Adequate Food Supply

There is no experimental evidence on the effect of planktonic food supply on coral growth rates. Obviously, oxidation of foodstuffs would supply free energy necessary for active transport of calcium ions as well as supply metabolic carbon dioxide. The only direct evidence that demonstrates the incorporation of metabolic carbon, derived from the oxidation of food, into coral skeletal carbonate is that of Pearse and Muscatine (1971), who fed ^{14}C-labelled protein to corals and subsequently found ^{14}C in the carbonate fraction of the skeleton.

Muscatine and Cernichiari (1969), having incubated *P. damicornis* in $NaH^{14}CO_3$ in the light, found photosynthetically fixed ^{14}C in the carbonate fraction of the skeleton. In view of the observation referred to above and the fact that zooxanthellae release photosynthate to their coral hosts, it is possible that some of the released ^{14}C products were oxidized to CO_2 by the coral, and this became incorporated as skeletal carbonate.

C. Effect of Light on Calcification Rates

Light is the environmental factor whose effect on calcification has been most studied. Goreau (1961) showed that corals with zooxanthellae calcified more rapidly in the light than in the dark, while both calcified more rapidly than corals from which the algal symbionts had been previously removed by prolonged dark treatment.

Two different hypotheses have been put forward to explain the role of zooxanthellae in the process of calcium deposition in corals. According to Goreau (1961) the algae act catalytically by removing the end products of the reaction producing calcium carbonate (see Eqn. 2), and may also remove animal metabolic wastes, thus enhancing the overall metabolic efficiency of the calicoblastic cells. On the other hand, Wainwright (1963) suggests that the rate of synthesis of the organic matrix may directly control the rate of deposition of the skeleton. Muscatine and Cernichiari (1969) inferred that the rate of supply of reduced carbon by the algae, used in matrix synthesis by the animal, may indirectly control calcification rates (Muscatine, 1968).

The physico-chemical reactions involved in the precipitation of calcium carbonate may be expressed as shown in Eqns. (1) to (4). The dissociation of H_2CO_3 in Eqns. (3) and (4) is thought to occur in the coral tissues under the influence of carbonic anhydrase.

$$Ca^{2+} + 2HCO_3^- \rightleftarrows Ca(HCO_3)_2 \qquad (1)$$

$$Ca(HCO_3)_2 \rightleftarrows CaCO_3 + H_2CO_3 \qquad (2)$$

$$H_2CO_3 \overset{c.a.}{\rightleftarrows} H^+ + HCO_3^- \qquad (3)$$

$$H_2CO_3 \overset{c.a.}{\rightleftarrows} H_2O + CO_2 \qquad (4)$$

As stated before, the zooxanthellae in corals are confined to the gastroderm, while the process of calcification is the function of the calicoblastic epidermis. There are therefore membrane and mesogleal boundaries separating the calicoblastic epidermis from the zooxanthellae. It must therefore be assumed, under the Goreau scheme, that photosynthetic utilization of bicarbonate or carbon dioxide establishes gradients of bicarbonate concentrations between the calicoblastic epidermis and the carbonate deposited in the skeleton.

Goreau's scheme assumes a partitioning between the two bicarbonate pools, i.e. the pool serving as a reactant (Eqn. 1) and that generated as the reaction product (Eqns. 3 and 4). If the reaction product were carbon dioxide, this distinction could be made in theory.

Carbon dioxide would however only be generated under acid conditions, whereas the precipitation of calcium carbonate must occur under alkaline conditions.

According to the Wainwright-Muscatine hypothesis, the algae release photosynthetic products which the coral uses to synthesize the organic matrix. The rate of supply of substrates by the algae and the rate of synthesis of the matrix then regulates the rate of deposition of the carbonate skeleton. In support of this, Muscatine and Cernichiari (1969) reported the incorporation of ^{14}C into the organic matrix of *P. damicornis* after animals were incubated in $NaH^{14}CO_3$. However, until the stoichiometric relationship between matrix synthesis and calcium deposition is established, the significance of the matrix in the control of calcium deposition remains unclear. Finally, it remains to be established whether the "matrix" of corals is a reality, i.e. whether the organic component in coral skeletons is analogous to that found in mammalian bone or is some form of organic contaminant in the skeleton.

Zooxanthellae are also thought to enhance the process of calcification in corals by removing carbonate crystal poisons and by removing animal metabolic wastes. Simkiss (1964) postulated that photosynthesis by the zooxanthellae may remove inorganic or organic phosphates, both carbonate crystal poisons, from the calcifying environment. The removal of nitrogenous waste was proposed by Yonge (1931).

Yamazato (1966) showed that in the coral *F. scutaria* there was an inverse relationship between phosphate concentration and ^{45}Ca deposition, both in the light and dark, but the inhibitory effect of phosphate was more pronounced in the dark. Lewis and Smith (1971) have shown that ammonium ion depresses photosynthesis in zooxanthellae, but increases fixation of carbon dioxide into released amino acids. Thus, zooxanthellae may remove metabolic nitrogenous waste while converting the nitrogen to a form potentially useful to the animal.

D. Zooxanthellae and Coral Nutrient Supply

The role of zooxanthellae in the nutrition of corals is currently a subject of controversy. That the algae can release photosynthetic products to the animal *in vivo* has been established (von Holt and von Holt, 1968; Muscatine and Cernichiari, 1969; Trench, 1969, 1970, 1971, b, c; Lewis and Smith, 1971). However, it is not known how much of the metabolic requirements of the animal is satisfied by the contributions of the algae. Coles (1969) has suggested that corals may not be able to derive all the carbon necessary for their respiratory metabolism from zooplankton available to them in the reef, and may therefore depend to varying extents on the photosynthetic products released by the zooxanthellae. They may however, derive most of the nitrogen and phosphate through an exogenous food supply.

If corals are dependent on their zooxanthellae for a supply of reduced carbon, why do they "expel" the algae under "adverse" conditions? (Goreau, 1964). Is there any other source of organic matter available to corals, other than that provided by the zooxanthellae and zooplankton? Corals may be able to absorb dissolved organic matter from solution (Stephens, 1962). Corals, and a variety of reef-dwelling invertebrates harbouring photosynthetic endosymbionts possess the morphological characteristics (a microvillated epidermis) usually associated with absorption, (see Goreau and Philpott, 1956; Goreau et al., 1971, 1973; Fankbonner, 1971; Trench and Gooday, 1973; Trench et al., 1973). Corals may also feed on organic detritus and bacteria (Goreau, personal communication). The contribution of each of these sources of food to the overall nutrition of the animals is difficult to quantify, but until estimate is made, the conclusion that corals are wholly autotrophic is unwarranted. Instead, it might be better to view these reef-dwelling invertebrates as nutritionally plastic, occupying several trophic levels simultaneously.

5. Synopsis

In this brief review of calcification in corals, I have attempted to point out the integrated roles of the ecological approach and the physiological and experimental approaches in understanding the process of calcification in corals. The mechanism of calcification in corals can be regarded as the "coral reef problem" of today. Although our knowledge of the possible role of zooxanthellae in coral biology, as well as some of the possible ecological factors that may influence the process of calcification has increased over the past few decades, in the final analysis, understanding the mechanism of calcification will be based on a clearer understanding of the functioning of the calicoblastic epidermis and of the chemical micro-environment between the calicoblastic cells and the deposited calcium carbonate. Heider (1881) and Ogilvie (1896) suggested that the calicoblastic cells themselves became calcified out of existence in the process of skeleton formation. Hayashi (1937) implied that calcification was extracellular, and Bryan and Hill (1941) suggested extracellular deposition of calcium carbonate on a colloidal gel matrix (mucopolysaccharide template of Goreau, 1961).

Since calcification is an extracellular process, how then is it controlled by the cells depositing the calcium carbonate? Is the function of the calicoblastic epidermis to "secrete" limestone or to modify and regulate the chemical micro-environment associated with the zones of carbonate crystal growth thereby improving the physico-chemical conditions essential to the precipitation of calcium carbonate? What is the composition of the "interstitial space" between the calicoblast cells and the carbonate skeleton? What properties of the matrix, if any, regulate crystal orientation, responsible in the long run for the species specific architechture of the skeleton? How are these patterns modified under the influence of varying environmental conditions experienced in the coral reef?

It is hoped that such questions will be answered when modern methods of biochemical and geochemical analysis, and careful field observations are brought to bear on what is probably one of the most fascinating problems in tropical marine biology today.

Acknowledgements

I would like to thank many of my colleagues whose free discussions on the problems of coral biology crystallized many of the questions posed in this brief review. I thank the Science Research Council, U.K., for support.

References

Bassett, C. A. L. (1965). *Calcified Tissues*, p. 78 (Eds. H. Fleish, H. J. J. Blackwood and M. Owen). Springer-Verlag, Berlin.

Bryan, W. H. and Hill, D. (1941). *Proc. Roy. Soc. Queensland* **52,** 78.

Clausen, C. (1968). *Experimental Coelenterate Biology* (Eds. H. Lenhoff, L. Muscatine and L. Davis). University of Hawaii Press.

Cloud, P. E., Jr. (1959). Geology of Saipan, Mariana Islands: Submarine topography and shoal water ecology. *Prof. Pap. U.S. geol. surv.* **280-K,** 361–445.

Coles, S. L. (1969). *Limnol. and Oceanogr.* **14,** 949.

Digby, P. S. B. (1968). *Symp. zool. Soc. Lond.* **22,** 93.

Fankbonner, P. V. (1971). Intracellular digestion of symbiotic zoanthellae by host amoebocytes in giant clams (Bivalvia: Tridacnidae) with a note on the nutritional role of the hypertrophied siphonal epidermis. *Biol. Bull. Mar. Biol. Woods Hole* **141,** 222.

Goreau, T. F. (1959a). The ecology of Jamaican coral reefs. I. Species composition and zonation. *Ecology* **40,** 67–90.

Goreau, T. F. (1959b). The physiology of skeletal formation in corals. I. A method for measuring the rate of calcium deposition in coral under different conditions. *Biol. Bull. Woods Hole* **116,** 59–75.

Goreau, T. F. (1961a). Problems of growth and calcium depositions in reef corals. *Endeavour* **20,** 32–39.

Goreau, T. F. (1961b). *The Biology of Hydra*, p. 269. (Eds. H. Lenhoff and W. F. Loomis). University of Miami Press.

Goreau, T. F. (1963). Calcium carbonate deposition by coralline algae and hermatypic corals in relation to their roles as reef builders. *Ann. N.Y. Acad. Sci.* **109,** 127 0 167.

Goreau, T. F. (1964). Mass expulsion of zoanthellae from Jamaica reef communities after Hurricane Flora. *Science, N.Y.* **145,** 383–386.

Goreau, T. F. and Philpott, D. E. (1956). Electromicrographic studies of flagellated epithelia in madreporian corals. *Exp. Cell. Res.* **10,** 552–556.

Goreau, T. F., Goreau, N. I. and Yonge, C. M. (1971). *Biol. Bull. Mar. Biol. Lab. Woods Hole* **141**, 247.

Goreau, T. F., Goreau, N. I. and Yonge, C. M. (1973). On the utilization of photosynthetic products from zoanthellae and of a dissolved amino acid in *Tridacna maxima f. elongata* (Mollusca: Bivalvia), *J. Zool. Lond.* **169**, 417–454.

Hayashi, K. (1937). *Palao. Trop: Biol. Sta. Studies* **1**, 169.

von Holt, C. and von Holt, M. (1968). *Comp. Biochem. Physiol.* **24**, 83.

Heider, A. (1881). Die Gattung *Cladocera*, Ehrenb. *Sitzungsber der K. Akad. der Wiss. Wien.*

Hyman, L. H. (1940). *The Invertebrates*, Vol. 1. Academic Press, New York and London.

Lewis, D. H. and Smith, D. C. (1971). The autotrophic nutrition of symbiotic marine coelenterates with special reference to hermatypic corals. I. Movement of photosynthetic products between the symbionts. *Proc. Roy. Soc. B.* **178**, 111–129.

Lowenstam, H. (1964). *J. Geol.* **62**, 284.

Muscatine, L. (1968). *Experimental Coelenterate Biology* (Eds. H. Lenhoff, L. Muscatine and L. Davis). University of Hawaii Press.

Muscatine, L. and Cernichiari, E. (1969). *Biol. Bull. Woods Hole* **137**, 506.

Ogilvie, M. (1896). *Phil. Trans. Roy. Soc. Lond.* (B) **187**, 83.

Pearse, V. B. and Muscatine, L. (1971). Role of symbiotic algae (Zoanthellae) in coral calcification. *Biol. Bull.* **141**, 350 363.

Shinn, E. A. (1966). *J. Paleont.* **40**, 233.

Simkiss, K. (1964). Phosphates as crystal poisons of calcification. *Biol. Rev.* **39**, 487–505.

Stephens, G. C. (1962). *Biol. Bull. Woods Hole* **123**, 648.

Stoddart, D. R. (1962). Three Caribbean atolls: Turneffe Islands, Lighthouse reef, Glovers reef, British Honduras. *Atoll Res. Bull.* **87**, 1–151.

Stoddart, D. R. (1969). Ecology and morphology of recent coral reefs. *Biol. Revs.* **44**, 433–498.

Thompson, T. G. and Chow, T. J. (1955). *Pap. Mar. Biol. and Oceanogr. Deep Sea Res. Suppl.* **3**, 20.

Trench, R. K. (1969). Ph.D. Thesis, U.C.L.A., Los Angeles, California.

Trench, R. K. (1970). Synthesis of a mucous cuticle by a zoanthid. *Nature (Lond.)* **227**, 1155–1156.

Trench, R. K. (1971a). The physiology and biochemistry of zoanthellae symbiotic with marine coelenterates. I. The assimilation of photosynthetic products of zoanthellae by two marine coelenterates. *Proc. Roy. Soc. B.* **177**, 225–235.

Trench, R. K. (1971b). The physiology and biochemistry of zoanthellae symbiotic with marine coelenterates. II. Liberation of fixed ^{14}C by zoanthellae *in vitro*. *Proc. Roy. Soc. B.* **177**, 237–250.

Trench, R. K. (1971c). The physiology and biochemistry of zoanthellae symbiotic with marine coelenterates. III. The effect of homogenates of host tissues on the excretion of photosynthetic products *in vitro* by zoanthellae from two marine coelenterates. *Proc. Roy. Soc. B.* **177**, 251–264.

Trench, R. K. and Gooday, G. W. (1973). Incorporation of ^{3}H-leucine into protein by animal tissues and by endosymbiotic chloroplasts in *Elysia viridis*. Montagu. *Comp. Biochem. Physiol.* **44A**, 321–330.

Trench, R. K., Boyle, J. E. and Smith, D. C. (1973). The association between chloroplasts of *Codium fragile* and the mollusc *Elysia viridis*. II. Chloroplast ultrastructure and photosynthetic carbon fixation in *E. viridis*. *Proc. Roy. Soc. B* **184**, 63–81.

Wainwright, J. W. (1963). Skeletal organization of the coral, *Pocillopora damicornis*. *Q. J. Microsc. Sci.* **104**, 169–183.

Wells, J. W. (1956). *Treatise on invertebrate Paleontology, Coelenterates* (Ed. R. C. Moore). University of Kansas Press.

Wells, J. W. (1957). Coral reefs, *in Treatise on marine ecology and paleo-ecology. I. Ecology* (Ed. J. W. Hedgpath). *Geol. Soc. Amer.* **67**, 609–631.

Yamazato, K. (1966). Ph.D. Thesis, University of Hawaii.

Yonge, C. M. (1931). Studies on the physiology of corals. III. Assimilation and Excretion. *Sci. Rep. Gt. Barrier Reef. Exped.* **1**, 83–91.

Yonge, C. M. (1963). The Biology of Coral reefs. *Adv. Mar. Biol.* **1**, 209–260.

Young, S. D. (1968). *Experimental Coelenterate Biology* (Eds. H. Lenhoff, L. Muscatine and L. Davis). University of Hawaii Press.

Archaeological Evidence for Eustatic Sea Level Change and Earth Movements in South West Turkey

N. C. FLEMMING
and
N. M. G. CZARTORYSKA

National Institute of Oceanography, Wormley, Surrey

1. Introduction 395
2. Field Methods and Data 396
3. Analysis 398
4. Discussion 399
References 403

1. Introduction

Allan (1966) uses bathymetric and gravimetric data to show that the arc of islands from the Peloponnese through Crete and Rhodes to South West Turkey is similar in many ways to a classical island arc feature, and stresses the importance of this arc as the southern boundary of the Aegean basin. McKenzie (1970) interprets the structure of the Aegean in terms of plate tectonics and indicates that the Cretan arc represents a plate boundary. The boundary is proposed as undergoing right shear movement and compression. Flemming (1968a, b) reported a broad tectonic doming of the Peloponnese associated with an actively folding anticlynal ridge trending south from the Elos peninsula towards Crete. The present work is concerned with a similar investigation of the eastern end of the island arc, with the purpose of revealing earth movements on the Turkish coast, and thus giving a more complete understanding of the plate boundary.

Negris (1904) and Haffemann (1960) conducted surveys of the coastal ruins of the Aegean in order to detect relative changes of land and sea level. Negris (1904) attributed the observed vertical movements entirely to tectonic factors, while Haffemann (1960) attributed them to eustatic change. Before commencing field work in this area we also had the benefit of studying the field notebooks of D. Blackman, University of Bristol Department of Classics, who had located a number of coastal ruins in the area.

2. Field Methods and Data

The sites investigated are shown in Fig. 1 and a typical ruin is shown in Fig. 2. The methods used to identify ruins, estimate dates, and derive relative local changes of level are described in Flemming (1969, pp. 6–13). Literary data on sites was gained from the Admiralty Pilot,

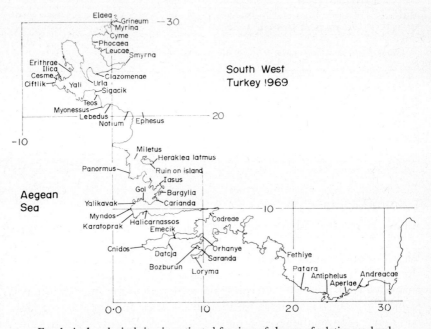

FIG. 1. Archaeological sites investigated for signs of change of relative sea level.

Admiralty charts, Bean and Cook (1957), Stark (1956), Bean (1966), and from archaeological reports on individual sites. The reliability of an estimate of change of level varies considerably from site to site, depending on the nature of the remains and their degree of preservation. It was considered undesirable to use a simple weighting procedure as this would inevitably be subjective, and might lead to reinforcement of preconceived hypotheses.

Each site was therefore allocated a uniform weight of 10 units, and in the event of a highly probable accurate estimate of displacement all points would be allotted to this estimate. For less reliable estimates, weighting points were loaded in successive 25 cm bands above and below the best estimate of displacement so as to generate a probability histogram for vertical displacement of a single site. Such a histogram is shown for site 23 in Fig. 3. In the majority of cases a total of 10 units was allocated, but in three cases where the overall reliability was low, the overall weighting was reduced.

The histograms are not in general symmetrical. In some cases ruins originally constructed on land are found underwater indicating a minimum necessary relative change of level, but in the absence of

FIG. 2. Submerged Roman harbour fortification at the entrance to the harbour of Myndos.

other evidence the change might have been considerably larger. Alternatively, ruins such as a fish tank, water channel, or breakwater, may dictate a maximum possible relative change of level, but set no minimum. Thus histograms of estimates and their probabilities may be highly skew. In other cases a low probability may be allocated equally to a wide range of values.

The area was divided into a 2 cm rectangular grid superimposed on a 1 : 500 000 mercator projection, and the position of sites recorded in terms of X and Y coordinates with reference to an arbitrary datum zero. Age of sites was estimated from the literature, supported or modified by field observations. Estimates are expressed in thousands of years to the nearest hundred years, but in many cases an uncertainty of dating of the order of 300 years must be accepted. This is due to uncertainty as to date of foundation, date of last occupation, and submergence of the smaller sites. Sites with ruins of many ages produce two or more estimates of displacement and date.

The data are summarized in Table 1. The weighting values and total weights are shown in the right hand columns. The weighting numbers listed for each site show the weight allocated to each 25 cm band on either side of the best estimate. All positive values indicate submergence of the land relative to the sea, or a relative rise in sea level,

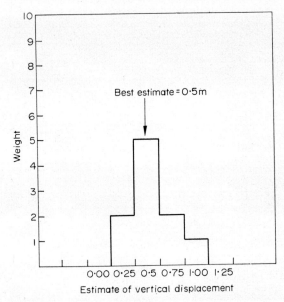

Fig. 3. Diagram illustrating the allocation of probability weighting to different estimates of the relative change of level at site 23, Karatoprak.

thus negative divergence from the best estimate indicates less submergence, and positive divergence indicates more submergence.

3. Analysis

Figure 4 shows the best estimates of displacement for each site plotted against age of site, together with the best fit least squares third degree curve. The mean displacement of sites and the variation in displacement both increase with age. This is compatible with earth movements of randomly varying rate combined with some eustatic change of sea level, but presentation of the data in this form provides no means of separating the factors.

The magnitude of displacement and the mean rate of displacement were then correlated with coordinate positions, taking into account weighting. The best fit was obtained with a fourth degree surface in terms of rate of displacement, indicating that earth movements are either continuous and gradual, or, if discontinuous, of short periodicity compared with the time span of the observations, that is about 2000 years. This analysis indicates a strong correlation with geographical location, but does not exclude the possibility of a small eustatic change. further statistical analysis designed to separate the eustatic and tectonic factors and quantify them will be published elsewhere.

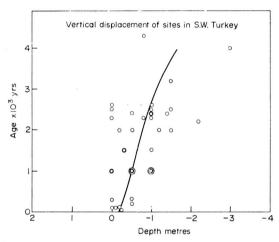

FIG. 4. Plot of the vertical displacement of sites against the age of the sites. The best fit third degree curve indicates the mean displacement increasing with age; this could be accounted for either by eustatic change or by net submergence of the coast over the whole area. The wide variation in rate of displacement of sites indicates that earth movements are of general occurrence.

This preliminary analysis indicates that the dominant features of earth movement are a general seaward depression combined with active anticlynal fold patterns leading west by the Cesme peninsula and South West across the Cnidian peninsula towards Rhodes (see Fig. 5). A surprising feature is the apparent increasing depression landwards, suggesting that the present coast is a fold axis, but this may be a spurious effect resulting from the surface-fitting programme having no control points of data inland.

4. Discussion

The bathymetric map produced by Giermann (1960) shows the continuous nature of the island arc from Greece to Turkey, with a

TABLE 1. Age, position, and estimate of vertical displacement for sites in the area shown in Fig. 1. Names followed by Roman numerals indicate estimates based on ruins of different ages at the same site. The weighting numbers in each case define a probability histogram of the type shown in Fig. 3.

Name	Age ×10³ yrs	Coordinates X	Y	Best Estimate E	Depth in Metres Weighting in 25 cm intervals −75	−50	−25	E	+25	+50	+75	Total Weight
Antiphellus I	2.2	23.60	0.85	2.2	—	—	—	8	1	1	—	10
Antiphellus II	1.0	23.60	0.85	1.0	—	—	1	7	1	1	—	10
Andriake	1.0	27.50	1.60	1.0	—	1	2	4	2	1	—	10
Bargylia	2.0	5.10	11.60	0.2	—	—	1	7	2	—	—	10
Bozburun	1.0	9.20	5.80	1.0	—	3	3	4	—	—	—	10
Caryanda I	0.3	5.05	10.90	0.0	—	—	—	8	2	—	—	10
Caryanda II	1.5	5.05	10.90	0.3	—	—	—	7	3	—	—	10
Cedreae	1.5	10.70	9.45	0.3	—	—	3	6	1	—	—	10
Cesme	0.1	−6.10	23.85	0.2	—	—	—	5	—	—	—	5
Ciftlik	0.1	−6.30	23.55	0.1	—	1	1	5	2	—	—	10
Clazomenae	2.3	−1.85	24.50	1.0	—	1	1	4	2	2	—	10
Cnidos	2.3	3.25	5.90	0.0	—	1	1	7	1	—	—	10
Cyme	2.4	−0.45	28.85	1.0	—	1	2	4	2	1	—	10
Datcja	2.5	6.15	6.20	0.0	—	—	1	8	1	—	—	10
Elaea	2.3	0.50	30.85	0.0	—	—	3	3	2	—	—	8
Erithrae I	0.05	−4.55	24.65	0.25	—	—	5	5	—	—	—	10

Site										
Erithrae II	0.2	−4.55	24.65	0.5	—	—	5	5	—	—	10
Erithrae III	2.4	−4.55	24.65	1.40	—	—	2	6	1	—	10
Halicarnassos	2.4	3.75	9.75	1.0	—	1	1	6	2	1	10
Heraklea Latmus	2.5	4.55	14.95	1.5	—	—	1	1	1	1	5
Iasus I	1.0	5.25	12.35	0.0	—	1	—	7	3	—	10
Iasus II	2.0	5.25	12.35	0.5	—	—	3	6	—	—	10
Ilica	2.4	−6.70	23.9	0.5	—	2	2	2	2	2	10
Karatoprak	1.0	2.25	9.40	0.5	—	—	2	5	2	1	10
Loryma	1.5	9.05	4.65	1.0	—	1	3	6	—	—	10
Miletus I	2.0	2.45	15.30	1.5	—	2	3	5	2	1	10
Miletus II	3.2	2.45	15.30	1.50	—	—	2	5	1	—	10
Myndos	2.0	2.05	10.0	1.2	—	1	1	8	2	1	10
Myonessus	2.6	−1.15	20.85	1.0	—	—	2	4	1	—	10
Orhaniye	1.0	10.05	6.80	0.0	—	1	—	5	2	1	10
Panormus	2.6	1.95	13.90	0.0	—	—	2	4	5	—	10
Saranda	1.0	9.60	5.70	0.5	—	1	5	5	2	—	10
Sigacik I	2.3	−1.85	22.50	0.8	—	—	2	7	—	—	10
Sigacik II	0.1	−1.85	22.50	0.0	2	—	—	5	5	1	10
Smyrna I	2.5	1.50	25.65	1.0	—	—	—	6	3	—	10
Old Smyrna	4.0	1.50	25.65	3.0	—	2	3	3	—	1	10
Teos	4.3	−1.75	22.30	0.8	—	1	2	7	1	—	10
Urla Beach	0.3	−2.65	23.9	0.5	—	—	3	6	—	—	10
Yali	1.0	−4.95	24.10	0.5	—	—	—	6	3	1	10

succession of sills at depths of the order of 500 m separating depths to north and south of the order of 2000 m. Allan (1966) and McKenzie (1970) show that the seismicity associated with this ridge is distributed so as to leave an aseismic zone or plate in the southern Aegean. Although lateral movements have been postulated by McKenzie (1970), no assessment has been made of the vertical movements to be expected at the plate boundary.

The structural geology of the highly indented coast of Turkey is not

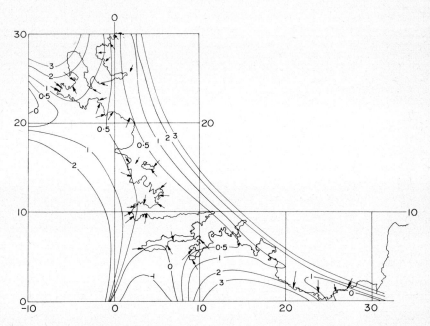

FIG. 5. Contour diagram relating rate of displacement of sites to geographical location. The figures represent relative submergence in metres per millenium. The arrows mark sites from which data were obtained; the names of the sites are given in Fig. 1.

known in detail, but there are a very large number of visible recently active faults. Thus the present contours of mean rate of earth movement can only be taken as very much smoothed averages both in time and place, indicating broad regional trends. Nevertheless, it is significant that the most marked feature of the contour pattern is the anticlynal fold trending west of south, directly on the axis of Rhodes, Karpathos, and Crete. This is compatible with the concept of an actively folding ridge along the axis of the island arc. The anticlynal pattern near the Cesme peninsula is less expected, but may be associated with the shallow plateau bordering Chios, and the sill at the southern margin of the northern basin of the Aegean, where there is a marked change in topography and relief.

The landward depression indicated by the present study may, as already noted, be a mathematical fiction. However, Ambraseys (1970) states that the main earthquake zone in historical times in this area was back from the coast in an arc from Izmir to Antalya, while McKenzie (1970) suggests that the eastern plate boundary of the Aegean is also set well back from the coast, though on a more north-south line. These two observations suggest that the coastal block between Izmir and Antalya is in a very unstable position, and it is just possible that the landward depression is genuine. This proposal is supported by the observations of Wendel (1968) who suggests that the extreme volumes of sediment accumulated in the estuary of the Meander and other rivers can only be explained if there has been ponding due to a landward tilting of the coast.

References

Allan, T. D. (1966). Recent geophysical studies in the Aegean and Eastern Mediterranean by the R/V Aragonese, *NATO Subcommittee on Oceanographic Research, Technical Report* **18**, p. 19–26.

Ambraseys, N. N. (1970). Some characteristic features of the Anatolean fault zone, *Tectonophysics* **9**, 143–165.

Bean, G. E. (1966). *Aegean Turkey*, 288 pp. Ernest Benn, London.

Bean, G. E. and Cook, J. M. (1957). The Carian Coast, *Annual of the British School at Athens* **52**.

Flemming, N. C. (1968a). Archaeological evidence for sea level changes in the Mediterranean, *Underwater Association Report*, 9–12.

Flemming, N. C. (1968b). Holocene earth movements and eustatic sea level changes in the Peloponnese, *Nature, Lond.* **217**, 1031–2.

Flemming, N. C. (1969). Archaeological evidence for eustatic change of sea level and earth movements in the western Mediterranean during the last 2000 years: Geological Society of America, Special Paper 109, 125 pp.

Giermann, G. (1960). Bathymetric chart of the Aegean, *Musée Oceanographique de Monaco*.

Haffemann, D. (1960). Ansteig des Meerespiegels in Gesichtlicher Zeit, *Die Umschau* **60**, n. 7, 193–196.

McKenzie, D. P. (1970). Plate Tectonics of the Mediterranean Region, *Nature* **226**, 239–243.

Negris, P. (1904). Vestiges antiques submergés, *Athenischer Mitteilungen* **29**, 340–363.

Stark, F. (1956). *The Lycian Shore*. John Murray, London.

Wendel, C. A. (1968). Tilting or silting? Which ruined ancient Aegean Harbours? *Archaeology* **22**, 322–4.

Cape Andreas Expedition, 1969

J. N. GREEN

Research Laboratory for Archaeology and History of Art, Oxford

1. Introduction 405
2. Methods 405
3. Results 407
4. Conclusion 411

1. Introduction

In the summer of 1969, an underwater archaeological survey was carried out at Cape Andreas, Cyprus. The object of the survey was to locate and investigate archaeological wrecks and artifacts in the area. Cape Andreas was selected as it had been visited briefly in 1967 by the author while working on an expedition in collaboration with Pennsylvania University. Within a period of a few days at the site on that occasion, three wrecks were located and it was felt that this area merited a more thorough investigation.

2. Methods

As there were no maps available of the area, a large scale map was constructed from aerial photographs. This map was taken to Cyprus and served as the basis for all the survey work carried out. The first phase of the work was to establish the extent of the 50 m contour, as this was to be the limit of the divers' search. An echo sounder survey was carried out and a complete hydrographic chart of the area constructed (Fig. 1).

The divers' search was based on a modified version of the "swimline" technique devised by Lt.-Comm., John Gratton. The swimline was guided by a jackstay stretched from the anchor block of a buoy marking the 50 m contour to the shore. The end diver on the swimline followed the jackstay and kept the swimline at right angles to it. The horizontal visibility was about 40 m, so the divers were stationed at 20 m intervals for good visual overlap. Figure 2 shows the arrangement for a swimline. A group of four divers surveyed one side of the jackstay and a second group of four surveyed the other side. Twenty-six swimlines

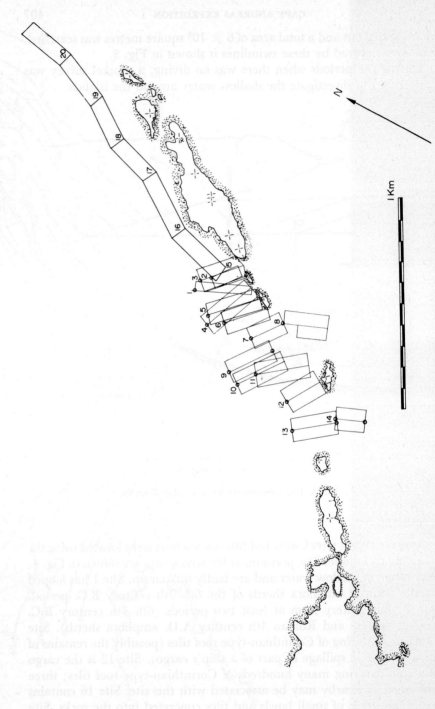

FIG. 3. The areas searched by the swimline technique.

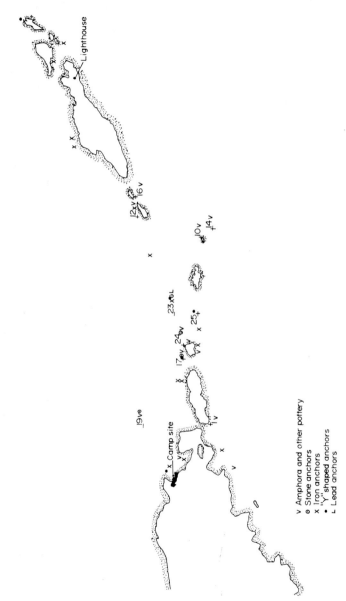

FIG. 4. The wreck sites loacted at Cape Andreas.

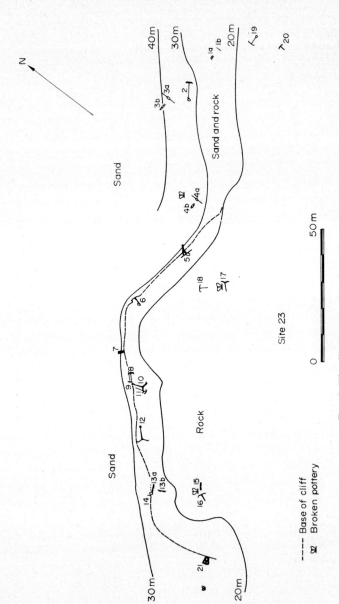

FIG. 5. The distribution of anchors at site 23.

19 is possibly two wrecks; pottery from this site was investigated in 1967 and indicated that it was of the 4th century B.C., but again looped handle fragments indicate an earlier date. A small stone anchor is clearly associated with this site. Sites 17 and 24, also located in 1967, contain Byzantine 7th century A.D. pottery with small "hour glass" amphoras and terracotta boxes (possibly sarcophagi).

The outline above indicates the variety of material found on the wreck sites. However, much more work is required on the material before a definitive report can be produced.

Of the fifty-six anchors located and coordinated on the map, only a few were investigated due to lack of compressed air and the relatively short time available. The most interesting anchors were located at site 23 where a total of twenty-one were found in an area approximately 150 m by 50 m (see Fig. 5). Seven were of lead, three being "Roman" type while the others showed the remains of lead stocks. There was only one stone anchor, all the rest being of iron. Three of the iron anchors were "Y" shaped, without any sign of a stock. Only a few anchors of this type have previously been located; three other examples of this type were also found at Cape Andreas. Two large iron anchors with shanks about 3 m long were found lodged in the base of the cliff. Of the twenty one anchors located at this site, there was only time to draw eleven.

It is possible that this site represents an anchorage, but in view of the close proximity of reefs, rocks and islands, it is not a particularly good one. Very little pottery was found in this area, whereas the two obvious anchorages to the North and South of the Cape were found to be littered with sherds over large areas. In view of this it is possible that the anchors at site 23 were lost by ships that were dragging their anchors in a northerly gale. Because of the very deep water to the North of the Cape, ships could only hold when the anchors reached the first rock outcrops at the base of the cliff at site 23. Either the ropes then chaffed away or the anchors were fouled and were abandoned. More work on the site is required before definite conclusions can be drawn.

4. Conclusion

The results of the 1969 survey indicate that, if underwater surveys are carried out in a systematic manner, a considerable amount of archaeological information can be obtained. Even though only a limited amount of work was done on the wreck sites and anchors, their positions are accurately known from the survey and this will allow rapid relocation on a future occasion. Furthermore, the precise extent of the divers' search is known and continuation of the work can be carried out rapidly and efficiently at any future date.

Subject Index

A

Absorbent chemicals in habitat, 256
Absorber, soda-lime charcoal, 256
Absorption of acoustic energy, 50
   light, 144–146, 148, 149, 150, 155, 160–162, 204, 366
   organic matter, 391
Acanthaster planci, 303, 306–308, 313–319, 325
   behaviour, 311, 313–318, 325
   distribution, 315
   feeding, 306, 307, 313–318
   habitat, 319
   nocturnal behaviour, 311, 313–318, 325
   population density, 313, 317
      dynamics, 303
Acclimatization, 243, 244
Accommodation of human eye, 158
Accumulators, 218, 258
Acidic vapour in habitat, 256
Acoustic baffle, 24, 101, 103, 131
   beacon, 116
   carrier, 153, 154
   controlled environment, 11, 24, 44
   energy, 4, 50, 51, 83, 98, 99, 101
   impedance, 57, 83, 98, 99, 265
   power, 85
   pressure, 85, 128, 262
   product, 4
   radiation, 131
   segments, 4
   shifts, 6–8, 13, 15–22, 24, 40, 51
   systems, 2, 41, 53, 55, 57–78
   transparency, 83
   tube, 7
   wave, 83, 179, 265
Acoustics, principles of, 82
Acropora, 310, 317
   A. cervicornis, 387
   A. palmata, 384, 385, 387
Acuity, auditory, 88, 105, 131
   visual, 160, 297
Adaptation to colour, 199, 206, 207

Adrenal cortex, 218
Adrenalin, 235
Aegean, 395, 402, 403
Aerial photography, 267, 405
   of radio pill, 226, 229, 230
"After drop", 234
Age of *Laminaria hyperborea*, 370–372
   of ruins, 396–398, 400, 401
Aggregates of sperulites, 384
Ahermatypic coral, 381, 382
Air cell, in theodolite, 270
Alcohol and visual attention, 210
Alcyonarians, 381
   feeding, 333
Alcyonium digitatum, 329, 333
Algae, 323, 324, 329, 332, 335, 348, 349, 353–355, 357, 358, 360, 362, 365, 366, 367, 369–378, 381, 382, 384, 387–391
   calcareous, 358
   Caulerpa prolifera, 360
   cementing, 384
   encrusting, 384
   epiphytic, 366
   kelp (see *Laminaria*), 323, 324, 349, 369, 378
   Lithothamnion, 324, 329
   Macrocystis, 349, 378
   planktonic, 348, 349, 353, 354
   Saccorhiza polyschides, 371
   symbiotic, 381, 382, 387–391
   Udotea, 365
   Ulva, 349
   unicellular, 329
Alveolar CO_2, 237, 241–243, 249
   pressure, 240
Amphora sherds, 407, 409, 411
Amplifier, 22, 32, 57, 66, 68, 89, 106, 157, 158, 165, 245
   Ithaca, 106
   Marantz, 32
   noise, 165
   video, 157

Anchors, 407, 409–411
Andriake, 396, 400
Anechoic space, 87, 105
  tank, 105
Angiosperms, 335–345, 347–355, 357–67
  Cymodocea nodosa, 360, 363, 366, 367
Anticlynal fold, 395, 399, 402
Antinomy trisulphide photoconductor, 155, 156
Antipatharians, 387
Antiphellus, 396, 400
Anxiety, 195, 197, 209, 210, 213, 217–225, 235
Aperiodic noise, 3
  signal, 40
  wave, 4
Apertures, 159, 168, 169
Aquanauts, 11, 12, 14–16, 28
Aragonite, 384, 388
Arouser, 210, 214
Arsnine, 257
Articulation, 13–16
  errors, 13
  normal, 15
Articulators, 4, 41, 54, 55
Articulatory movement, 4
Artifacts, 405, 407, 409–411
Aseismic zone, 402
Aspect ratio, 281
Attention and narcosis, 209–216
  profile, 210
Attenuation of light, 161, 175, 337, 340, 342–344, 349
  of sound, 83, 87, 88, 90, 91, 96, 101
  coefficient, 149, 161, 162, 171–173, 180, 338, 349
  length, 148
Attenuator, 89, 260–262
  recording, 89, 91
Attitude of submersible, 285, 291–296, 298
Audio frequency spectrometer, 24
Audiogram, 87
Audiometer, 90, 91, 258, 260–262
  Béséky, 90
  Rudmose (ARJ-4), 91
Auditory mechanism, 83, 101, 133
  perception, 81, 83, 101
  sensitivity, 86, 87, 88–100, 263, 264
  system, 81–83
  threshold, 100
Autotrophic coral, 391

Avon Rubber Company, 254

B

Back scatter, 151, 164, 165, 168, 171–174, 178–184
Background light, 156, 168, 173
  noise, 81, 123
Bacteria, 391
Baffle, acoustic, 24, 101, 103, 131
Balanophyllia regia, 328, 329
Ballast, 255
  tanks, 281
Bargylia, 396, 400
Barnacle, 321, 327–331, 333
Bathymetric data, 395
  map, 399
Batteries, 218, 229, 256–258, 277, 279, 280, 282, 286, 290, 291, 296
  lead acid, 256, 258, 279
  mallory cell, 229
  nickel-cadmium, 282
Beat frequency oscillator, 25
Békésy technique, 88
Bending of light, 146
Benthos, quantitative measurement, 285–298
  stereophotography, 301
Bicarbonate, 353, 389
Bilabial, 13, 14
Binaural localization, 103, 104, 128
Binocular vision, 294
Binoculars, 159
Binomial test, 193, 194, 196, 202
Bioassay of macrophytes, 348
Biocoenoses, 299–302
  depth, 299
  distribution, 300, 301
  growth, 303
  population dynamics, 301
  productivity, 301
  quantitative measurement, 299–302
Biomass *Laminaria hyperborea*, 373
  macrophytes, 336, 342
  Posidonia oceanica, 361
Biomass maxima, 342–344
Biota, 384
  epilithic, 320, 324, 326, 329, 332, 333
Blood pressure, 218
  samples, 220, 225–227
Bone conduction of sound, 87, 93, 97–99, 103, 122, 130–132, 265
  oscillator, 132

sensitivity, 98
threshold, 97, 98
Boredom, 262
Boring organisms on *Carryophyllia*, 321, 326, 328, 329
  starfish, 305
Bozburun, 396, 400
Bradycardia, 249
Brittle stars, 304
Bryozoans, 366
Bubble on theodolite, 272
  noise, 41, 57, 69, 91, 108, 261
Buffer solution, 348, 351, 354
Bugg Springs field facility, 42, 43, 88
Buoy, fixed, 304, 307, 308
  marker, 373
  towed, 134, 286
Buoyancy of diver, 191, 194, 197
  habitat, 254
  submersible, 281
  theodolite, 271
Buoyancy controls, 282

C

Calcification, 381–392
Calcite, 381
Calcium, 353, 384, 390
  ions, 388
Calcium carbonate, 381–392
Calice, 321, 322, 328–331
Calipers, 321
Camera, photographic, 141, 152–154, 162, 164, 170, 287, 288, 290
  aperture, 159
  exposure, 154
  lens, 142, 177
  shutter, 154
  stereoscopic, 300
  television, 45, 46, 67–69, 141, 143, 149, 154, 162, 170, 176, 177, 184
Cape Andreas, 405–412
Carbohydrate, 362
Carbon, 353, 360, 362, 365–367, 372–374, 378, 388–391
  14, 348, 358, 360, 362, 373, 388, 390
  ^{14}C sodium bicarbonate, 348, 359, 372–374, 377, 378, 388, 390
  fixation, 348, 354, 360, 362, 364, 367, 374, 388, 390
Carbon dioxide and coral skeletons, 388–390

habitats, 248, 256–259
  and macrophytes, 353
Carbon monoxide, 257
Carbonate, 384, 388, 389
Carbonic anhydrase, 389
Cardiac rhythm, 223
Carrier wave, 66
Cartilage, 83, 84, 131
Caryanda, 396, 400
Caryophyllia alaskensis, 320
  C. clavus, 320
  C. smithi, 319–334
    density, 322
    depth, 322
    distribution, 320, 322–325
    ecology, 319–334
    feeding, 332–333
    growth, 326–328, 331, 333
    gut content, 323, 332
    larval settlement, 327
    morphology, 322–327
    mortality, 326–329
    nematocysts, 330
    ova, 327
    planulae, 325, 327
    population, 321
      density, 321, 324–327, 333
    reproduction, 327, 333
    size, 321, 326, 327
    skeleton, 321, 325–329
    sperm, 327
    taxonomy, 320
Caulerpa prolifera, 360
Cds cell, 200
Cedreae, 396, 400
Cells, Cds, 200
  coral, 383, 389, 391
  fuel, 257, 280
  kerr, 179
Central nervous system, 131
Centre of buoyancy, 280, 281, 296
  gravity, 256, 280, 281, 296
  pressure, 281
  revolution, 281
Centres of calcification, 384
Centrifugal force, 280
Cesme, 396, 400
Characeae, 335
Charophyta, 341
Chart recorder, 289, 348
Charting, 267
Chemical indicator tubes, 257

Chlorophyll, 375
Chromatography, 362
Ciftlik, 396, 400
Ciliary mechanism, 325
Clazomenae, 396, 400
Cliona, 329
Cnidos, 396, 400
Cochlea, 83, 84, 86, 98, 99, 130, 131
Codes, international, 2
   morse, 2, 41
Coenocyathus dohrni, 320
Coenothecalia, 381
Cold, 244, 259, 262, 282
Collimating, 268, 270, 273
Collimation axis, 268, 273
Colonization, 297, 340–342, 371
Colour adaptation, 199, 206, 207
   apparent, 201, 202
   classification, 206
   chart, 199–202, 204, 207
   contrast, 273
   difference, 142
   estimation, 202
   fluorescent, 272, 273
   laser, 180
   loss, 140, 142, 143
   perception, 199, 201, 202–204, 206
   vision, 199–207
   water, 140, 145, 149–151, 155, 156, 180, 199, 203, 204, 206
Communication, 1–137
   by codes, 2, 41
   by facial expression, 2, 41
   by gestures, 2, 41
   by speech, 1–80
   by writing, 2, 41, 272, 307, 314
Communication sciences laboratory, 36, 49, 106
   systems, 2, 41, 57–78
      acoustic, 2, 41, 53, 55, 57–78
      amplitude modulated, 66, 68, 71, 74
      Aquaphone, 66, 70, 73
      Aquasonics, 66
      Aquasonics 420, 53, 55, 70, 71, 73, 75, 76
      Aquasonics 811, 70, 73
      Bendix Divercom, 66, 70, 71, 73
      Bendix Watercom, 66, 70–73
      British Buddy line, 66, 70
      DUCS, 218, 226, 227
      Erus, 70, 73
      hard line, 66, 71, 74

NAS, 70
PQC 2, 70, 72, 73
PCQ-3, 70, 72, 73
PQC-1a, 70, 72, 73
Raytheon yack-yack, 66, 70, 73
Scubacom, 73
Sea-Tel, 66, 70
Subcom, 66, 71, 73
Compass, 133, 314
   gyro, 286, 288, 291
   heading, 288, 290, 294, 297
   ships, 286, 288
Compensation point, 372, 377, 378
Competition for food, 353
   for habitat, 324, 326, 340, 371, 386, 387
Compression, plate tektonics, 395
   sound, 82
Computer, 161, 169, 183, 245, 246, 297
   card punch, 48, 69, 90
   key punch, 47, 69, 90, 106, 108
Condensation, 245, 259
Condenser, 21
   microphone, 265
Cone of vision, 291–294
Consonant, 4, 6, 8, 9, 13, 24, 40
   amplitude, 8
   bilabial, 13, 14
   dental, 13, 14, 16
   distortion, 13, 16, 40
   energy, 8, 9, 21, 24, 40
   enhancement, 9
   errors, 14
   glottal, 13, 14
   palatal, 13, 14, 16
   suppression, 9
Consonant-vowel amplitude, 9
   ratio, 8, 10, 21, 22
Contact lenses, 274
Continental shelf, 50, 282
Contraction rates, 239
Contrast, 142, 143, 159, 168, 169, 172, 174, 177, 182, 183
   apparent target, 165, 168
   colour, 273
   inherent target, 164, 165, 167, 168, 273
   loss, 164, 165, 172
   output, 156
   perception, 157
   ratio, 165, 172, 174–177, 179, 180
   reproduction, 156, 157

SUBJECT INDEX 417

variation, 172
Convective acceleration, 238
Convolution processing, 17, 19, 22, 23
Coral, 303, 304, 306, 307, 310, 315, 316
   acropora, 310, 317
   Acropora cervicornis, 387
   A. palmata, 384, 385, 387
   ahermatypic, 381, 382
   autotrophic, 391
   blue, 381
   Caryophyllia alaskensis, 320
     C. clavus, 320
     C. smithi, 319–334
   Coenocyanthus dohrni, 320
   Coenothecalia, 381
   Devonshire cup, 319–334
   distribution, 384–387
   elkhorn, 385
   food, 388, 390–392
   Fungia scutaria, 383, 388, 390
   gastroderm, 382, 389
   gorgonians, 386, 387
   growing point, 387
   growth, 384–392
   hermatypic, 381, 382, 384, 388
   hydro, 381
   madreporarian, 319, 381
   Millepora complanata, 385, 386
   Milleporina, 381
   Montastrea annularis, 385–387
   morphology, 382–384
   Mussa angulosa, 383
   Mycetophyllia lamarkana, 383
   Pocillopora damicornis, 388, 390
   polyps, 306, 319, 321, 325, 327, 330–333, 382, 383
   porites, 310
   reef, 306–311, 381, 382, 384–388, 391
   profile, 305
   respiration, 390
   scleractinian, 319, 381, 382, 384
   skeleton, 381–392
   soft, 329, 333, 381
   spicule, 381
   stinging, 381
   stony, 381
   table, 314–318
   tentacles, 382
Coronal head plane, 116
Corynactis viridis, 322, 329
Course corrections, 289, 290
Cover of species, 330, 302, 340, 341

Cross hairs, etched, 268
   wire, 270, 273
Crown-of-thorns, 303, 306, 308, 313–318
Crystals, in coral, 390, 391
   in lasers, 180, 181
Cues, 4, 50, 102, 134, 135, 191–195, 197
   acoustic, 82, 106
   auditory, 119, 129, 130
   binaural, 122, 128
   depth, 191, 192, 194–197
   intensity, 101, 131, 132
   pressure, 191, 197
   rope, 192–194
   seabed, 191
   tactile, 113, 123, 128–130
   temporal, 101, 132
   time, 132
   visual, 82, 191, 192, 194, 197
Current of water, 42, 114, 144, 147, 186, 281, 285, 286, 288–291, 297, 325, 326, 333, 387
Current meter, 288, 289
Cyme, 396, 400
Cymodocea nodosa, 360, 363, 366, 367

D

Datcja, 396, 400
Dating, 396–398
Decompression, 219, 227, 259, 271, 314
   chamber, 58, 59, 62
   depth, 191, 193
   time, 191, 201
Deflection of sound, 101
Demodulator, 66
Density of species, 300, 301, 322, 373, 374
   of water, 82, 278
Dental, 13, 14, 16
Depressant, 210
Depression—sea level changes, 399, 403
Depth, estimation, 191–198
   gauge, 191–195, 200, 287, 340
Design of habitat, 253–255
Destructive interference, 52
Detector illuminance, 166
   tubes, 258
Detritus, 391
Devonshire cup coral, 319–334
Dexterity, manual, 209, 219, 221, 223, 224, 226
Diadema setosum, 306, 310
DICORS, 44–46, 53, 55, 56, 67–69, 75, 89,

DICORS, (continued)
90, 94, 105, 114
  mini, 46, 67, 68, 75, 114
"Diffuse masking", 184
Digital coding, 17, 19–22
Dinophyceae, 382
Discharge lamps, 151, 152
Displacement, 396–402
Diver Auditory Localization System, 105–107, 109, 110
  Communication Research System, 44–46, 53, 55, 56, 67–69, 75, 89, 90, 94, 105, 114
  efficiency, 217, 225
  Transport Vehicle, 277–283
    design, 277–282
    uses, 282, 283
Dolphin, 132
  noises, 50, 110, 111
Drag, 277–279
Duke University hyperbaric facility, 11, 40, 99, 100
Duty cycle function, 179
Dynamic pressure, 277, 281
  stability, 280
Dysprosium, 151

E

Ear, dolphin, 132
  human, 83–86, 88, 91, 96–98, 101, 103, 110, 114, 125, 128, 131, 264
    external, 83, 84, 86, 87, 91, 93–99, 130–132, 265
Ear plugs, 94, 95
Eardrums, 83, 84, 86, 87, 94, 98, 192
Earphones, 15, 39, 87, 98, 102, 130
Earth movements, 395–403
Earthquake zone, 403
Echinoderms, 303–311, 313–318, 328, 358
  abundance, 303
  Acanthaster planci, 303, 306, 308, 313–318
  behaviour, 303, 309, 311, 313–318
  brittle stars, 304
  depth, 305, 308
  Diadema setosum, 306, 310
  Echinometra mathei, 305
  Echinostrephus molare, 305
  Echinothrix diadema, 310
  ecology, 303, 304, 311
  Eucidaris metularia, 304
  habitat, 306, 311

Heterocentrotus mammilatus, 306, 310
Paracentrotus lividus, 360, 366
population, 303–311
  density, 304, 305, 307
  level, 303
  size, 303
Tripneustes gratilla, 304
Echinometra mathei, 305
Echinostrephus molare, 305
Echinothrix diadema, 310
Echo sounder, 297, 298
  survey, 405
Ecological gradient, 302
  evidence, 381–392
Ecosystem, 382, 384
Eel grass, 305
Effort, 239, 241
Elaea, 396, 400
Electrocardiography, 218, 220, 221, 223
  machine, 221
Electro-magnetic radiation, 144
Electro-optical imaging tubes, 143, 152–154
  receivers, 141, 156, 165, 173, 179
  systems, 141, 143, 150, 170, 185
Electrodes, 220, 221, 257
Electron, 167
  beam, 153, 154
  charge, 166
  image intensifier, 153
  multiplication, 153, 158
  noise, 165
  scanning, 178
Elevators, 281
Endocrine organs, 218
Energy, acoustic, 4, 50, 51, 83, 98, 99, 101
  high frequency, 9
  image intensifiers, 159
  laser, 152
  low frequency, 9
  speech, 40, 49, 51
Enzymes, 388
Epibiota, 322, 327–329, 332
Epidermis of coral, 382
  of starfish, 310, 311
  calicoblastic, 382, 389, 391
  microvillated, 391
Epiglottis, 3
Epilimnion, 340, 341
Epilithic biota, 320, 324, 326, 329, 332, 333
Epiphyte, 358, 359, 366, 367

SUBJECT INDEX

Equal pressure point, 240, 241
Ergometer, 246, 247
Erithrae, 396, 400, 401
Ethanol, 360, 362, 373
Eucidaris metularia, 304
Eustachian tube, 84
Eustatic sea level changes, 395–403
Expiratory flow, 240, 243
Expired gas, 245
   minute volume, 239, 241
Explosives, 133
Exposure, 154
Eye, fish, 156
   human, 141, 143, 146, 147, 149, 152–160, 162, 165, 170, 176, 185, 269
      accommodation, 158
      acuity, 160
      colour vision, 201, 206
      spectral sensitivity, 156, 203, 206
Eyepiece of theodolite, 268–270, 273

F

Fabric stress in habitat, 254, 255
Facial expressions, 2, 41
Faults, sea level changes, 402
Feeding, *Acanthaster planci*, 306, 307, 313–318
   Alcyonium, 333
   Caryophyllia smithi, 332, 333
   Paracentrotus lividus, 360, 366
Filament lamps, 150, 151, 156
Film, photographic, 153, 155–159, 165, 170, 176, 180, 181, 185, 204, 302
   cine, 170
   polaroid type, 180, 181
Filter, chance, 338
   infra-red, 200
   monochromatic wratten, 200
   polarizing, 182, 183
   red, 314
   spectral, 145, 156
Fish farming, 283
Flash, bulb, 152
   electronic, 152, 300
Florida aquanaut research expedition, 67, 76
Fluid in body, 218
   ear, 84, 99
Fluorimetric analysis, 220
Focal plane, 184, 268
Focusing of theodolite, 270, 271
Fontinalis antipyretica, 340

Food, *Caryophyllia smithi*, 332, 333
   corals, 388, 390–392
Food chain, 384
Force of sound, 98
Formant, 4, 8, 9, 10, 13
   frequencies, 6–8, 13, 16, 17, 19, 21
   frequency shift, 6–8, 15, 16, 19–21, 24, 40
   positions, 19
   restoring vocoder, 19, 20
   transitions, 4, 8, 24, 40
Fourier components, 183, 184
   synthesis, 183
Frequency counter, 106
   dependant filter, 184
   domain processing, 17–19, 24
   doubling, 180
   subtraction, 17, 18
Fricative, 5, 13, 14, 16, 54
Frond of *Laminaria hyperborea*, 370–378
Fundamental frequency, 10, 15, 18, 55, 56
   indicator, 15
   recorder, 15
Fungia scutaria, 383, 388, 390

G

Gametophyte of *Lammaria hyperborea*, 369, 374
Gapmeter, 258
Gas analysis, 244, 258
   density, 5, 237, 238
   exchange, 238, 240
   flow, 238, 240, 241, 244–246
   flow proportional counter, 373
   meter, 244
   mixtures, HeO_2, 1, 7–37, 55, 78, 100, 242, 244, 249, 250
Gastroderm of coral, 382, 389
Gastropods, 328
Geometric projection, 291
Geometric isolation, 143, 174–177, 179, 180
Gestures, 2, 41
Gigartina sp., 349
Glands, adrenal, 218
   pituitary, 218
Glide, 5, 14, 110–112, 114
Glottal, 13, 14
   period, 21
   pulse, 21
Gorgonians, 386, 387

Graphic level recorder, 25, 89, 106
Graticule, 268, 270, 273
Gravimetric data, 395
Grazing of *Laminaria hyperborea*, 371, 373, 374
  of macrophytes, 344
  of *Posidonia oceanica*, 366

H

Habitats, divers, 11, 28, 31, 32, 40, 56, 77, 243, 253–266, 283
Halicarnassos, 396, 401
Halo, 160, 161
Haptera, 369, 370
Hard line communication system, 66, 71, 74
Hearing, 81–137, 259–266
  directional, 260–266
  level, 87
  loss, 92, 99
  mechanism, 88, 101, 133
  perception, 81, 83, 101
  sensitivity, 86–100, 263, 264
Heart rate, 218, 223, 249
Heat loss, 231, 234
Heating of diver vehicle, 282
Helmets, 4, 57, 77
  advanced, 60, 63
  Kirby Morgan, 58, 60, 63, 64, 71
Heraklea latmus, 396, 401
Hermatypic coral, 381, 382, 384, 388
Heterocentrotus mammilatus, 306, 310
Hiatella arctica, 329
Holdfast, 369–373
Hoods, 96, 97, 114, 119, 122, 123, 128–130, 269
Hoplangia durotrix, 329
Hormones, 218, 220–226, 235
  noradrenalin, 235
  plasma cortisol, 218, 220–226, 235
Horse power, 279
Humidity in habitat, 256, 265
Hull of submersible, 277–281
Hydrogen in habitat, 256, 257
  in submersible, 279
Hydrogen peroxide, 280
Hydrographic chart, 405
  surveys, 267, 405
Hydroids, 329
Hydrophone, 45, 46, 51, 52, 66, 68, 89, 97, 106, 258, 260, 261
Hydroplanes, 278, 281

Hyperbaric facility, 10, 11, 27, 40, 43, 99, 100
Hypothermia, 119

I

Iasus, 396, 401
Ilica, 396, 401
Illuminance, 149, 159, 166, 167
Image, brightness, 159
  contrast, 142–144, 156, 157, 174, 184
  converter tube, 181
  detail reproduction, 183
  detection, 158, 167, 168
  dissector, 178
  high frequency component, 184
  illuminance, 166, 175
  information, 183
  inspection, 167, 168
  intensifiers, 143, 153, 154, 159, 160, 181
  intensity, 142–144, 162, 169
  irradiance, 157, 167, 175
  low frequency component, 184
  motion, 170
  noise, 184
  orthicon, 153, 170, 181
  perception, 273
  processing technique, 143, 183–185
  quality, 154, 183, 184
  recognition, 167, 168
  resolution, 142–144, 154, 158, 162, 163
  scanning, 152
  storage, 184–185
  transmission, 141, 162, 163
Immersion heater, 256
Inclination, submersible, 293
Inclinometer, 287, 288, 290
Incubation, 360, 373
  jar, 372
Inertial navigation system, 297
Inflatable habitat, 253–258, 275
Infra-red, 150, 151, 155
Inner ear, 131
Inquilinism, 331
Inspiratory effort, 237
  flow, 243
Integration in respiration, 246
Interference patterns in speech, 51, 52
Intra-frame correlation, 184
Ionic conduction field, 257
Irradiance, 160, 164–166, 168, 169, 171, 172, 174, 178, 337, 340, 342, 348, 374–376

meter, 149
Island arc, 395, 399, 402
Isocon, 153, 159, 160
Isotope, 348

K
Karatoprak, 396, 401

L
Lambertian reflector, 172
Lambertson diving apparatus, 104
Lamina disc cutter, 372
Laminar flow in respiration, 238, 245
Laminaria, 324, 333, 349, 369–378
  Laminaria hyperborea, 369–378
    age, 370–372
    density, 373, 374
    depth, 373–375
    frond, 370–378
    gametophyte, 369, 374
    growth, 369–378
    haptera, 369, 370
    holdfast, 369, 373
    meristem, 369, 370, 371
    morphology, 369, 370
    productivity, 371–374, 378
    respiration, 372, 374, 375
    specific leaf area, 372, 375
    sporophyte, 369
    stipe, 369–373, 375, 377, 378
  Laminaria saccharina, 371
Larynx, 3
Laser, 143, 152, 161, 177–181, 184, 267, 274, 275
  argon gas, 152, 179
  KDP crystal, 180
  liquid, 152
  lithium niobate crystal, 181
  neodynium doped glass rod, 152, 180
  ruby, 152
  solid state, 152
  yttrium aluminium garnet rod, 152, 181
Laterization, 102, 122
Leaf, freshwater macrophytes, 336, 337, 347, 348, 350, 351
  Posidonia oceanica, 357–362, 364, 366, 367
  area, 348–350, 362
  morphology, 336, 337
  scars, 358, 361, 362

tissue density, 362, 365
Learning effects, 119, 218, 262
Lens, 142, 146, 168, 169, 170, 177, 184, 270–272, 274
  transmission, 159
  wide angle, 142
  zoom, 177
Leptopsammia pruvoti, 329
Level, 268, 272
  surveys, 268
Levelling staff, 274
Life support module, 256–259
  system, 256
Lift-drag ratio, 281
Light, absorption, 144–146, 149, 150, 155, 160–162, 204, 366
  amplification, 153
  angular distribution, 147, 149, 163, 172
  asymptotic distribution, 149
  attenuation, 149
  beam, 148, 160–162, 185, 199
  detection and ranging, 179
  energy, 150, 349, 366, 374
  intensity, 142, 143, 145, 148, 149, 158, 159, 161, 164, 179, 180, 185, 191, 197, 201, 202, 207, 317, 347, 350, 351, 371–376, 387
  meter, 200, 373
    integrating, 372
  monochromatic, 148, 149
  numbers, 200
  polarized, 143, 144, 182, 183
  polychromatic, 148
  pulse, 151, 179, 180, 185, 274
  quality, 349–351
  rays, 161
  reflected, 163–165, 169–173, 181, 182
  scattered, 143, 144, 146–149, 160–164, 168, 171–174, 176, 178, 181–183, 186, 269
    angular distribution of, 147–149
    function, 162, 172
    multiple scattering, 161, 163, 171, 172
    selective absorption, 143, 145, 149–51
  transmission, 145, 155, 156, 160–162
  white, 210, 211
Limen, 99
Limestone, 381–392
Limit of maximum performance, 241
Lincoln index, 309
Lips, 3, 5

Liquid monofuel, 280
Lithothamnion, 324, 329
Littoral shelf, 343
Litorella uniflora, 335
Load on motor, 280
Loch water, 335–345, 347, 354
Lock, 11, 288, 289
Loryma, 396, 401
Loudness, 4
Low pressure oxygen line, 257
Luminance, scene, 158–160
Luminous efficiency, 151
Lungs, 3, 238–241

M

Macrocystis, 349, 378
Macrophytes, freshwater, 335–345, 347–354
  depth, 335–343, 350, 351
  distribution, 335, 342–344
  growth, 343, 344, 347
  Potamogeton crispus, 338, 340
  P. filliformis, 337
  P. gramineus, 337
  P. natans, 337
  P. obtusifolius, 336–340, 347, 353, 354
  P. perfoliatus, 337, 338, 340, 347, 350, 351, 353, 354
  P. polygonifolius, 336, 337, 340
  P. praelongus, 335, 337, 339, 344, 347–352
  P. sps., 341, 344
  Potamogetonaceae, 357
  productivity, 344, 347–354
  respiration, 337
  shading, 338
  shoot extension, 340
  specific leaf area, 336–340, 350
  zonation, 335–345, 347
Madrepore, 319, 381
Magnesium, 353
Magnetometer, 282
Malleus, 84
Mammals, 104, 116, 121, 122
Mannitol, 378
Manometry, 348, 362
Matrix, 384, 389–391
  chitinous, 384
  proteinaceous, 384
Map construction, 405
Mapping, 282, 314, 316
Marine charts, 267

Marking and recapture, 310, 311
Masks, 2, 4, 26, 28, 29, 32, 41, 55, 61–64, 70, 72, 310
  Desco, 60, 63, 70
  M.D.L., 26, 29, 32, 35, 39, 61, 62, 64, 69
  Normalair, 310
  Scott, 26, 29, 32, 35, 39, 59, 61–63, 70
  U.S. divers, 61, 63, 70
Mass density of seawater, 277, 278
Maximum breathing effort, 237, 241, 242
  voluntary ventilation, 237–240, 242
Measurement of lack to fit, 268
Memory, 220, 221, 226
  bank, 246
Mercator projection, 397
Mercury arc, 151
Meristem, 370, 371
Meristematic zones, 369
Mesoglea, 382
Mesogleal boundary, 389
Metabolism, 218
Method of limits, 87
Metrology, 152
Microphone, 15, 23–32, 34, 35, 38–40, 55, 57–64, 66, 69, 71, 258, 287, 290, 308
  advanced, 63
  aquaphone, 59, 60, 62, 63
  aquasonics, 55, 58–63
  Bendix, 59, 60, 62, 63
  CSL, 64
  dynamagnetic, 59, 60, 62, 63
  Electrovoice 664, 15, 27–32, 34, 35, 38, 39
  IRPI, 27, 29, 30
  LTV, 31, 58, 60, 64
  MDL, 29, 32, 59, 61, 62, 64, 69
  NCSL, 31
  NSRDL, 60
  Roanwell, 27, 28, 30, 32, 35, 39
  S/GP, 27, 30
  U.S. Navy, 27, 30, 31
  Yack-Yack, 59, 60, 62, 64
Microstereo-comparator, 300
Middle ear, 83, 86, 87, 91–93, 98, 99, 103, 131
  activity, 84
  system, 93, 94, 97, 98, 265
Midsagittal head plane, 116
Mie theory, 146
Miletus, 396, 401
Millepora complanata, 385, 386

SUBJECT INDEX

Milleporina, 381
Minimum audible angle, 102, 116–125, 128, 129, 131
　field, 86, 87
　pressure, 87, 88
Minute ventilation, 247
　volume, 239
Mirrors, 168
　rotating, 178
Modulation transfer function, 163, 165, 176, 184
Modulator, 18, 66
Monia squama, 329
Monastrea annularis, 385–387
Mooring of habitat, 255, 256
Morphological adaptation
　Laminaria hyperborea, 375
　macrophytes, 336–337, 347, 350
Mosses, 335, 340
Motor, behaviour, 124
　electric, 280
　efficiency, 277, 279, 280
Mouth, coral, 382
　man, 3
Mouthcups, 42, 58, 77
Multiple sound source, 105–108
Muscular effort, 240
Musculature, 384
Mussa angulosa, 383
Muzzle, 4, 42, 55, 57–64, 70, 71, 73, 74, 77
　bencom, 70
　bioengionics, 59, 62, 63, 69, 71, 73
　　nautilus, 55, 57, 58, 71, 73, 74
　M.D.L., 59
　nautilus, 70
　raytheon, 59, 61, 62, 33, 70
Mycetophyllia lamarkana, 383
Myndos, 396, 401
Myonessus, 396, 401

N

Nasal cavity, 3, 6
　speech, 14, 15
　voice quality, 5, 6
National Oceanic and Atmosphere Administration—
　man undersea science and technology, 76
　manned undersea activities programme, 78
Naupilius, 329, 330, 332

Navigation, 112, 116, 122, 130, 132–135, 142, 282, 286–288, 297
Nematocysts, 330
Nitella opaca, 341
Nitrogen, 5, 7, 8, 257, 259, 390, 391
　mixture, 259
　narcosis, 191, 192, 209–224, 271
　effects, 5, 209–216
Nitrogenous waste in corals, 390
Nocturnal behaviour, *Acanthaster planci*, 311, 313, 318
Noise, 15, 28, 31, 32, 50, 101, 103, 108, 115, 165, 209, 210, 261–265
　ambient, 4, 10, 31, 42, 50, 52, 55, 68, 87, 114, 122, 264, 265
　amplifier, 165
　background, 81, 123
　biological, 50, 110, 111, 114, 116, 117, 264, 265
　current, 166, 167
　dolphin, 50, 110, 111, 114, 116, 117
　electron, 165
　fish, 50
　propellor, 50
　snapping shrimp, 50, 264, 265
　thermal, 27, 50, 101, 102, 106, 114, 115, 119, 129, 130
　white, 199, 121, 123
Noradrenalin, 235
Nose, 3
Nutrients, 342, 343, 348, 351–355, 390–392
Nylon-neoprene fabric, 254

O

Objective in theodolite, 270
Octave filter bank, 261, 263
Odometer, 286, 287, 289, 297
　wheel, 288, 295, 297
Old Smyrna, 401
Optical efficiency, 152
　frequency doubling, 152
　system, 267, 270
Oral cavity, 3, 42
　disc, 382
Orhanive, 396, 401
Orthicon, 153, 159, 160, 170, 181
Oscillator, 18, 21, 25, 52, 89, 90, 106, 132, 229, 261
Oscilloscope, 106, 245
Ossicles, 83, 84, 86, 309

Ossicular inertia, 131
Oxidation, 388
Oxygen, 7, 43, 257, 259, 280, 325, 348
  consumption, 239, 241, 242
  meter, 258
Ova, 327
Oval window, 84, 86

P

Pachymatisma, 329
Palatal, 13, 14, 16
Palate, 3
Panormus, 396, 401
Paracentrotus lividus, 360, 366
Paracyathus inornatus, 320
Parallex, 271, 273
Paramagnetic meter, 257
Patch counts, 306, 307, 311
  reef, 309, 313
Peak flow valve, 241
Pedal disc, 382
Pendulum inclinometer, 290
Periodic signals, 40
  sound, 82
  tone, 3
  wave, 4
Periphyton, 344
Pharyngeal cavity, 3
Pharynx, 3
Phase distortion, 52
Phi phenomenon, 134, 135
Phonene, 5, 7, 13–16, 54, 110–112, 114
Phonetics, 15
Phoronis hippocrepia, 329
Phosphate, 348, 351–354, 390, 391
Phosphor screen, 153
Photic zone, 340
Photo current, 166
Photo detector, 158, 178
Photo-electric converting surface, 154
Photo-emissive surface, 153, 155, 157, 170
Photogrammetry, 267, 282, 299–301
Photographic processing, 156–157, 173, 183, 185
Photography, 141, 142, 149, 150, 153, 157, 165, 204
Photosynthesis, 353, 358, 362, 365, 366, 369–378, 388–390
Photosynthesis rate, 337, 358–360, 362, 363, 365, 366, 367, 372
  organ, 348

Phototactic, 318
Phytoplankton, 348, 349, 353, 354
Pieso-electric effects, 387
Pigment, 374, 375
Pigmentation, 367
Pinnae, 83, 84, 96, 101
Pitch, 280, 281, 288, 290, 291, 293, 297
  angles, 294–297
  moments, 281
  plane, 281
  repetition rate, 21
Placopecten magellanicus, 285
Planck's constant, 190
Plane table, 275
Planing board, 307, 308
Plankton, 286, 348, 349, 353, 354, 388
Planulae of *Caryophyllia smithi*, 325, 327
Plasma cortisol, 218, 220–226, 235
Plate boundary, 395, 402, 403
  tectonics, 395, 399, 402
Plethysmograph, 244
Pleural pressure, 239, 241
Plumb bob, 269, 271
Pneumatic drilling, 301
Pneumotachograph, 244, 245
Pocillopora damicornis, 388, 390
Polyp expansion, 332, 333
Polyzoa, 324, 329
Ponding, 403
Porites, 310
Porpoise, 121
Posidonia meadows, 357, 360, 362, 365, 366, 367
  P. oceanica, 357–367
    depth, 361, 363–365, 366
    distribution, 357
    ecology, 357
    growth, 357–367
    morphology, 359
    productivity, 358
    respiration, 362, 365, 366, 367
    rhizomes, 361, 362
    shading, 366
    shoot, 358, 360, 363, 364
  P. sps., 362
Posidoniaceae, 357
Potamilla reniformis, 329
Potamogeton crispus, 338, 340
  P. filliformis, 337
  P. graminseus, 337
  P. natans, 337
  P. obtusifolius, 336–340, 347, 353, 354

P. perfoliatus, 337, 338, 340, 347, 350, 351, 353, 354
P. polygonifolius, 336, 337, 340
P. praelongus, 335, 337, 339, 344, 347, 348, 349–352
P. sps., 341, 344
Potamogetonaceae, 357
Potassium of dihydrogen phosphate, 152
Power of vehicle, 278, 280
Pottery, 410, 411
Pre-palatal, 14
Preservations of specimens, 321, 360, 373
Pressure, acoustic, 85, 128, 262
  alveolar, 240
  chamber, 28, 29, 31, 32, 100, 209–212, 215, 217, 223, 224, 243–248
  cues, 191, 197
  differential, 255, 280
  drop, 238, 245
  in ear, 83, 86, 91
  effects, 2, 4–11, 13, 16, 25, 28, 32, 37, 40, 41, 43, 52, 54, 82, 83, 87, 88, 92, 93, 99, 100
  head, 238, 239, 241
  pleural, 239–241
  sensitive radio pill, 229
  static, 87
Pressure-flow relationship, 241
Probe, 229, 230, 348
Projector, 32, 52, 66, 89, 104–106, 117, 118, 125, 126
Propellor, 281
  efficiency, 279, 280
  noise, 50
  position, 280
  variable pitch, 280
Propulsion, 278, 280
Protein, 381, 388
Protractor, 288, 290
Puerto Rico International Undersea Laboratory, 56
Pulmonary function, 237
Pulse generator, 245, 246
  train, 111, 112, 114, 116, 262
Pulses, laser, 152, 180, 181
  light, 151, 179, 180, 185, 274
  sound, 101, 106, 110, 119, 129, 135
  time, 286, 289, 290
Pupil of eye, 170
Pure tone sensitivity, 92
Pyrgoma anglicum, 321, 327–331, 333
  distribution, 329, 330

naupilius, 329, 330, 332
population, 329, 330, 331

Q

Quadrats, 304, 305, 310, 311, 321, 358, 362, 372, 373
Quantum efficiency, 165, 170
"Quiet" conditions, 261, 265

R

Radar, 286
Radial artery, 223
Radiance transfer, 273
Radiation, 152, 165
  detectors, 165
Radio pill, 226, 227, 229–233
Radioactivity, 348, 360, 373, 374
Radioisotope, 359, 360, 373
Range of vehicle, 277
  gating, 143, 152, 179–181
Rarefactions, 82
Rate of ascent, 194
  descent, 192, 194
Recall, 220, 221
Recorder, graphic level, 25
  polar, 125
Red light, 314
Reflection of laser light, 181
  coefficient, 169
  total, 172
  total internal, 51
Reflectivity, 169, 172
Reflectors, 104, 105, 169, 177
Refraction, 144, 146
  multiple, 146
Refractive deterioration, 163, 291
  index, 146, 183, 270, 291
Remainder variance, 212
Resonance, 7, 42
Resonant characteristics, 4, 7, 83, 84
Resonation of vocal cavities, 3
Resonators, 3
Respiration and anxiety, 218
  coral, 390
  depth, 237–240, 242, 243, 248, 249
  Laminaria hyperborea, 372, 374, 375
  macrophytes, 337
  measurement, 237–250
  Posidonia oceanica, 362, 365–367
Respiratory function, 237, 243, 244, 249
  gas, 245
  loss, 362, 366

Respiratory function, *continued*
  mechanism, 3
  minute volume, 239, 241
  muscles, 239
Retina, 170
Reverberation chamber, 51
  tank, 104
Reynold's number, 278
Rhizomes of *Posidonia oceanica*, 357, 358, 362
  growth, 361, 362
  leaf scars, 358, 361
Rib cage, 3
"Ringrose" meter, 257
Roll, 280, 281, 288, 290, 291, 294–297
  angles, 294, 296
  stability, 280
Roof tiles, 407
Rooting depth, 335–337
Rotameter, 245, 246
Rotary Blower, 246
Royal Naval Physiological Laboratory, 244
Rudder, 281
  control, 281

S

Saccorhiza polyschides, 371
Salinity, 123, 146
Saranda, 396, 401
Saturation, diving, 1, 7–41, 77, 243–244, 247, 250, 258, 259
  levels, film, 156
Scallops, 285, 288, 289, 295–297
  Plactopecten magellanicus, 285
  population density, 285, 296–298
Scan lines, 158, 160
Scanning, 109, 110, 154, 157, 176–179
  beam, 153, 154, 181
  electron, 153, 154, 178
  laser, 177
  linear, 154, 176
  speeds, 182
  television, 153, 154, 167, 177, 183
Scatter meters, 186
Scleractinia, 319, 381, 382, 384
Sea bed, 50–52, 65, 191, 218, 282, 286, 290–297, 299–302, 321, 361, 362
  mapping, 282
  sand, 147
  silt, 147
  slope, 340

Sea conditions, 51, 52, 65, 68, 186, 270–272, 275
Sea level changes, 395–403
Sea urchins, 305, 360
Seal, 104, 121
Sealab, 11, 12, 24
Sealion, 104, 121
Seams in habitat, 255
Search, 142, 169, 185, 405, 407, 411
  open, 304–306, 308, 310
  swimline, 407, 408
Sediment, 403
Sedimentation, 299, 324, 325
Semi-circular canals, 84
Sentance comprehension, 219, 221, 223, 224, 226
Serum caps, 359
  ports, 373
Servometer, 125
Shoot, 358, 360, 363, 364
  extension, 340
Shutter, 154
Sigacik, 396, 401
Signal to noise ratio, 40, 103, 128, 154, 158, 165–168, 170, 178, 180
Signal processing, 183
  routing, 17, 19, 23
Sills, 402
Silting, 324, 325, 333
Sinus cavity, 265
Sinusoids, 51, 90, 101, 106, 112, 163
Skin, 83
  blood flow, 234
Skull, 98, 99, 132
Smyrna, 396, 401
Snapping shrimp, 50, 264, 265
Soda-lime absorber, 256
Soft palate, 3, 6
Sonar, 50, 104, 133
  domes, 125
  doppler, 297
  side-scan, 282
Sonic beacon, 297
Sound, amplitude, 4, 98, 101, 109, 260, 262
  barrier, 83
  definition, 82
  diffraction, 101
  duration, 122
  energy, 4, 50, 51, 83, 98, 99, 101
  intensity, 83, 86, 102, 112–114, 119, 123, 128–130, 131, 260

SUBJECT INDEX

level meter, 15, 56, 258, 260, 261
localization, 83, 98, 100–132
localization, acuity, 116, 123
 response, 104, 109, 110, 112, 117, 126, 127
 perceived, 4
 perception, 98, 101, 102
 pressure field, 263
  level, 40, 51, 52, 83, 88, 91, 92, 106, 110, 111, 114, 117, 129, 264, 265
 propogation, 49–52, 83
 segment, 4
 transmission characteristic, 7
 velocity, 265
 wave, 265
Sounds, 82
 aperiodic, 3
 clicks, 122
 high percussion, 103
 periodic, 3, 82
South-west Turkey, 395–403
Spatial frequencies, 163, 165, 184
 orientation, 82, 101, 135
 pattern, 300, 302
Speaking intensity, 10, 15
 rate, 15, 54–56
Species composition, 300
Spectral absorption, 149, 199
 distribution, 203
 filter, 145, 156
 intensity, 335, 351
 lighting, 201
 response, 154–156, 301
 sensitivity, 154–156, 200, 203, 206
Spectrometer, 200
Spectrophotometer, 201, 202, 374
Spectroradiometer, 338, 348, 349
Spectrum speech, 4, 6, 52
 light, 144, 149, 151, 184, 200, 204, 206, 207
Speech, 1–80
 behaviour, 10
 correction, 10, 12–14, 16, 32, 33
 degradation, 4–8, 10, 50, 52, 57
 discrimination, 92, 93
 distortion, 2, 4, 7, 10, 11, 12, 14, 16, 24, 31, 33, 40, 52, 54, 55, 57
 errors, 10
 feedback, 56, 57
 filtered, 49, 50
 helium, 7–40
 intelligibility, 1–77

loudness, 4
mechanism, 2–4, 9
over-articulate, 54, 55
perception, 4, 9, 17, 24
propogation, 49–52
reception threshold, 92, 93
scuba, 41–76
spectral envelope, 18, 19
spectrum, 4, 6
transitions, 4, 8, 9, 16, 24, 40
transmission laboratory, Sweden, 22
unscramblers
 DYCOR, 20, 22, 36–38
 HELLE, 20, 22, 36–38
 HRBS, 25, 27, 28, 29
 IEC, 18, 36, 38
 IRPI, 20, 22, 27, 30, 33, 34, 36–38
 NASL, 18, 19, 25–29, 33, 34, 36
 RAYTHEON, 21, 22, 27, 29, 36, 38
 RELA, 20, 22, 36–38
 rotating head, 21
 Singer/GP, 20, 21, 22, 26, 27, 29, 30, 36–38
 Westinghouse, 20, 22, 25–29, 33, 34, 36
unscrambling, 16–41
 frequency domain processing, 17–19, 24
 time domain processing, 17, 19–22, 24
 vocoder, 17–20
waves, 4
Speed of current, 277
 vehicle, 277–280, 286, 295, 297
Sperm, 327
Spicule, 381
Spines of starfish, 308, 309
Spirometer, 244
Sponges, 324, 325, 329, 366, 386, 387
Sporophyte of *Laminaria hyperborea*, 369
Stability of vehicle, 280–281
Stabilizing fins, 278
Stapes, 86
"Starting effect", 194, 195, 197
Static margin, 281
Steering, 281
Stereography, 299–302
Sternum, 220
Stibine, 257
Stipe of *Laminaria hyperborea*, 369–373, 375, 377, 378
Stoichiometric relationship, 390

Stoneworts, 335
Stop, 5, 13, 14
Straw gauge, 244
Stress, 56, 209, 210, 215, 220, 223–225
  cold, 223–225, 235
  fabric, 254, 255
Stress-strain relationship, 255
Strontium, 384
Submergence, 397–399, 402
Submersible, 141, 277–283, 285–298
  "Cubmarine", 285, 289
  height off bottom, 289–297
  "Shelf Diver", 285–297
Substratum, 147, 340, 343, 344, 357, 358, 360, 369, 371
Sucrose, 362
Surf zone, 384–387
Survey, 152, 186, 255, 267, 268, 271, 272, 296, 405, 411
  grid, 267
  string line, 267
  tape, 267, 273
Suspended particles, 199, 340
  scattering by. 143, 144, 146–148, 164, 168, 173, 182
Switches, 47–49, 66, 68, 90, 91, 108, 119, 179, 180, 200, 211, 261, 289
  photocell, 48
  reed, 48, 49
  toggle, 47, 48
Sympathetic nervous system, 218
Syringes, 220, 359, 373

T

Tags on *Acanthaster planci*, 308, 309, 314, 316
Tagging of *Acanthaster planci*, 308, 309, 314
Tape recorder, 15, 19, 25–27, 32, 33, 53, 55, 66, 68, 106, 245, 287, 290
  manipulation, 17, 19
  playback, 19
Teeth, 3
Telephone, 257, 258
Telescope, 268–270, 272, 274
Telethermometer, 228
Television, camera, 45, 46, 67–69, 141, 143, 149, 153, 154, 156, 157, 159, 160, 162, 165, 167, 170, 173, 176, 177, 179, 184
  isocon, 153, 159, 160
  lines, 154, 157, 159, 176
  monitor, 45, 46, 56, 68, 69, 90, 106, 143, 153, 156, 158, 160, 176, 178
  orthocon, 153, 170, 181
  scanning, 153, 154, 177, 183
  scan lines, 157, 158, 160, 167
  scans, 167
  tube, 143, 150, 152, 153, 154, 157–159, 162, 167, 170, 175, 177, 185
  videcon, 150, 153–156, 158, 159, 170, 175, 177
Tektite, 43, 74, 114
Temperature, air, 259
  body, 224–235
  changes, 229, 231
  coefficient of resistance, 227, 228
  oral, 224
  profile of loans, 341
  rise in habitat, 256
  thermistor probes, 226–231
  variation, 245
  water, 42, 118, 123, 146, 192, 195, 218, 226, 231, 259, 365, 372, 388
Tentacles of coral, 382
Teos, 396, 401
Thalassia, 305
Thallium, 151, 175
Theodolite, 267–275
  vernier, 269, 270
Thermistor, 227, 228
Thermocline, 2, 43, 134, 192, 194–197, 340, 341
Thermometer, 372
Thoracic airway, 240, 241
Thrust, 277, 279, 280
Thrusters, 281
Thunderflash, 133
Tidal range, 218, 286
  speed, 286
  stream, 325, 330, 333
  volume, water, 324
    air, 246, 248, 249
Tilting, 403
Time dependant isolation, 179–181, 182
  domain processing, 17, 19–22, 24
Tirfor pulling machine, 256
Tone, 260–263
  level, 261
  pure, 92, 93
  wave pattern, 261
Tongue, 3
Torch, 199–202, 206, 207, 314
Towing of diver, 307, 309

SUBJECT INDEX

Trachea, 3
Training effects, man, 109, 113, 114, 119, 121, 127, 128
  animals, 121
Transducers, 28, 30, 40, 45, 46, 49, 52, 55, 65, 66, 90, 108, 114, 119, 129, 134, 245, 260, 265, 287, 290
Transects, 286, 306–308, 312, 313, 335, 384, 385, 387
Transformer, 89
Translocation, 374, 377, 378
Transmissometer, 186
Trim, 281
Trim tank, 281
Tripheustes gratilla, 305
"Tunnel effect", 291
Tunicates, 325
Turbulence, 3, 299, 340, 343, 344, 376
Turbulent flow, 238, 241
Tympanic cavity, 83, 84
  membrane, 83, 84, 87, 94, 192
Tympanum, 83, 84, 86, 94, 98

U

Udotea, 365, 366
Ulva, 349
Umbilical cable, 253, 282
U.S. Deep Submergence Rescue Vehicle, 143
  Naval Applied Sciences Laboratory, 11, 18, 43, 51, 107
  Naval Coastal Systems Laboratory, 15, 33, 38, 44
  Naval Research Laboratory Underwater Sound Reference Division, 42, 43, 88, 105, 129
  Navy Experimental Diving Unit, 11, 24, 27, 33, 43
    hyperbaric facility, 10, 27
  Navy Manned Undersea Activities programme, 78
  Navy Mine Defence Laboratory, 107
Urinary excretion products, 218
Urine, 226, 227
Urla, 396, 401

V

Variance ratio, 212
Vasoconstriction, 231
Vehicle, diver transport, 277–283
Velar, 14
Velocity of light, 144, 179, 181, 190
  sound, 5, 23, 24, 51, 82, 83
Vent valve, 255
Ventilation, 237, 239, 241–243, 248
  minute, 248, 249
Ventilatory capacity, 237
Vernier, 270–272, 274
  magnifiers, 271
  micrometer, 274
  scales, 274
Verruca stroemia, 329
Viewing, 139–190
  clear water, 141, 142, 150, 161, 163, 192, 333
  close range, 150, 155, 156
  coastal water, 145, 147, 149, 172, 186
  long range, 141, 142–144, 150, 155, 156
  oceanic water, 141, 145, 147, 149, 161, 162, 172
  short range, 141, 142, 151, 164, 180
  turbid water, 141, 142, 143, 150, 162–164, 172, 177, 182, 194, 323, 324, 326, 332, 333
Viewport of submersible, 289–294, 297
Viscosity of air, 238
  seawater, 278
Visibility, 186, 192, 195, 218, 226, 273, 275, 282, 286, 297, 302, 311
Vision, colour, 199–207
  photopic, 149, 150, 158, 159
  scotopic, 150, 159
Visual acuity, 297
  attention, 209–216
  field, 209, 210, 213, 215, 216, 285, 291–297
  location, 141
  mechanism, 102
  perception, 185
  range, 273
Vital capacity, 243
Vocal cavities, 3, 4, 6
  walls, 6
  folds, 3, 4, 9
  intensity, 15, 54, 55, 56
  tract, 3, 4, 7, 9, 23, 28, 41
Vocoder, 17–20
Voltmeter, 106
Vowel amplitude, 8
  distortion, 14, 16, 40
  energy, 21, 24, 40
  formants, 9, 10, 13, 16
Vowels, 4, 6–9, 13, 40

W

Wake turbulence, 278
Wastes and poisons, 389, 390
Water clarity, 141
  loch, 335–345, 347, 354
  movement, 324–326, 330
  oceanic, 141, 145, 149, 161, 162, 172, 321
  pure water, 145, 146
  samples, 360, 373
  vapour, 256
"Water Mate" connector, 220, 228
Wave action, 370, 372, 375
  guide, 51, 52
  motion, 259, 271, 326
  pattern, 262
  sound, 82
Westinghouse Corporation Ocean Simulation Facility, 11
  oceans research and engineering centre 15, 24, 27
Word lists, 5, 11, 15, 26, 27, 29, 31, 32, 50, 52–60, 62–65, 67–69, 71, 73
Work, 1, 74–77, 237, 239–243, 246–249, 258
Worms, 325, 329
Wrecks, investigation, 405, 407, 409, 411
  location, 405, 407, 411

Y

Yali, 396, 401
Yaw, 286–291, 294–297
  protractor, 287–289
Yellow substance, 145, 147

Z

Zooplankton, 390, 391
Zooxanthellae, 382, 387–391